Deep Learning for Natural Language Processing

Deep learning is becoming increasingly important in a technology-dominated world. However, the building of computational models that accurately represent linguistic structures is complex, as it involves an in-depth knowledge of neural networks and the understanding of advanced mathematical concepts such as calculus and statistics. This book makes these complexities accessible to those from a humanities and social sciences background by providing a clear introduction to deep learning for natural language processing. It covers both theoretical and practical aspects and assumes minimal knowledge of machine learning, explaining the theory behind natural language in an easy-to-read way. It includes pseudo code for the simpler algorithms discussed and actual Python code for the more complicated architectures, using modern deep learning libraries such as PyTorch and Hugging Face. Providing the necessary theoretical foundation and practical tools, this book will enable readers to immediately begin building real-world, practical natural language processing systems.

MIHAI SURDEANU is an associate professor in the computer science department at the University of Arizona. He works in both academia and industry on natural language processing systems that process and extract meaning from natural language.

MARCO A. VALENZUELA-ESCÁRCEGA is a research scientist in the computer science department at the University of Arizona. He has worked on natural language processing projects in both industry and academia.

Deep Learning for Natural Language Processing

A Gentle Introduction

Mihai Surdeanu

University of Arizona

Marco Antonio Valenzuela-Escárcega

University of Arizona

Shaftesbury Road, Cambridge CB2 8EA, United Kingdom

One Liberty Plaza, 20th Floor, New York, NY 10006, USA

477 Williamstown Road, Port Melbourne, VIC 3207, Australia

314–321, 3rd Floor, Plot 3, Splendor Forum, Jasola District Centre,
New Delhi – 110025, India

103 Penang Road, #05–06/07, Visioncrest Commercial, Singapore 238467

Cambridge University Press is part of Cambridge University Press & Assessment,
a department of the University of Cambridge.

We share the University's mission to contribute to society through the pursuit of
education, learning and research at the highest international levels of excellence.

www.cambridge.org
Information on this title: www.cambridge.org/9781316515662

DOI: 10.1017/9781009026222

First published 2024

A catalogue record for this publication is available from the British Library

*A Cataloging-in-Publication data record for this book is available from the Library of
Congress*

ISBN 978-1-316-51566-2 Hardback
ISBN 978-1-009-01265-2 Paperback

Contents

Figures

Tables

Preface

Upon encountering this publication, one might ask the obvious question, "Why do we need another deep learning and natural language processing book?" Several excellent ones have been published, covering both theoretical and practical aspects of deep learning and its application to language processing. However, from our experience teaching courses on natural language processing, we argue that, despite their excellent quality, most of these books do not target their most likely readers. The intended reader of this book is one who is skilled in a domain other than machine learning and natural language processing and whose work relies, at least partially, on the automated analysis of large amounts of data, especially textual data. Such experts may include social scientists, political scientists, biomedical scientists, and even computer scientists and computational linguists with limited exposure to machine learning.

Existing deep learning and natural language processing books generally fall into two camps. The first camp focuses on the theoretical foundations of deep learning. This is certainly useful to the aforementioned readers, as one should understand the theoretical aspects of a tool before using it. However, these books tend to assume the typical background of a machine learning researcher and, as a consequence, we have often seen students who do not have this background rapidly get lost in such material. To mitigate this issue, the second type of book that exists today focuses on the machine learning practitioner – that is, on how to use deep learning software, with minimal attention paid to the theoretical aspects. We argue that focusing on practical aspects is similarly necessary but not sufficient. Considering that deep learning frameworks and libraries have become fairly complex, the chance of misusing them due to theoretical misunderstandings is high. We have commonly seen this problem in our courses too.

This book therefore aims to bridge the theoretical and practical aspects of deep learning for natural language processing. We cover the necessary theoretical background and assume minimal machine learning background from the reader. Our aim is that anyone who took introductory linear algebra and calculus courses will be able to follow the theoretical material. To address practical aspects, this book includes pseudocode for the simpler algorithms discussed and actual Python code for the more complicated architectures. The code should be understandable to anyone who has taken a Python programming course. After reading this book, we expect that the reader will have the necessary foundation to immediately begin building real-world, practical natural language processing systems, and to expand their knowledge by reading research publications on these topics.

1   Introduction

Machine learning (ML) has become a pervasive part of our lives. For example, Pedro Domingos, an ML faculty member at the University of Washington, discusses a typical day in the life of a twenty-first-century person, showing how she is accompanied by ML applications throughout the day from early in the morning (e.g., waking up to music that the machine matched to their preferences) to late at night (e.g., taking a drug designed by a biomedical researcher with the help of a robot scientist) (Domingos, 2015). Of all approaches in ML, *deep learning* has seen explosive success in the past decade, and today it is ubiquitous in real-world applications of ML. At a high level, deep learning is a subfield of ML that focuses on artificial neural networks, which were "inspired by information processing and distributed communication nodes in biological systems."[1]

Natural language processing (NLP) is an important interdisciplinary field that lies at the intersection of linguistics, computer science, and ML. In general, NLP deals with programming computers to process and analyze large amounts of natural language data.[2] As an example of its usefulness, consider that PubMed, a repository of biomedical publications built by the National Institutes of Health, has indexed more than one million research publications *per year* since 2010 (Vardakas et al., 2015).[3] Clearly, no human reader (or team of readers) can process so much material. We need machines to help us manage this vast amount of knowledge. As one example out of many, an interdisciplinary collaboration that included our research team showed that machine reading discovers an order of magnitude more protein signaling pathways in biomedical literature than exist today in humanly curated knowledge bases (Valenzuela-Escárcega et al., 2018).[4] Only 60–80% of these automatically discovered biomedical interactions are correct (a good motivation for *not* letting the machines work alone!). But, without NLP, all of these would

---

[1] https://en.wikipedia.org/wiki/Deep_learning.

[2] https://en.wikipedia.org/wiki/Natural_language_processing.

[3] www.ncbi.nlm.nih.gov/pubmed.

[4] Protein signaling pathways "govern basic activities of cells and coordinate multiple-cell actions." Errors in these pathways "may cause diseases such as cancer." See https://en.wikipedia.org/wiki/Cell_signaling.

remain "undiscovered public knowledge" (Swanson, 1986), limiting our ability to understand important diseases such as cancer. Other important and more common applications of NLP include web search, machine translation, and speech recognition, all of which have had a major impact in almost everyone's life.

Since approximately 2014, a "deep learning tsunami" has hit the field of NLP (Manning, 2015) to the point that, today, a majority of NLP publications use deep learning. For example, the percentage of deep learning publications at four top NLP conferences increased from under 40% in 2012 to 70% in 2017 (Young et al., 2018). There is good reason for this domination: deep learning systems are relatively easy to build (due to their modularity), and they perform better than many other ML methods.[5] For example, the site nlpprogress.com, which keeps track of state-of-the-art results in many NLP tasks, is dominated by results of deep learning approaches.

This book explains deep learning methods for NLP, aiming to cover both theoretical aspects (e.g., how do neural networks learn?) and practical ones (e.g., how do I build one for language applications?).

The goal of the book is to do this while assuming minimal technical background from the reader. The theoretical material in the book should be completely accessible to the reader who took linear algebra, calculus, and introduction to probability theory courses, or who is willing to do some independent work to catch up. From linear algebra, the most complicated notion used is matrix multiplication. From calculus, we use differentiation and partial differentiation. From probability theory, we use conditional probabilities and independent events. The code examples should be understandable to the reader who took a Python programming course.

Starting nearly from scratch aims to address the background of who we think will be the typical reader of this book: an expert in a discipline other than ML and NLP, but who needs ML and NLP for her job. There are many examples of such disciplines: the social scientist who needs to mine social media data, the political scientist who needs to process transcripts of political discourse, the business analyst who has to parse company financial reports at scale, the biomedical researcher who needs to extract cell signaling mechanisms from publications, and so forth. Further, we hope this book will also be useful to computer scientists and computational linguists who need to catch up with the deep learning wave. In general, this book aims to mitigate the impostor syndrome (Dickerson, 2019) that affects many of us in this era of rapid change in the field of ML and artificial intelligence (AI) (this author certainly has suffered and still suffers from it![6]).

---

[5] However, they are not perfect. See Section 1.3 for a discussion.

[6] Even the best of us suffer from it. Please see Kevin Knight's description of his personal experience involving tears (not of joy) in the introduction of this tutorial (Knight, 2009).

1.1   What This Book Covers

This book interleaves chapters that discuss the theoretical aspects of deep learning for NLP with chapters that focus on implementing the previously discussed theory. For the implementation chapters, we will use PyTorch, a deep learning library that is well suited for NLP applications.[7]

Chapter 2 begins the theory thread of the book by attempting to convince the reader that ML is easy. We use a children's book to introduce key ML concepts, including our first learning algorithm. From this example, we start building several basic neural networks. In the same chapter, we formalize the perceptron algorithm, the simplest neural network. In Chapter 3, we transform the perceptron into a logistic regression network, another simple neural network that is surprisingly effective for NLP. In Chapters 5 and 6, we generalize these algorithms into feed-forward neural networks, which operate over arbitrary combinations of artificial neurons.

The astute historian of deep learning will have observed that deep learning had an impact earlier on image processing than on NLP. For example, in 2012, researchers at the University of Toronto reported a massive improvement in image classification when using deep learning (Krizhevsky et al., 2012). However, it took more than two years to observe similar performance improvements in NLP. One explanation for this delay is that image processing starts from very low-level units of information (i.e., the pixels in the image), which are then hierarchically assembled into blocks that are more and more semantically meaningful (e.g., lines and circles, then eyes and ears, in the case of facial recognition). In contrast, NLP starts from words, which are packed with a lot more semantic information than pixels and, because of that, are harder to learn from. For example, the word *house* packs a lot of commonsense knowledge (e.g., houses generally have windows and doors and they provide shelter). Although this information is shared with other words (e.g., *building*), a learning algorithm that has seen *house* in its training data will not know how to handle the word *building* in a new text to which it is exposed after training.

Chapter 8 addresses this limitation. In it, we discuss word2vec, a method that transforms words into a numerical representation that captures (some) semantic knowledge. This technique is based on the observation that "you shall know a word by the company it keeps" (Firth, 1957) – that is, it learns these semantic representations from the context in which words appear in large collections of texts. Under this formalization, similar words such as *house* and *building* will have similar representations, which will improve the learning capability of our neural networks. An important limitation of word2vec is that it conflates all senses of a given word into a single numerical representation. That is, the word *bank* gets a single numerical representation regardless of whether its

---

[7] https://pytorch.org.

current context indicates a financial sense – for example, *Bank of London* – or a geological one – for example, *bank of the river*.

Chapter 10 introduces sequence models for processing text. For example, while the word *book* is syntactically ambiguous (i.e., it can be either a noun or a verb), the information that it is preceded by the determiner *the* in a text gives strong hints that this instance of it is a noun. In this chapter, we cover recurrent neural network architectures designed to model such sequences, including long short-term memory networks and conditional random fields.

This word2vec limitation is addressed in Chapter 12 with contextualized embeddings that are sensitive to a word's surroundings. These contextualized embeddings are built using transformer networks that rely on "attention," a mechanism that computes the representation of a word using a weighted average of the representations of the words in its context. These weights are learned and indicate how much "attention" each word should pay to each of its neighbors (hence the name).

Chapter 14 discusses encoder-decoder methods (i.e., methods tailored for NLP tasks that require the transformation of one text into another). The most common example of such a task is machine translation, for which the input is a sequence of words in one language, and the output is a sequence that captures the translation of the original text in a new language.

Chapter 16 shows how several NLP applications such as part-of-speech tagging, syntactic parsing, relation extraction, question answering, and machine translation can be robustly implemented using the neural architectures introduced previously.

As mentioned before, the theoretical discussion in these chapters is interleaved with chapters that discuss how to implement these notions in PyTorch. Chapter 4 shows an implementation of the perceptron and logistic regression algorithms introduced in Chapters 2 and 3 for a text classification application. Chapter 7 presents an implementation of the feed-forward neural network introduced in Chapters 5 and 6 for the same application. Chapter 9 enhances the previous implementation of a neural network with the continuous word representations introduced in Chapter 8.

Chapter 11 implements a part-of-speech tagger using the recurrent neural networks introduced in Chapter 10. Chapter 13 shows the implementation of a similar part-of-speech tagger using the contextualized embeddings generated by a transformer network. The same chapter also shows how to use transformer networks for text classification. Last, Chapter 15 implements a machine translation application using some of the encoder-decoder methods discussed in Chapter 14.

We recommend that the reader not familiar with the Python programming language first read Appendixes A and B for a brief overview of the programming language and pointers on how to handle international characters represented in Unicode in Python.

1.2   What This Book Does Not Cover

It is important to note that deep learning is only one of the many subfields of ML. In his book, Domingos provides an intuitive organization of these subfields into five "tribes" (Domingos, 2015):

Connectionists: This tribe focuses on ML methods that (shallowly) mimic the structure of the brain. The methods described in this book fall into this tribe.

Evolutionaries: The learning algorithms adopted by this group of approaches, also known as genetic algorithms, focus on the "survival of the fittest." That is, these algorithms "mutate" the "DNA" (or parameters) of the models to be learned, and preserve the generations that perform the best.

Symbolists: The symbolists rely on inducing logic rules that explain the data in the task at hand. For example, a part-of-speech tagging system in this camp may learn a rule such as if the previous word is *the*, then the next word is a noun.

Bayesians: The Bayesians use probabilistic models such as Bayesian networks. All these methods are driven by Bayes's rule, which describes the probability of an event.

Analogizers: The analogizers' methods are motivated by the observation that "you are what you resemble." For example, a new email is classified as spam because it uses content similar to other emails previously classified as such.

It is beyond the goal of this book to explain these other tribes in detail. For a more general description of ML, the interested reader should look to other sources such as Domingos's book or Hal Daumé III's excellent *Course in Machine Learning*.[8]

Even from the connectionist tribe, we focus only on neural methods that are relevant for fundamental language processing and that we hope serve as a solid stepping stone toward research in NLP.[9] Other important, more advanced topics are not discussed. These include: domain adaptation, reinforcement learning, dialog systems, and methods that process multimodal data such as text and images.

1.3   Deep Learning Is Not Perfect

While deep learning has pushed the performance of many ML applications beyond what we thought possible just 10 years ago, it is certainly not perfect. Gary Marcus and Ernest Davis provide a thoughtful criticism of deep learning in their book, *Rebooting AI* (Marcus and Davis, 2019). Their key arguments are:

---

[8] http://ciml.info.

[9] Most methods discussed in this book are certainly useful and commonly used outside of NLP as well.

Deep learning is opaque: While deep learning methods often learn well, it is unclear *what* is learned – that is, what the connections between the network neurons encode. This is dangerous, as biases and bugs may exist in the models learned, and they may be discovered only too late, when these systems are deployed in important real-world applications such as medical diagnoses or self-driving cars.

Deep learning is brittle: It has been repeatedly shown both in the ML literature and in actual applications that deep learning systems (and for that matter most other ML approaches) have difficulty adapting to new scenarios they have not seen during training. For example, self-driving cars that were trained in regular traffic on US highways or large streets do not know how to react to unexpected situation such as a firetruck stopped on a highway.[10]

Deep learning has no common sense: An illustrative example of this limitation is that object recognition classifiers based on deep learning tend to confuse objects when they are rotated in three-dimensional space – for example, an overturned bus in the snow is confused with a snowplow. This happens because deep learning systems lack the commonsense knowledge that some object features are inherent properties of the category itself regardless of the object position – for example, a school bus in the United States usually has a yellow roof, while some features are just contingent associations – for example, snow tends to be present around snowplows. (Most) humans naturally use common sense, which means that we do generalize better to novel instances, especially when they are outliers.

All the issues Marcus and Davis raised remain largely unsolved today.

1.4   Mathematical Notations

While we try to rely on plain language as much as possible in this book, mathematical formalisms cannot (and should not) be avoided. Where mathematical notations are necessary, we rely on the following conventions:

- We use lowercase characters such as x to represent scalar values, which will generally have integer or real values.
- We use bold lowercase characters such as $\mathbf{x}$ to represent arrays (or vectors) of scalar values, and x_i to indicate the scalar element at position i in this vector. Unless specified otherwise, we consider all vectors to be column vectors during operations such as multiplication, even though we show them in text as horizontal. We use $[\mathbf{x}; \mathbf{y}]$ to indicate vector concatenation. For example, if $\mathbf{x} = (1, 2)$ and $\mathbf{y} = (3, 4)$, then $[\mathbf{x}; \mathbf{y}] = (1, 2, 3, 4)$.
- We use bold uppercase characters such as $\mathbf{X}$ to indicate matrices of scalar values. Similarly, x_{ij} points to the scalar element in the matrix at row i

---

[10] www.teslarati.com/tesla-model-s-firetruck-crash-details.

and column j. $\mathbf{x}_i$ indicates the vector corresponding to the entire row i in matrix $\mathbf{X}$.

- We collectively refer to matrices of arbitrary dimensions as *tensors*. By and large, in this book, tensors will have one dimension (i.e. vectors) or two (matrices). Occasionally, we will run into tensors with three dimensions.
- A word with an arrow on top refers to the *distributional representation* or *embedding vector* corresponding to that word. For example, $\vec{queen}$ indicates the embedding vector for the word *queen*.

2   The Perceptron

This chapter covers the perceptron, the simplest neural network architecture. In general, neural networks are ML architectures loosely inspired by the structure of biological brains. The perceptron is the simplest example of such architectures: it contains a single artificial neuron.

The perceptron will form the building block for the more complicated architectures discussed later in the book. However, rather than starting directly with the discussion of this algorithm, we will start with something simpler: a children's book and some fundamental observations about ML. From these, we will formalize our first ML algorithm, the perceptron. In the following chapters, we will improve upon the perceptron with logistic regression (Chapter 3), and deeper feed-forward neural networks (Chapter 5).

2.1   Machine Learning Is Easy

Machine learning is easy. To convince you of this, let us read a children's story (Donaldson and Scheffler, 2008). The story starts with a little monkey that lost her mom in the jungle (Figure 2.1). Luckily, a butterfly offers to help and collects some information about the mother from the little monkey (Figure 2.2). As a result, the butterfly leads the monkey to an elephant. The monkey explains that her mom is neither grey nor big, and does not have a trunk. Instead, her mom has a "tail that coils around trees." Their journey through the jungle continues until, after many mistakes (e.g., snake, spider), the pair end up eventually finding the monkey's mom and the family is happily reunited.

In addition to the exciting story that kept at least a toddler and this parent glued to its pages, this book introduces several fundamental observations about (machine) learning.

First, *objects are described by their properties*, also known in ML terminology as *features*. For example, we know that several features apply to the monkey mom: isBig, hasTail, hasColor, numberOfLimbs, etc. These features have values, which may be Boolean (true or false), a discrete value from a fixed set, or a number. For example, the values for the aforementioned features are: false, true, brown (out of multiple possible colors), and 4. As we will see soon, it is preferable to convert these values into numbers because

Figure 2.1 A wonderful children's book that introduces the fundamentals of machine learning: *Where's My Mom?*, by Julia Donaldson and Axel Scheffler © Julia Donaldson 2000, illustrations copyright © Axel Scheffler 2000

US credit: Illustrations by Axel Scheffler, copyright © 2000 by Axel Scheffler; from WHERE'S MY MOM? by Julia Donaldson. Used by permission of Dial Books for Young Readers, an imprint of Penguin Young Readers Group, a division of Penguin Random House LLC. All rights reserved.

UK credit: From Monkey Puzzle, first published in 2000 by Macmillan Children's Books an imprint of Pan Macmillan. Reproduced by permission of Macmillan Publishers International Limited.

most ML can be reduced to numeric operations such as addition and multiplication. For this reason, Boolean features are converted to 0 for false and 1 for true. Features that take discrete values are converted to Boolean features by enumerating over the possible values in the set. For example, the color feature is converted into a set of Boolean features such as hasColorBrown with the value true (or 1), hasColorRed with the value false (or 0), and so forth.

Second, *objects are assigned a discrete label*, which the learning algorithm or *classifier* (the butterfly has this role in our story) will learn how to assign to new objects. For example, in our story, we have two labels: isMyMom and isNotMyMom. When there are two labels to be assigned such as in our story, we

Table 2.1 An example of a possible feature matrix **X** (left table) and a label vector **y** (right table) for three animals in our story: elephant, snake, and monkey

isBig	hasTail	hasTrunk	hasColor Brown	numberOf Limbs	Label
1	1	1	0	4	isNotMyMom
0	1	0	0	0	isNotMyMom
0	1	0	1	4	isMyMom

Little monkey: "I've lost my mom!"

"Hush, little monkey, don't you cry. I'll help you find her," said butterfly. "Let's have a think, How big is she?"

"She's big!" said the monkey. "Bigger than me."

"Bigger than you? Then I've seen your mom. Come, little monkey, come, come, come."

"No, no, no! That's an elephant."

Figure 2.2 The butterfly tries to help the little monkey find her mom, but fails initially (Donaldson and Scheffler, 2008)

call the problem at hand a *binary classification problem*. When there are more than two labels, the problem becomes a *multiclass classification task*. Sometimes, the labels are continuous numeric values, in which case the problem at hand is called a *regression task*. An example of such a regression problem would be learning to forecast the price of a house on the real estate market from its properties – for example, number of bedrooms and year it was built. However, in NLP, most tasks are classification problems (we will see some simple ones in this chapter and more complex ones starting with Chapter 10).

To formalize what we know so far, we can organize the examples the classifier has seen (also called a training dataset) into a matrix of features **X** and a vector of labels **y**. Each example seen by the classifier takes a row in **X**, with each of the features occupying a different column. Each y_i is the label of the corresponding example x_i. Table 2.1 shows an example of a possible matrix **X** and label vector **y** for three animals in our story.

The third observation is that a good learning algorithm *aggregates its decisions over multiple examples with different features*. In our story, the butterfly learns that some features are positively associated with the mom (i.e., she is likely to have them), while some are negatively associated with her. For example, from the animals the butterfly sees in the story, it learns that the mom is likely to have a tail, fur, and four limbs, and she is not big, does not have a trunk, and her color is not grey. We will see soon that this is exactly the intuition behind the simplest neural network, the perceptron.

Last, learning algorithms produce incorrect classifications when not exposed to sufficient data. This situation is called *overfitting*, and it is more formally defined as when an algorithm performs well in training (e.g., once the butterfly sees the snake, it will reliably classify it as not the mom when it sees in the future), but poorly on unseen data (e.g., knowing that the elephant is not the mom did not help much with the classification of the snake). To detect overfitting early, ML problems typically divide their data into three partitions: (a) a training partition from which the classifier learns, (b) a development partition that is used for the *internal* validation of the trained classifier – that is, if it performs poorly on this dataset, the classifier has likely overfitted – and (c) a testing partition that is used *only* for the final, formal evaluation. Machine learning developers typically alternate between training (on the training partition) and validating what is being learned (on the development partition) until acceptable performance is observed. Once this is reached, the resulting classifier is evaluated (ideally once) on the testing partition.

2.2   Use Case: Text Classification

In the rest of this chapter, we will begin to leave the story of the little monkey behind us and change to a related NLP problem, *text classification*, in which a classifier is trained to assign a label to a text. This is an important and common NLP task. For example, email providers use binary text classification to classify emails into spam or not spam. Data mining companies use multiclass classification to detect how customers feel about a product – for example, like, dislike, or neutral. Search engines use multiclass classification to detect the language a document is written in before processing it.

Throughout the next few chapters, we will focus on text classification for simplicity. We will consider only two labels for the next few chapters, and we will generalize the algorithms discussed to multiclass classification (i.e., more than two labels) in Chapter 6. After we discuss sequence models (Chapter 10), we will introduce more complex NLP tasks such as part-of-speech tagging and syntactic parsing.

For now, we will extract simple features from the texts to be classified. That is, we will simply use the frequencies of words in a text as its features. More formally, the matrix **X**, which stores the entire dataset, will have as many columns as words in the vocabulary. Each cell x_{ij} corresponds to the number of times the word at column j occurs in the example stored at row i. For example, the text *This is a great great buy* will produce a feature corresponding to the word *buy* with value 1, one for the word *great* with value 2, and so forth,

Table 2.2 Example output of a hypothetical classifier on five evaluation examples and two labels: positive (+) and negative (−). The "Gold" column indicates the correct labels for the five texts; the "Predicted" column indicates the classifier's predictions

	Gold	Predicted
1	+	+
2	+	−
3	−	−
4	−	+
5	+	+

while the features corresponding to all the other words in the vocabulary that do not occur in this document receive a value of 0. This feature design strategy is often referred to as a *bag of words* because it ignores all the syntactic structure of the text, and treats the text simply as a collection of independent words. We will revisit this simplification in Chapter 10, where we will start to model sequences of words.

2.3 Evaluation Measures for Text Classification

The simplest evaluation measure for text classification is *accuracy*, defined as the proportion of evaluation examples that are correctly classified. For example, the accuracy of the hypothetical classifier shown in Table 2.2 is $3/5 = 60\%$ because the classifier was incorrect on two examples (rows 2 and 4).

Using the four possible outcomes for binary classification summarized in the matrix shown in Table 2.3, which is commonly referred to as a confusion matrix, accuracy can be more formally defined as:

$$\text{Accuracy} = \frac{TP + TN}{TP + FN + FP + TN}. \tag{2.1}$$

For example, for the classifier output shown in Table 2.2, TP = 2 (rows 1 and 5), TN = 1 (row 3), FP = 1 (row 4), and FN = 1 (row 2).

While accuracy is obviously useful, it is not always informative. In problems where the two labels are heavily unbalanced – that is, one is much more frequent than the other, and we care more about the less frequent label – a classifier that is not very useful may have a high accuracy score. For example, assume we build a classifier that identifies high-urgency Medicaid

Table 2.3 Confusion matrix showing the four possible
outcomes in binary classification, where $+$ indicates the
positive label and $-$ indicates the negative label

	Classifier predicted $+$	Classifier predicted $-$
Gold label is $+$	True positive (TP)	False negative (FN)
Gold label is $-$	False positive (FP)	True negative (TN)

applications– that is, applications must be reviewed quickly due to the patient's
medical condition.[1] The vast majority of applications are not high-urgency,
which means they can be handled through the usual review process. In this
example, the positive class is assigned to the high-urgency applications. If a
classifier labels all applications as negative (i.e., not high-urgency), its accuracy will be high because the TN count dominates the accuracy score. For
example, say that out of 1,000 applications only 1 is positive. Our classifier's accuracy is then: $\frac{0+999}{0+1+0+999} = 0.999$, or 99.9%. This high accuracy is
obviously misleading in any real-world application of the classifier.

For such unbalanced scenarios, two other scores that focus on class of interest (say, the positive class) are commonly used: precision and recall. Precision
(P) is the proportion of correct positive examples out of all positives predicted
by the classifier. Recall (R) is the proportion of correct positive examples out
of all positive examples in the evaluation dataset. More formally:

$$P = \frac{TP}{TP + FP}, \tag{2.2}$$

$$R = \frac{TP}{TP + FN}. \tag{2.3}$$

For example, both the precision and recall of this classifier are 0 because $TP = 0$ in its output. On the other hand, a classifier that predicts two positives, out of
which only one is incorrect, will have a precision of $1/2 = 0.5$ and a recall of
$1/1 = 1$, which are clearly more informative of the desired behavior.

Often it helps to summarize the performance of a classifier using a single
number. The F_1 score achieves this as the harmonic mean of precision and
recall:

$$F_1 = \frac{2PR}{P + R}. \tag{2.4}$$

For instance, the F_1 score for the previous example is: $F_1 = \frac{2 \times 0.5 \times 1}{0.5 + 1} = 0.67$.
A reasonable question to ask here is why not use instead the simpler arithme-
tic mean between precision and recall ($\frac{P+R}{2}$) to generate this overall score?

---

[1] Medicaid is a federal and state program in the United States that helps with medical costs for
some people with limited income and resources.

The reason for choosing the more complicated harmonic mean is that this formula is harder to game. For example, consider a classifier that labels everything as positive. Clearly, this would be useless in our example of classifying high-urgency Medicaid applications. This classifier would have a recall of 1 (because it did identify all the high-urgency applications) and a precision of approximately 0 (because everything else in the set of 1,000 applications is also labeled as high-urgency). The simpler arithmetic mean of the two scores is approximately 0.5, which is an unreasonably high score for a classifier that has zero usefulness in practice. In contrast, the F_1 score of this classifier is approximately 0, which is more indicative of the classifier's overall performance. In general, the F_1 score penalizes situations where the precision and recall values are far apart from each other.

A more general form of the F_1 score is:

$$F_\beta = (1 + \beta^2)\frac{PR}{(\beta^2 P) + R}, \tag{2.5}$$

where β is a positive real value, which indicates that recall is β times more important than precision. This generalized formula allows one to compute a single overall score for situations when precision and recall are not treated equally. For example, in the high-urgency Medicaid example, we may decide that recall is more important than precision. That is, we are willing to inspect more incorrect candidates for high-urgency processing as long as we do not miss the true positives. If we set $\beta = 10$ to indicate that we value recall as being 10 times more important than precision, the classifier in our example ($P = 0.5$ and $R = 1$) has a $F_{\beta=10}$ score of: $F_{\beta=10} = 101\frac{0.5 \times 1}{(100 \times 0.5)+1} = 0.99$, which is much closer to the classifier's recall value (the important measure here) than the F_1 score.

We will revisit these measures in Chapter 3, where we will generalize them to multiclass classification – that is, to situations where the classifier must produce more than two labels – and in Chapter 4, where we will implement and evaluate multiple text classification algorithms.

2.4   The Perceptron

Now that we understand our first NLP task, text classification, let us introduce our first classification algorithm, the *perceptron*. The perceptron was invented by McCulloch and Pitts (1943) and first implemented by Rosenblatt (1958). Its aim was to mimic binary decisions made by a single neuron. Figure 2.3 shows a depiction of a biological neuron, and Rosenblatt's computational simplification, the perceptron.[2] As the figure suggests, the perceptron is the simplest

---

[2] By BruceBlaus – Own work, CC BY 3.0, https://commons.wikimedia.org/w/index .php?curid=28761830.

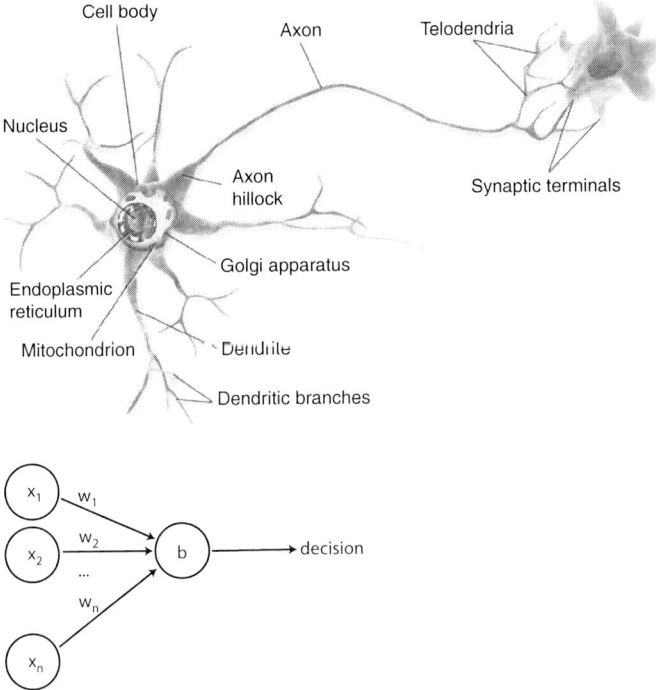

Figure 2.3 A depiction of a biological neuron, which captures input stimuli through its dendrites and produces an activation along its axon and synaptic terminals (left), and its computational simplification, the perceptron (right)

possible artificial neural network. We will generalize from this single-neuron architecture to networks with an arbitrary number of neurons in Chapter 5.

The perceptron has one input for each feature of an example $\mathbf{x}$, and produces an output that corresponds to the label predicted for $\mathbf{x}$. Importantly, the perceptron has a real-value weight vector $\mathbf{w}$, with one weight w_i for each input connection i. Thus, the size of $\mathbf{w}$ is equal to the number of features, or the number of columns in $\mathbf{X}$. Further, the perceptron also has a bias term, b, that is scalar (we will explain why this is needed later in this section). The perceptron outputs a binary decision, let's say Yes or No (e.g., Yes, the text encoded in $\mathbf{x}$ contains a positive review for a product, or No, the review is negative), based on the decision function described in Algorithm 1. The $\mathbf{w} \cdot \mathbf{x}$ component of the decision function is called the *dot product* of the vectors $\mathbf{w}$ and $\mathbf{x}$. Formally, the dot product of two vectors $\mathbf{x}$ and $\mathbf{y}$ is defined as:

$$\mathbf{x} \cdot \mathbf{y} = \sum_{i=1}^{n} x_i y_i, \tag{2.6}$$

Algorithm 1 The decision function of the perceptron

---

1 **if w · x** + b > 0 **then**
2 | return Yes
3 **else**
4 | return No
5 **end**

---

where n indicates the size of the two vectors. In words, the dot product of two vectors, **x** and **y**, is found by adding (Σ), the values found by multiplying each element of **x** with the corresponding value of **y**. In the case of the perceptron, the dot product of **x** and **w** is the weighted sum of the feature values in **x**, where each feature value x_i is weighted by w_i. If this sum (offset by the bias term b, which we will discuss later) is positive, then the decision is Yes. If it is negative, the decision is No.

Sidebar 2.1   The dot product in linear algebra

In linear algebra, the dot product of two vectors **x** and **y** is equivalent to $\mathbf{x}^T\mathbf{y}$, where T is the transpose operation. However, in this book, we rely on the dot product notation for simplicity.

Sidebar 2.2   The sign function in the perceptron

The decision function listed in Algorithm 1 is often shown as sign($\mathbf{w} \cdot \mathbf{x} + b$), where the + sign is used to represent one class and the − sign the other.

There is an immediate parallel between this decision function and the story of the little monkey. If we consider the Yes class to be isMyMom, then we would like the weights of the features that belong to the mom (e.g., hasColorBrown) to have positive values, so the dot product between **w** and the **x** vector corresponding to the mom turns out positive, and the features specific to other animals (e.g., hasTrunk) to receive negative weights, so the corresponding decision is negative. Similarly, if the task to be learned is review classification, we would like positive words (e.g., *good, great*) to have positive weights in **w**, and negative words (e.g., *bad, horrible*) to have negative weights.

In general, we call the aggregation of a learning algorithm or classifier and its learned parameters (**w** and b for the perceptron) a *model*. All classifiers aim to learn these parameters to optimize their predictions over the examples in the training dataset.

The key contribution of the perceptron is a simple algorithm that learns these weights (and bias term) from the given training dataset. This algorithm is

---

Algorithm 2 Perceptron learning algorithm

---

1 $\mathbf{w} = 0$
2 $b = 0$
3 **while** *not converged* **do**
4   **for** *each training example $\mathbf{x}_i$ in X* **do**
5   d = decision($\mathbf{x}_i$, $\mathbf{w}$, b)
6   **if** $d == y_i$ **then**
7   continue
8   **else if** $y_i ==$ *Yes* **and** $d ==$ *No* **then**
9            $b = b + 1$
10            $\mathbf{w} - \mathbf{w} \mid \mathbf{x}_l$
11   **else if** $y_i ==$ *No* **and** $d ==$ *Yes* **then**
12            $b = b - 1$
13            $\mathbf{w} = \mathbf{w} - \mathbf{x}_i$
14   **end**
15 **end**

---

summarized in Algorithm 2. Let us dissect this algorithm next. The algorithm starts by initializing the weights and bias term with zeros. Note that lines of pseudocode that assign values to a vector such as line 1 in the algorithm ($\mathbf{w} = 0$) assign this scalar value to *all* the elements of the vector. For example, the operation in line 1 initializes all the elements of the weight vector with zeros.

Lines 3 and 4 indicate that the learning algorithm may traverse the training dataset more than once. As we will see in the following example, sometimes this repeated exposure to training examples is necessary to learn meaningful weights. Informally, we say that the algorithm *converged* when there are no more changes to the weight vector (we will define convergence more formally later in this section). In practice, on real-world tasks, it is possible that true convergence is not reached, so, commonly, line 3 of the algorithm is written to limit the number of traversals of the training dataset (or *epochs*) to a fixed number.

Line 5 applies the decision function in Algorithm 1 to the current training example. Lines 6 and 7 indicate that the perceptron simply skips over training examples that it already knows how to classify – that is, its decision d is equal to the correct label y_i. This is intuitive: if the perceptron has already learned how to classify an example, there is limited benefit in learning it again. In fact, the opposite might happen: the perceptron weights may become too tailored for the particular examples seen in the training dataset, which will cause it to overfit. Lines 8–10 address the situation when the correct label of the current training example $\mathbf{x}_i$ is Yes, but the prediction according to the current weights and bias is No. In this situation, we would intuitively want the weights and bias to have higher values such that the overall dot product plus the bias is more likely to be positive. To move toward this goal, the perceptron simply *adds* the feature values in $\mathbf{x}_i$ to the weight vector $\mathbf{w}$, and adds 1 to the bias. Similarly, when the perceptron makes an incorrect prediction for the label No (lines 11–13), it decreases the value of the weights and bias by *subtracting* $\mathbf{x}_i$ from $\mathbf{w}$ and subtracting 1 from b.

Sidebar 2.3   Error-driven learning

The class of algorithms such as the perceptron that focus on "hard" examples in training – that is, examples for which they make incorrect predictions at a given point in time – are said to perform *error-driven learning*.

Figure 2.4 shows an intuitive visualization of this learning process.[3] In this figure, for simplicity, we are ignoring the bias term and assume that the perceptron decision is driven solely by the dot product $\mathbf{x} \cdot \mathbf{w}$. Figure 2.4 (a) shows the weight vector $\mathbf{w}$ in a simple two-dimensional space, which would correspond to a problem that is represented using only two features.[4] In addition of $\mathbf{w}$, the figure also shows the *decision boundary* of the perceptron as a dashed line that is perpendicular on $\mathbf{w}$. The figure indicates that all the vectors that lie on the same side of the decision boundary with $\mathbf{w}$ are assigned the label Yes, and all the vectors on the other side receive the decision No. Vectors that lie exactly on the decision boundary (i.e., their decision function has a value of 0) receive the label No according to Algorithm 1. In the transition from (a) to (b), the figure also shows that redrawing the boundary changes the decision for $\mathbf{x}$.

Why is the decision boundary line perpendicular on $\mathbf{w}$, and why are the labels so nicely assigned? To answer these questions, we need to introduce a new formula that measures the cosine of the angle between two vectors, *cos*:

$$cos(\mathbf{x}, \mathbf{y}) = \frac{\mathbf{x} \cdot \mathbf{y}}{||\mathbf{x}||||\mathbf{y}||}, \tag{2.7}$$

where $||\mathbf{x}||$ indicates the length of vector $\mathbf{x}$, that is, the distance between the origin and the tip of the vector's arrow – measured with a generalization of Pythagoras's theorem:[5] $||\mathbf{x}|| = \sqrt{\sum_{i=1}^{N} x_i^2}$. The cosine similarity, which ranges between -1 and 1, is widely used in the field of information retrieval to measure the similarity of two vectors (Schütze et al., 2008). That is, two perfectly

---

[3] This visualization was introduced by Schütze et al. (2008).

[4] This simplification is useful for visualization, but it is highly unrealistic for real-world NLP applications, where the number of features is often proportional with the size of a language's vocabulary – that is, in the hundreds of thousands.

[5] Pythagoras' theorem states that the square of the hypotenuse, c, of a right triangle is equal to the sum of the squares of the other two sides, a and b, or, equivalently: $c = \sqrt{a + b}$. In our context, c is the length of a vector with coordinates a and b in a two-dimensional space.

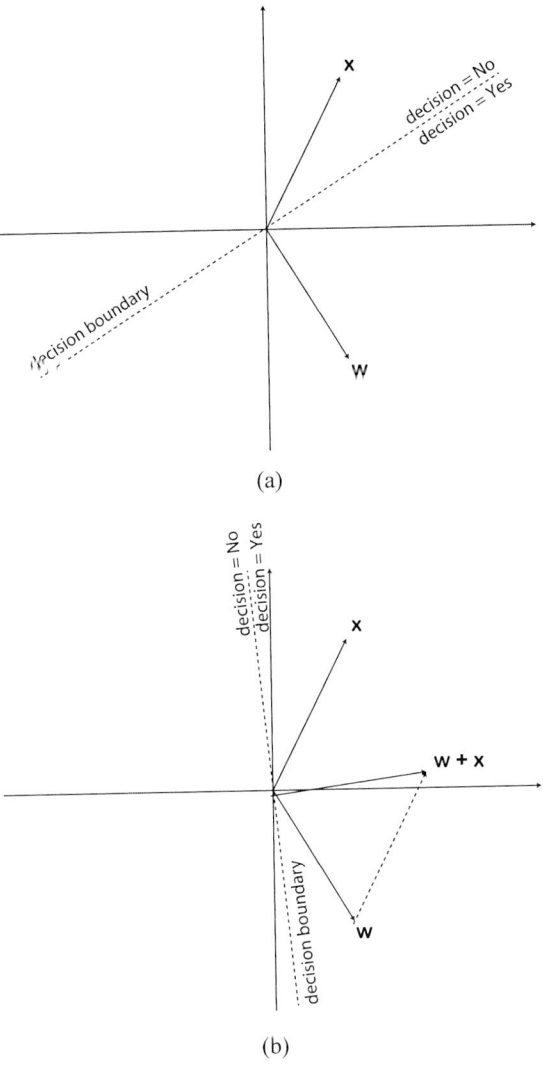

Figure 2.4 Visualization of the perceptron learning algorithm: (a) incorrect classification of the vector **x** with the label Yes, for a given weight vector **w**; and (b) **x** lies on the correct side of the decision boundary after **x** is added to **w**

similar vectors will have an angle of 0° between them, which has the largest possible cosine value of 1. Two "opposite" vectors have an angle of 180° between them, which has a cosine of −1. We will extensively use the cosine similarity formula starting with the next chapter. But, for now, we will simply

Table 2.4 The feature matrix **X** (left table) and label
vector **y** (right table) for a review classification training
dataset with three examples

#	good	excellent	bad	horrible	boring	Label
#1	1	1	1	0	0	Positive
#2	0	0	1	1	0	Negative
#3	0	0	1	0	1	Negative

observe that the cosine similarity value has the same sign with the dot product
of the two vectors (because the length of a vector is always positive). Because
vectors on the same side of the decision boundary with **w** have an angle with
w in the interval $[-90°, 90°]$, the corresponding cosine (and, thus, dot product
value) will be positive, which yields a Yes decision. Similarly, vectors on the
other side of the decision boundary will receive a No decision.

Sidebar 2.4 Hyperplanes and perceptron convergence

In a one-dimensional feature space, the decision boundary for the perceptron
is a dot. As shown in Figure 2.4, in a two-dimensional space, the decision
boundary is a line. In a three-dimensional space, the decision boundary is a
plane. In general, for an n-dimensional space, the decision boundary of the
perceptron is a *hyperplane*. Classifiers such as the perceptron whose decision
boundary is a hyperplane – that is, it is driven by a linear equation in **w** (see
Algorithm 1) – are called *linear classifiers*.

If such a hyperplane that separates the labels of the examples in the training dataset exists, it is guaranteed that the perceptron will find it, or will find
another hyperplane with similar separating properties (Block, 1962; Novikoff,
1963). We say that the learning algorithm has *converged* when such a hyperplane is found, which means that all examples in the training data are correctly
classified.

Figure 2.4 (a) shows that, at that point in time, the training example **x**
with label Yes lies on the incorrect side of the decision boundary. Figure 2.4
shows how the decision boundary is adjusted after **x** is added to **w** (line 10 in
Algorithm 2). After this adjustment, **x** is on the correct side of the decision
boundary.

To convince ourselves that the perceptron is indeed learning a meaningful
decision boundary, let us go trace the learning algorithm on a slightly more
realistic example. Table 2.4 shows the matrix **X** and label vector **y** for a training
dataset that contains three examples for a product review classification task. In
this example, we assume that our vocabulary has only the five words shown

Table 2.5 The perceptron learning process
for the dataset shown in Table 2.4, for one
pass over the training data. Both **w** and b are
initialized with 0s

Example seen: #1
$\mathbf{x} \cdot \mathbf{w} + b = 0$
Decision = Negative
Update (add): $\mathbf{w} = (1, 1, 1, 0, 0)$, $b = 1$

Example seen: #2
$\mathbf{x} \cdot \mathbf{w} + b = 2$
Decision = Positive
Update (subtract): $\mathbf{w} = (1, 1, 0, -1, 0)$, $b = 0$

Example seen: #3
$\mathbf{x} \cdot \mathbf{w} + b = 0$
Decision = Negative
Update: none

---

in **X** – for instance, the first data point in this dataset is a positive review that contains the words *good*, *excellent*, and *bad*.

Table 2.5 traces the learning algorithm as it iterates through the training examples. For example, because the decision function produces the incorrect decision for the first example (No), this example is added to **w**. Similarly, the second example is subtracted from **w**. The third example is correctly classified (barely), so no update is necessary. After just one pass over this training dataset, also called an *epoch*, the perceptron has converged. We will let the reader convince themself that all training examples are now correctly classified. The final weights indicate that the perceptron has learned several useful things. First, it learned that *good* and *excellent* are associated with the Yes class, and it has assigned positive weights to them. Second, it learned that *bad* is not to be trusted because it appears in both positive and negative reviews, and, thus, it assigned it a weight of 0. Last, it learned to assign a negative weight to *horrible*. However, it is not perfect: it did not assign a nonzero weight to *boring* because of the barely correct prediction made on example #3. There are other bigger problems here. We discuss them in Section 2.7.

This example as well as Figure 2.4 seem to suggest that the perceptron learns just fine without a bias term. So why do we need it? To convince ourselves that the bias term is useful, let us walk through another simple example, shown in Table 2.6. The perceptron needs four epochs – that is, four passes over this training dataset – to converge. The final parameters are: $\mathbf{w} = (2)$ and $b = -4$. We encourage the reader to trace the learning algorithm through this dataset on their own as well. These parameters indicate that the hyperplane for

Table 2.6 The feature matrix **X** (left table) and label vector **y** (right table) for a review classification training dataset with four examples. In this example, the only feature available is the *total* number of positive words in a review

#	Number of positive words	Label
#1	1	Negative
#2	10	Positive
#3	2	Negative
#4	20	Positive

this perceptron, which is a dot in this one-dimensional feature space, is at 2 (because the final inequation for the positive decision is $2x - 4 > 0$). That is, in order to receive a Yes decision, the feature of the corresponding example must have a value > 2 – that is, the review must have at least three positive words. This is intuitive, as the training dataset contains negative reviews that contain one or two positive words. What this shows is that the bias term allows the perceptron to shift its decision boundary *away* from the origin. It is easy to see that, without a bias term, the perceptron would not be able to learn anything meaningful, as the decision boundary will always be in the origin. In practice, the bias term tends to be more useful for problems that are modeled with few features. In real-world NLP tasks that are high-dimensional, learning algorithms usually find good decision boundaries even without a bias term (because there are many more options to choose from).

Sidebar 2.5   Implementations of the bias term

Some ML software packages implement the bias term as an additional feature in **x** that is always active – that is, it has a value of 1 for all examples in **X**. This simplifies the math a bit – that is, instead of computing $\mathbf{x} \cdot \mathbf{w} + b$, we now have to compute just $\mathbf{x} \cdot \mathbf{w}$. It is easy to see that modeling the bias as an always-active feature has the same functionality as the explicit bias term in Algorithm 2. In this book, we will maintain an explicit bias term for clarity.

2.5   Voting Perceptron

As we saw in the previous examples, the perceptron learns well, but it is not perfect. Often, a very simple strategy to improve the quality of classifier is to use an *ensemble model*. One such ensemble strategy is to *vote* between the decisions of multiple learning algorithms. For example, Figure 2.5 shows a

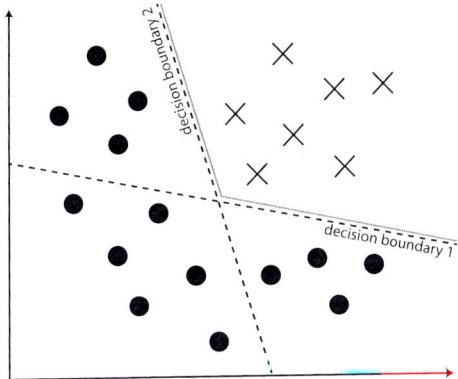

Figure 2.5 An example of a binary classification task and a voting
perceptron that aggregates two imperfect perceptrons. The voting
algorithm classifies correctly all the data points by requiring two
votes for the × class to yield a × decision. The decision boundary
of the voting perceptron is shown with a continuous line

visualization of such a voting perceptron, which aggregates two individual
perceptrons by requiring that both classifiers label an example as × before
issuing the × label.[6]

The figure highlights two important facts. First, the voting perceptron performs better than either individual classifier. In general, ensemble models that
aggregate models that are sufficiently different from each other tend to perform better than the individual (or base) classifiers that are part of the ensemble
(Dieterich, 2000). This observation holds for people too! It has been repeatedly shown that crowds reach better decisions than individuals. For example,
in 1907, Sir Francis Galton observed that while no individual could correctly
guess the weight of an ox at a fair, averaging the weights predicted by *all* individuals came within a pound or two of the real weight of the animal (Young,
2009). Second, the voting perceptron is a *nonlinear classifier* – that is, its
decision boundary is no longer a line (or a hyperplane in *n* dimensions) – in
Figure 2.5, the nonlinear decision boundary for the voting perceptron is shown
with a continuous line.

While the voting approach is an easy way to produce a nonlinear classifier that improves on the basic perceptron, it has drawbacks. First, we need
to produce several individual perceptron classifiers. This can be achieved in
at least two distinct ways. For example, instead of initializing the **w** and *b*
parameters with 0s (lines 1 and 2 in Algorithm 2), we initialize them with random numbers (typically small numbers centered around 0). For every different

---

[6] This example was adapted from Erwin Chan's Ling 539 course at the University of Arizona.

set of initial values in **w** and *b*, the resulting perceptron will end up with a different decision boundary and thus a different classifier. The drawback of this strategy is that the training procedure must be repeated for each individual perceptron. A second strategy for producing multiple individual perceptron that avoids this training overhead is to keep track of all **w**s and *b*s that are produced during the training of a single perceptron. That is, before changing the *b* and **w** parameters in Algorithm 2 (lines 9 and 12), we store the current values (before the change) in a list. This means that at the end of the training procedure, this list will contain as many individual perceptrons as the number of updates performed in training. We can even sort these individual classifiers by their perceived quality: the more iterations a specific *b* and **w** combination "survived" in training, the better the quality of this classifier is likely to be. This indicator of quality can be used to assign weights to the "votes" given to the individual classifiers, or to filter out base models of low quality (e.g., remove all classifiers that survived fewer than 10 training examples).

The second drawback of the voting perceptron is its runtime overhead at evaluation. When the voting perceptron is applied on a new, unseen example, it must apply all its individual classifiers before voting. Thus the voting perceptron is N times slower than the individual perceptron, where N is the number of individual classifiers used. To mitigate this drawback, we will need the average perceptron, discussed next.

2.6   Average Perceptron

The average perceptron is a simplification of the voting perceptron we discussed previously. The simplification consists in that, instead of keeping track of *all* **w** and *b* parameters created during the perceptron updates like the voting algorithm, these parameters are averaged into a *single* model, say **avgW** and *avgB*. This algorithm, which is summarized in Algorithm 3, has a constant runtime overhead for computing the average model – that is, the only additional overhead compared to the regular perceptron are the additions in lines 12–14 and 18–20, and the divisions in lines 25 and 26. Further, the additional memory overhead is also constant, as it maintains a single extra weight vector (**totalW**) and a single bias term (*totalB*) during training. After training, the average perceptron uses a decision function different from the one used during training. This function has a similar shape to the one listed in Algorithm 1, but uses **avgW** and *avgB* instead.

Despite its simplicity, the average perceptron tends to perform well in practice, usually outperforming the regular perceptron, and approaching the performance of the voting perceptron. But why is the performance of the average perceptron so good? After all, it remains a linear classifier just like the regular perceptron, so it must have the same limitations. The high-level explanation is that the average perceptron does a better job than the regular

---

Algorithm 3 Average perceptron learning algorithm

---

1   $\mathbf{w} = 0$

2   b = 0

3   numbertotalOfUpdates = 0

4   **totalW** = 0

5   totalB = 0

6   **while** *not converged* **do**

7   **for** *each training example* $\mathbf{x}_i$ **in X do**

8   d = decision($\mathbf{x}_i$, $\mathbf{w}$, b)

9   **if** $d == y_i$ **then**

10   continue

11   **else if** $y_i == $ *Yes* **and** $d == $ *No* **then**

12   numberOfUpdates = numberOfUpdates + 1

13   **totalW** = **totalW** + **w**

14   totalB = totalB + b

15         $\mathbf{w} = \mathbf{w} + \mathbf{x}_i$

16         $b = b + 1$

17   **else if** $y_i == $ *No* **and** $d == $ *Yes* **then**

18   numberOfUpdates = numberOfUpdates + 1

19   **totalW** = **totalW** + **w**

20   totalB = totalB + b

21         $\mathbf{w} = \mathbf{w} - \mathbf{x}_i$

22         $b = b - 1$

23   **end**

24   **end**

25   avgB = totalB/numberOfUpdates

26   **avgW** = **totalW**/numberOfUpdates

---

perceptron at controlling for *noise*. Kahneman et al. (2021) define noise as unwanted variability in decision-making. Note that noise is a common occurrence in both human and machine decisions. For example, Kahneman et al. (2021) report that judges assign more lenient sentences if the outside weather is nice, or if their favorite football team won their match the prior weekend. Clearly, these decisions should not depend on such extraneous factors.

Similarly, in the ML space, the regular perceptron may be exposed to such noisy, unreliable features during training. When this happens, these features will receive weight values in the perceptron model (the $\mathbf{w}$ vector) that are all over the place, sometimes positive and sometimes negative. All these values are averaged in the average vector, and thus the average weight value for these unreliable features will tend to be squished to (or close to) zero. The effect of this squishing is that the decision function of the average perceptron will tend to not rely on these features (because their contribution to the dot product in the decision function will be minimal). This differs from the regular perceptron, which does not benefit from this averaging process that reduces the weights of unimportant features. In general, this process of squishing the weights of features that are not important is called *regularization*. We will see other regularization strategies in Chapter 6.

2.7   Drawbacks of the Perceptron

The perceptron algorithm and its variants are simple, easy to customize for other tasks beyond text classification, and they perform fairly well (especially in the voting and average form). However, they also have important drawbacks. We discuss these drawbacks here, and we will spend a good part of this book discussing solutions that address them.

The first obvious limitation of the perceptron is that, as discussed in this chapter, it is a linear classifier. Yes, the voting perceptron removes this constraint, but it comes at the cost of maintaining multiple individual perceptrons. Ideally, we would like to have the ability to learn a single classifier that captures a nonlinear decision boundary. This ability is important, as many tasks require such a decision boundary. A simple example of such a task was discussed by Minsky and Papert as early as 1969: the perceptron cannot learn the XOR function (Minsky and Papert, 1969). To remind ourselves, the XOR function takes two binary variables – that is, numbers that can take only one of two values – 0 (which stands for False) or 1 (or True), and outputs 1 when exactly one of these values is 1, and 0 otherwise. A visualization of the XOR is shown in Figure 2.6. It is immediately obvious that there is no linear decision boundary that separates the dark circles from the clear ones. More importantly in our context, language is beautiful, complex, and ambiguous, which means that, usually, we cannot model tasks that are driven by language using methods of limited power such as linear classifiers. We will address this important limitation in Chapter 5, where we will introduce neural networks that can learn

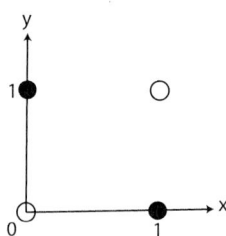

Figure 2.6 Visualization of the XOR function operating over two variables, x and y. The dark circles indicate that the XOR output is 1; the clear circles stand for 0

nonlinear decision boundaries by combining multiple layers of "neurons" into a single network.

A second more subtle but very important limitation of the perceptron is that it has no "smooth" updates during training – that is, its updates are the same regardless of how incorrect the current model is. This is caused by the decision function of the perceptron (Algorithm 1), which relies solely on the *sign* of the dot product. That is, it does not matter how large (or small) the value of the dot product is; when the sign is incorrect, the update is the same: adding or subtracting the *entire* example x_i from the current weight vector (lines 10 and 13 in Algorithm 2). This causes the perceptron to be a slow learner because it jumps around good solutions. One University of Arizona student called this instability "Tony Hawk-ing the data."[7] On data that are linearly separable, the perceptron will eventually converge (Novikoff, 1963). However, real-world datasets do not come with this guarantee of linear separation, which means that this "Tony Hawk-ing" situation may yield a perceptron that is far from acceptable. What we would like to have is a classifier that updates its model *proportionally* with the errors it makes: a small mistake causes a small update, while a large one yields a large update. This is exactly what the logistic regression does. We detail this in the next chapter.

The third drawback of the perceptron, as we covered it so far, is that it relies on hand-crafted features that must be designed and implemented by the ML developer. For example, in the text classification use case introduced in Section 2.2, we mentioned that we rely on features that are simply the words in each text to be classified. Unfortunately, in real-world NLP applications, feature design gets complicated very quickly. For example, if the task to be learned is review classification, we should probably capture negation. Certainly the phrase *great* should be modeled differently than *not great*. Further, maybe we should investigate the syntactic structure of the text to be classified. For example, reviews typically contain multiple clauses, whose sentiment must be composed into an overall classification for the entire review. For instance, the review *The wait was long, but the food was fantastic* contains two clauses: *The wait was long* and *but the food was fantastic*, each one capturing a different sentiment, which must be assembled into an overall sentiment toward the corresponding restaurant. Further, most words in any language tend to be very infrequent (Zipf, 1932), which means that a lot of the hard work we might invest in feature design might not generalize enough. That is, suppose that the reviews included in a review classification training dataset contain the word *great* but not the word *fantastic*, a fairly similar word in this context. Then any ML algorithm that uses features that rely on explicit words will correctly learn how to associate *great* with a specific sentiment, but will not know what to

---

[7] Tony Hawk is an American skateboarder, famous for his half-pipe skills. See https://en.wikipedia.org/wiki/Tony_Hawk.

do when they see the word *fantastic*. Chapter 8 addresses this limitation. We will discuss methods to transform words into a numerical representation that captures (some) semantic knowledge. Under this representation, similar words such as *great* and *fantastic* will have similar forms, which will improve the generalization capability of our ML algorithms.

Last, in this chapter, we focused on text classification applications such as review classification that require a simple ML classifier, which produces a single binary label for an input text – for example, positive versus negative review. However, many NLP applications require multiclass classification (i.e., more than two labels), and, crucially, produce *structured* output. For example, a part-of-speech tagger, which identifies which words are nouns, verbs, and so forth must produce the *sequence* of part of speech tags for a given sentence. Similarly, a syntactic parser identifies syntactic structures in a given sentence such as which phrase serves as a subject for a given verb. These structures are typically represented as *trees*. The type of ML algorithms that produce structures rather than individual labels are said to perform *structured learning*. We will begin discussing structured learning in Chapter 10.

2.8   Historical Background

The perceptron was invented by McCulloch and Pitts in 1943 (McCulloch and Pitts, 1943). Frank Rosenblatt provided a first software implementation in 1958 (Rosenblatt, 1958), and soon after, a hardware implementation as the "Mark I Perceptron," a machine built for image recognition. The Mark I Perceptron now resides at the Smithsonian Institution. Interestingly enough, at the time Rosenblatt was a research psychologist at the Cornell Aeronautical Laboratory, Warren McCulloch was a professor of psychiatry at the University of Illinois at Chicago, while Walter Pitts was an unofficial student of mathematics, logic, and biology. Computer science did not yet exist as a formal academic discipline. The first computer science department in the United States was established at Purdue University only in 1962.

Following the development of the perceptron, Rosenblatt stated: "Stories about the creation of machines having human qualities have long been a fascinating province in the realm of science fiction ... Yet we are about to witness the birth of such a machine – a machine capable of perceiving, recognizing and identifying its surroundings without any human training or control" (Lefkowitz, 2019). Needless to say, such statements were premature, especially considering the perceptron's limitations as a linear classifier – that is, it cannot learn simple nonlinear functions such as the XOR (Minsky and Papert, 1969). This discrepancy between claims and reality caused the first AI "winter" – that is, a period of several decades during which government funding for AI was drastically reduced. Some argue that Rosenblatt has been vindicated by the tremendous empirical achievements of today's neural networks (Lefkowitz,

2019), while others continue to argue that statements such as Rosenblatt's (and many other AI researchers') remain disconnected from what today's AI can actually do (Dreyfus, 1992; Marcus and Davis, 2019).

Nevertheless, regardless where one stands in this controversy, it is clear that the perceptron and its variants (see next section) made a tremendous contribution to ML and NLP, and paved the way for today's deep learning field (as we will see throughout the rest of this book).

2.9   References and Further Readings

The original perceptron papers were McCulloch and Pitts (1943) (theory) and Rosenblatt (1958) (first implementation). Block (1962) and Novikoff (1963) demonstrated the convergence of the perceptron training algorithm – that is, if a hyperplane that separates the labels of the examples in the training dataset exists, it is guaranteed that the perceptron will find it, or will find another hyperplane with similar separating properties.

Minsky and Papert (1969) demonstrated the limitations of the perceptron – that is, that it cannot learn nonlinear functions such as the XOR.

Despite its simplicity (or perhaps because of it), the perceptron has been widely used and extended for various problems in ML and NLP. For example, Duda et al. (1973) extended the original binary perceptron to multiclass classification. Crammer and Singer (2003) and Crammer et al. (2006) proposed a generalized multiclass setting for the perceptron and introduced several new training algorithms for it that have improved worst-case behavior. Collins (2002) introduced a variant of the perceptron adapted for sequence problems in NLP such as part-of-speech tagging. Collins and Roark (2004) extended this algorithm for syntactic parsing.

2.10   Summary

This chapter presented the perceptron, one of the simplest ML algorithms, which will serve as the building block for the neural networks explored throughout the rest of the book. We also discussed a couple of perceptron variants, starting with the voting perceptron, our first exposure to a nonlinear classifier. It was followed by the average perceptron, which introduced regularization – that is, reducing the importance of noisy information in the learned model.

3   Logistic Regression

As mentioned in the previous chapter, the perceptron does not perform smooth updates during training, which may slow learning down or cause it to miss good solutions entirely in real-world situations. In this chapter, we will discuss logistic regression (LR), an ML algorithm that elegantly addresses this problem.

3.1   The Logistic Regression Decision Function and Learning Algorithm

As we discussed, the lack of smooth updates in the training of the perceptron is caused by its reliance on a discrete decision function driven by the sign of the dot product. The first thing LR does is replace this decision function with a new, *continuous* function, which is:

$$\text{decision}(\mathbf{x}, \mathbf{w}, b) = \frac{1}{1 + e^{-(\mathbf{w} \cdot \mathbf{x} + b)}}. \tag{3.1}$$

The $\frac{1}{1+e^{-x}}$ function is known as the logistic function, hence the name of the algorithm. The logistic function belongs to a larger class of functions called sigmoid functions because they are characterized by an S-shaped curve. Figure 3.1 shows the curve of the logistic function. In practice, the name sigmoid (or σ) is often used instead of logistic, which is why the LR decision function is often summarized as: $\sigma(\mathbf{w} \cdot \mathbf{x} + b)$. For brevity, we will use the σ notation in our formulas as well.

Figure 3.1 shows that the logistic function has values that monotonically increase from 0 to 1. We will use this property to implement a better learning algorithm, which has "soft" updates that are proportional to how incorrect the current model is. To do this, we first arbitrarily associate one of the labels to be learned with the value 1, and the other with 0. For example, for the review classification task, we (arbitrarily) map the positive label to 1 and the negative label to 0. Intuitively, we would like to learn a decision function that produces values close to 1 for the positive label and values close to 0 for the negative one. The difference between the value produced by the decision function and

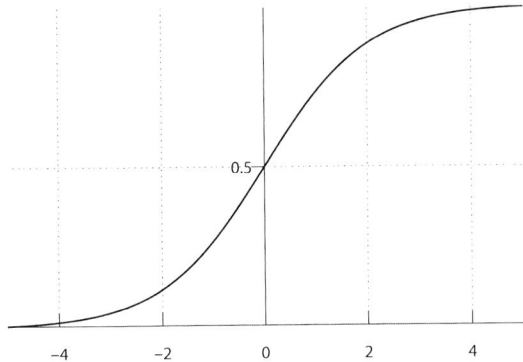

Figure 3.1 The logistic function

the gold value for a training example will quantify the algorithm's confusion at a given stage in the learning process.

Algorithm 4 lists the LR learning process that captures these intuitions. We will discuss later in this chapter how this algorithm was derived. For now, let us make sure that this algorithm does indeed do what we promised.

Note that the only new variable in this algorithm is α, known as the learning rate. The learning rate takes a positive value that adjusts up or down the values used during the update. We will revisit this idea later on in this chapter. For now, let us assume $\alpha = 1$.

It is easy to see that, at the extreme (i.e., when the prediction is perfectly correct or incorrect), this algorithm reduces to the perceptron learning algorithm. For example, when the prediction is perfectly correct (say, $y_i = 1$ for the class associated with 1), y_i is equal to d, which means that there is no weight or bias update in lines 6 and 7. This is similar to the perceptron (lines 6 and 7 in Algorithm 2). Further, when a prediction is perfectly incorrect, say, $y_i = 1$ (Yes) when $d = 0$ (No), this reduces to adding $\mathbf{x}_i$ to $\mathbf{w}$ and 1 to b (similar to the perceptron update, lines 8–10 in Algorithm 2). When $y_i = 0$ when $d = 1$, the algorithm reduces to subtracting $\mathbf{x}_i$ from $\mathbf{w}$ and 1 from b (similar to lines 11–13 in Algorithm 2).

The interesting behavior occurs in the majority of the situations when the LR decision is neither perfectly correct nor perfectly incorrect. In these situations, the LR performs a soft update that is proportional to how incorrect the current decision is, which is captured by $y_i - d$. That is, the more incorrect the decision is, the larger the update. This is exactly what we would like a good learning algorithm to do.

Once the algorithm finishes training, we would like to use the learned weights ($\mathbf{w}$ and b) to perform binary classification – for example, classify a text into a positive or negative review. For this, at prediction time we will convert the LR decision into a discrete output using a threshold τ, commonly set

---

Algorithm 4 Logistic regression learning algorithm

1   $\mathbf{w} = 0$
2   $b = 0$
3   **while** *not converged* **do**
4   **for** *each training example* $\mathbf{x}_i$ **in X do**
5   d = decision($\mathbf{x}_i$, $\mathbf{w}$, b)
6         $\mathbf{w} = \mathbf{w} + \alpha(y_i - d)\mathbf{x}_i$ // y_i is the correct label for example $\mathbf{x}_i$
7         $b = b + \alpha(y_i - d)$
8   **end**
9   **end**

---

to 0.5.[1] That is, if decision($\mathbf{x}, \mathbf{w}, b) \geq 0.5$, then the algorithm outputs one class (say, positive review); otherwise, it outputs the other class.

3.2   The Logistic Regression Cost Function

The next three sections of this chapter focus on deriving the LR learning algorithm shown in Algorithm 4. The reader who is averse to math, or is satisfied with the learning algorithm and the intuition behind it, may skip to Section 3.7. However, we encourage the reader to try to stay with us through this derivation. These sections introduce important concepts – that is, cost functions and gradient descent, which are necessary for a thorough understanding of the following chapters in this book. We will provide pointers to additional reading where more mathematical background may be needed.

The first observation that will help us formalize the training process for LR is that the LR decision function implements a conditional probability – that is, the probability of generating a specific label given a training example and the current weights. More formally, we can write:

$$p(y = 1|\mathbf{x}; \mathbf{w}, b) = \sigma(\mathbf{x}; \mathbf{w}, b), \tag{3.2}$$

The left term of this equation can be read as the probability of generating a label y equal to 1, given a training example $\mathbf{x}$ and model weights $\mathbf{w}$ and b (the vertical bar "|" in the conditional probability formula should be read as "given"). Intuitively, this probability is an indicator of confidence (the higher the better). That is, the probability approaches 1 when the model is confident that the label for $\mathbf{x}$ is 1, and 0 when not. Similarly, the probability of y being 0 is:

$$p(y = 0|\mathbf{x}; \mathbf{w}, b) = 1 - \sigma(\mathbf{x}; \mathbf{w}, b). \tag{3.3}$$

---

[1] Other values for this threshold are possible. For example, for applications where it is important to be conservative with predictions for class 1, τ would take values larger than 0.5.

These probabilities form a probability distribution – that is, the sum of probabilities over all possible labels equals 1. Note that while we aim to minimize the use of probability theory in this section, some of it is unavoidable. The reader who wants to brush up on probability theory may consult other material on this topic such as Griffiths (2008).

To simplify notations, because we now know that we estimate label probabilities, we change the notation for the two probabilities to: $p(1|\mathbf{x}; \mathbf{w}, b)$ and $p(0|\mathbf{x}; \mathbf{w}, b)$. Further, when it is obvious what the model weights are, we will skip them and use simply $p(1|\mathbf{x})$ and $p(0|\mathbf{x})$. Last, we generalize the two formulas to work for any of the two possible labels with the following formula:

$$p(y|\mathbf{x}) = (\sigma(\mathbf{x}; \mathbf{w}, b))^y (1 - \sigma(\mathbf{x}; \mathbf{w}, b))^{1-y}. \qquad (3.4)$$

It is trivial to verify that this formula reduces to one of the two equations, for $y = 1$ and $y = 0$.

Intuitively, we would like the LR training process to maximize the probability of the correct labels in the entire training dataset. This probability is called the *likelihood of the data* (L), and is formalized as:

$$L(\mathbf{w}, b) = p(\mathbf{y}|\mathbf{X}) \qquad (3.5)$$

$$= \Pi_{i=1}^{m} p(y_i|\mathbf{x}_i), \qquad (3.6)$$

where $\mathbf{y}$ is the vector containing all the correct labels for all training examples, $\mathbf{X}$ is the matrix that contains the vectors of features for all training examples, and m is the total number of examples in the training dataset. Note that the derivation into the product of individual probabilities is possible because we assume that the training examples are independent of each other, and the joint probability of multiple independent events is equal to the product of individual probabilities (Griffiths, 2008).

A common convention in ML is that instead of maximizing a function during learning, we instead aim to minimize a *cost* or *loss* function C, which captures the amount of errors in the model.[2] By definition, C must return only positive values. That is, C will return large values when the model does not perform well, and is 0 when the learned model is perfect. We write the LR cost function C in terms of likelihood L as:

$$C(\mathbf{w}, b) = -\log L(\mathbf{w}, b) \qquad (3.7)$$

$$= -\sum_{i=1}^{m} (y_i \log \sigma(\mathbf{x}_i; \mathbf{w}, b) + (1 - y_i) \log(1 - \sigma(\mathbf{x}_i; \mathbf{w}, b))). \qquad (3.8)$$

---

[2] Formally, the loss function operates on a single training example while the cost function considers all examples in the training dataset. However, this terminology has become more ambiguous in the literature. For this reason, we will use "loss" and "cost" interchangeably in this book. For example, in the theory chapters, we prefer to use "cost" because we tend to apply it to an entire training set (or a partition of it). In the coding chapters, we will use "loss" more frequently because it matches PyTorch's terminology.

Equation 3.7 is often referred to as the *negative log likelihood* of the data, a descriptive term that summarizes well the content of the equation. It is easy to see that C satisfies the constraints of a cost function, which are:

- First, the cost function must always return positive values. In our case, the logarithm of a number between 0 and 1 is negative; the negative sign in front of the sum turns the value of the sum into a positive number.
- Second, the cost function returns large values when the model makes many mistakes (i.e., the likelihood of the data is small), and approaches 0 when the model is correct (i.e., the likelihood approaches 1).

Thus, we can formalize the goal of the LR learning algorithm as minimizing the cost function. Next we will discuss how we do this efficiently.

3.3   Gradient Descent

The missing component that connects the cost function just introduced with the LR training algorithm (Algorithm 4) is gradient descent. Gradient descent is an iterative method that finds the parameters that minimize a given function. In our context, we will use gradient descent to find the LR parameters (**w** and b) that minimize the cost function C.

However, for illustration purposes, let us take a step away from the LR cost function and begin with a simpler example: let us assume we would like to minimize the function $f(x) = (x + 1)^2 + 1$, which is plotted in Figure 3.2. Clearly, the smallest value this function takes is 1, which is obtained when $x = -1$. Gradient descent finds this value by taking advantage of the function slope, or derivative of $f(x)$ with respect to x – that is, $\frac{d}{dx} f(x)$. Note: if the reader needs a refresher on what function derivatives are and how to compute them, now is a good time to study it. Any calculus textbook or even the Wikipedia page for function derivatives provides sufficient information for what we need in this book.[3]

One important observation about the slope of a function is that it indicates the function's direction of change. That is, if the derivative is negative, the function decreases; if it is positive, the function increases; and if it is zero, we have reached a local minimum or maximum for the function. Let us verify that is the case for our simple example. The derivative of our function $\frac{d}{dx}((x + 1)^2 + 1)$ is $2(x + 1)$, which has negative values when $x < -1$, positive values when $x > -1$, and is 0 when $x = -1$. Intuitively, gradient descent uses this observation to take small steps toward the function's minimum in the opposite direction indicated by the slope. More formally, gradient descent starts by initializing x with some random value – for example, $x = -3$ – and then repeatedly subtracts a quantity proportional to the derivative of x, until it *converges* – that is,

---

[3] https://en.wikipedia.org/wiki/Derivative.

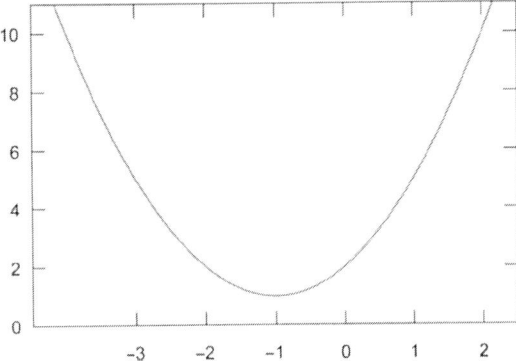

Figure 3.2 Plot of the function $f(x) = (x + 1)^2 + 1$

it reaches a derivative of 0 (or close enough so we can declare success). That is, we repeatedly compute:

$$x = x - \alpha \frac{d}{dx} f(x) \tag{3.9}$$

until convergence.

Sidebar 3.1   Partial derivative notation

In this book, we use the Leibniz notation for derivatives. That is, $\frac{d}{dx} f$ indicates the derivative of function f with respect to x – that is, the amount of change in f in response to an infinitesimal change in x. This notation is equivalent to the Lagrange notation (sometimes attributed to Newton) of $f'(x)$.

α in the equation is the same learning rate introduced before in this chapter. Let us set $\alpha = 0.1$ for this example. Thus, in the first gradient descent iteration, x changes to $x = -3 - 0.1 \times 2(-3 + 1) = -2.6$. In the second iteration, x becomes $x = -2.6 - 0.1 \times 2(-2.6 + 1) = -2.28$. And so on, until, after approximately 30 iterations, x approaches -1.001, a value practically identical to what we were looking for.

This simple example also highlights that the learning rate α must be positive (so we don't change the direction indicated by the slope), and small (so we do not "Tony Hawk" the data). To demonstrate the latter situation, consider when $\alpha = 1$. In this case, in the first iteration, x becomes 1, which means we already skipped over the value that yields the function's minimum ($x = -1$). Even worse, in the second iteration, x goes back to -3 and we are now in danger of entering an infinite loop! To mitigate this, α usually takes small positive values, say, between 0.00001 and 0.1. In Chapter 6, we will discuss other strategies to dynamically shrink the learning rate as the learning advances, so we further reduce our chance of missing the function's minimum.

The gradient descent algorithm generalizes to functions with multiple parameters: we simply update each parameter using its own partial derivative of the function to be minimized. For example, consider a new function that has two parameters, x_1 and x_2: $f(x_1, x_2) = (x_1 + 1)^2 + 3x_2 + 1$. For this function, in each gradient descent iteration, we perform the following updates:

$$x_1 = x_1 - \alpha \frac{d}{dx_1} f(x_1, x_2) = x_1 - 0.1(2x_1 + 2)$$

$$x_2 = x_2 - \alpha \frac{d}{dx_2} f(x_1, x_2) = x_2 - 0.1(3),$$

or, in general, for a function $f(\mathbf{x})$, we update each parameter x_i using the formula:

$$x_i = x_i - \alpha \frac{d}{dx_i} f(\mathbf{x}). \tag{3.10}$$

One obvious question that should arise at this moment is, why are we not simply solving the equation where the derivative equals 0, as we were taught in calculus? For instance, for the first simple example we looked at, $f(x) = (x + 1)^2 + 1$, zeroing the derivative yields immediately the exact solution $x = -1$. While this approach works well for functions with a single parameter or two, it becomes prohibitively expensive for functions with four or more parameters. Machine learning in general falls in the latter camp: it is very common that the functions we aim to minimize have thousands (or even millions) of parameters. In contrast, as we will see later, gradient descent provides a solution whose runtime is linear in the number of parameters times the number of training examples. Further, in some situations, training data are not available ahead of time, but, instead, are provided sequentially – that is, a few examples at a time. This type of ML is called *online learning*. For example, the reviews necessary to train a review classifier might not be available ahead of time, but come in over time as buyers review products. Gradient descent is well suited for online learning because it operates on one (or a few) training examples at a time.

It is important to note that gradient descent is not perfect. It does indeed work well for convex functions – that is, functions that have exactly one minimum and are differentiable at every point such as our simple example – but it does not perform so well in more complex situations. Consider the function shown in Figure 3.3.[4] This function has two minima (around $x = 3$ and $x = -2$). Because gradient descent is a "greedy" algorithm – that is, it commits to a solution relying only on local knowledge without understanding the bigger picture – it may end up finding a minimum that is not the best. For example, if

---

[4] This example of a function with multiple minima was taken from https://en.wikipedia.org/wiki/Derivative.

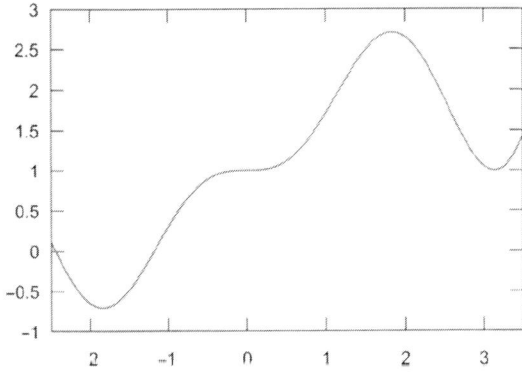

Figure 3.3 Plot of the function $f(x) = x \sin(x)^2 + 1$

x is initialized with 2.5, gradient descent will follow the negative slope at that position, and end up discovering the minimum around $x = 3$, which is not the best solution. However, despite this known limitation, gradient descent works surprisingly well in practice.

Now that we have a general strategy for finding the parameters that minimize a function, let us apply it to the problem we care about in this chapter – that is, finding the parameters $\mathbf{w}$ and b that minimize the cost function $C(\mathbf{w}, b)$ (Equation 3.8). A common source of confusion here is that the parameters of C are $\mathbf{w}$ and b, not $\mathbf{x}$ and y. For a given training example, $\mathbf{x}$ and y are known and constant. That is, we know the values of the features and the label for each given example in training, and all we have to do is compute $\mathbf{w}$ and b. Thus, the training process of LR reduces to repeatedly updating each w_j in $\mathbf{w}$ and b features by the corresponding partial derivative of C:

$$w_j = w_j - \alpha \frac{d}{dw_j} C(\mathbf{w}, b) \qquad (3.11)$$

$$b = b - \alpha \frac{d}{db} C(\mathbf{w}, b). \qquad (3.12)$$

Given a sufficient number of iterations and a learning rate α that is not too large, $\mathbf{w}$ and b are guaranteed to converge to the optimal values because the LR cost function is convex.[5]

However, one problem with this approach is that computing the two partial derivatives requires the inspection of *all* training examples (this is what the summation in Equation 3.8 indicates), which means that the learning algorithm

---

[5] Demonstrating that the LR cost function is convex is beyond the scope of this book. The interested reader may read other materials on this topic such as http://mathgotchas.blogspot.com/2011/10/why-is-error-function-minimized-in.html

---

Algorithm 5 Logistic regression learning algorithm using stochastic gradient descent

---

1 $\mathbf{w} = 0$

2 $b = 0$

3 **while** *not converged* **do**

4   **for** *each training example* x_i **in X do**

5   **for** *each* w_j *in* $\mathbf{w}$ **do**

6              $w_j = w_j - \alpha \frac{d}{dw_j} C_i(\mathbf{w}, b)$

7   **end**

8          $b = b - \alpha \frac{d}{db} C_i(\mathbf{w}, b)$

9   **end**

10 **end**

---

would have to do many passes over the training dataset before any meaningful changes are observed. Because of this, in practice, we do not compute C over the whole training data, but over a small number of examples at a time. This small group of examples is called a *minibatch*. In the simplest case, the size of the minibatch is 1 – that is, we update the $\mathbf{w}$ and b weights after seeing each individual example i, using a cost function computed for example i alone:[6]

$$C_i(\mathbf{w}, b) = -(y_i \log \sigma(\mathbf{x}_i; \mathbf{w}, b) + (1 - y_i) \log(1 - \sigma(\mathbf{x}_i; \mathbf{w}, b))). \qquad (3.13)$$

This simplified form of gradient descent is called *stochastic gradient descent* (SGD), where "stochastic" indicates that we work with a stochastic approximation (or an estimate) of C. Building from the previous three equations, we can write the LR training algorithm as shown in Algorithm 5. The reader will immediately see that this formulation of the algorithm is similar to Algorithm 4, which we introduced at the beginning of this chapter. In the next section, we will demonstrate that these two algorithms are indeed equivalent by computing the two partial derivatives $\frac{d}{dw_j} C_i(\mathbf{w}, b)$ and $\frac{d}{db} C_i(\mathbf{w}, b)$. Importantly, the runtime of this algorithm is linear in the number of parameters (lines 5 and 8) times the number of training examples (line 4), which makes this algorithm a practical solution for training on large datasets.

3.4   Deriving the Logistic Regression Update Rule

Here, we will compute the partial derivative of the cost function $C_i(\mathbf{w}, b)$ of an individual example i, with respect to each feature weight w_j and bias term b.

---

[6] Technically, C_i is a loss function because it applies to a single data point. However, we will continue to use the term "cost function" for readability.

For these operations, we will rely on several rules to compute the derivatives of a few necessary functions. These rules are listed in Table 3.1.

Let us start with the derivative of C with respect to one feature weight w_j:

$$\frac{d}{dw_j}C_i(\mathbf{w}, b) = \frac{d}{dw_j}(-y_i \log \sigma(\mathbf{x}_i; \mathbf{w}, b) - (1 - y_i) \log(1 - \sigma(\mathbf{x}_i; \mathbf{w}, b))).$$

Let us use σ_i to denote $\sigma(\mathbf{x}_i; \mathbf{w}, b)$, for simplicity:

$$= \frac{d}{dw_j}(-y_i \log \sigma_i - (1 - y_i) \log(1 - \sigma_i)).$$

Pulling out the y_i constants and then applying the chain rule on the two logarithms:

$$= -y_i \frac{d}{d\sigma_i} \log \sigma_i \frac{d}{dw_j}\sigma_i - (1 - y_i)\frac{d}{d(1-\sigma_i)} \log(1-\sigma_i)\frac{d}{dw_j}(1-\sigma_i).$$

After applying the derivative of the logarithm:

$$= -y_i \frac{1}{\sigma_i} \frac{d}{dw_j}\sigma_i - (1 - y_i)\frac{1}{1 - \sigma_i} \frac{d}{dw_j}(1 - \sigma_i).$$

After applying the chain rule on $\frac{d}{dw_j}(1 - \sigma_i)$:

$$= -y_i \frac{1}{\sigma_i} \frac{d}{dw_j}\sigma_i + (1 - y_i)\frac{1}{1 - \sigma_i} \frac{d}{dw_j}\sigma_i$$

$$= (-y_i\frac{1}{\sigma_i} + (1 - y_i)\frac{1}{1 - \sigma_i})\frac{d}{dw_j}\sigma_i$$

$$= \frac{-y_i(1 - \sigma_i) + (1 - y_i)\sigma_i}{\sigma_i(1 - \sigma_i)}\frac{d}{dw_j}\sigma_i$$

$$= \frac{\sigma_i - y_i}{\sigma_i(1 - \sigma_i)}\frac{d}{dw_j}\sigma_i.$$

After applying the chain rule on σ_i:

$$= \frac{\sigma_i - y_i}{\sigma_i(1 - \sigma_i)}\frac{d}{d(\mathbf{w} \cdot \mathbf{x}_i + b)}\sigma_i\frac{d}{dw_j}(\mathbf{w} \cdot \mathbf{x}_i + b).$$

After the derivative of the logistic function and then canceling numerator and denominator:

$$= \frac{\sigma_i - y_i}{\sigma_i(1 - \sigma_i)}\sigma_i(1 - \sigma_i)\frac{d}{dw_j}(\mathbf{w} \cdot \mathbf{x}_i + b)$$

$$= (\sigma_i - y_i)\frac{d}{dw_j}(\mathbf{w} \cdot \mathbf{x}_i + b).$$

Table 3.1 Rules of computation for a few functions necessary to derive the logistic regression update rules. In these formulas, f and g are functions, a and b are constants, and x is a variable

Description	Formula
Chain rule	$\frac{d}{dx}f(g(x)) = \frac{d}{dg(x)}f(g(x))\frac{d}{dx}g(x)$
Derivative of summation	$\frac{d}{dx}(af(x) + bg(x)) = a\frac{d}{dx}f(x) + b\frac{d}{dx}g(x)$
Derivative of natural logarithm	$\frac{d}{dx}\log(x) = \frac{1}{x}$
Derivative of logistic	$\frac{d}{dx}\sigma(x) = \frac{d}{dx}(\frac{1}{1+e^{-x}}) = -\frac{1}{(1+e^{-x})^2}(-e^{-x})$
	$= \sigma(x)(1 - \sigma(x))$
Derivative of dot product between vectors $\mathbf{x}$ and $\mathbf{a}$ with respect to x_i	$\frac{d}{dx_i}(\mathbf{x} \cdot \mathbf{a}) = a_i$

Last, after applying the derivative of the dot product:

$$= (\sigma_i - y_i)x_{ij}, \tag{3.14}$$

where x_{ij} is the value of feature j in the feature vector $\mathbf{x_i}$.

Following a similar process, we can compute the derivative of C_i with respect to the bias term as:

$$\frac{d}{db}C_i(\mathbf{w}, b) = \frac{d}{db}(-y_i \log \sigma(\mathbf{x}_i; \mathbf{w}, b) - (1 - y_i)\log(1 - \sigma(\mathbf{x}_i; \mathbf{w}, b))) = \sigma_i - y_i. \tag{3.15}$$

Knowing that σ_i is equivalent with decision($\mathbf{x}_i$, $\mathbf{w}$, b), we can immediately see that applying Equation 3.15 in line 8 of Algorithm 5 transforms the update of the bias into the form used in Algorithm 4 (line 7). Similarly, replacing the partial derivative in line 6 of Algorithm 5 with its explicit form from Equation 3.14 yields an update equivalent with the weight update used in Algorithm 4. The superficial difference between the two algorithms is that Algorithm 5 updates each feature weight w_j explicitly, whereas Algorithm 4 updates *all* weights at once by updating the entire vector $\mathbf{w}$. Needless to say, these two forms are equivalent. We prefer the explicit description in Algorithm 5 for clarity. But, in practice, one is more likely to implement Algorithm 4 because vector operations are efficiently implemented in most ML software libraries.

3.5   From Binary to Multiclass Classification

So far, we have discussed binary LR, where we learned a classifier for two labels (1 and 0), where the probability of predicting label 1 is computed as:

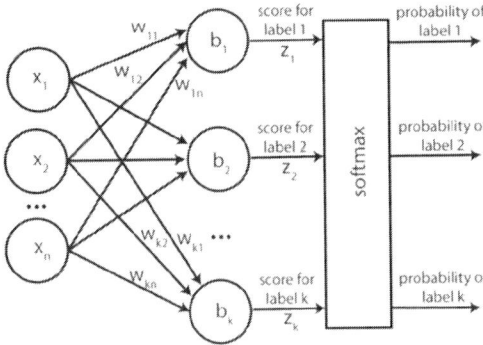

Figure 3.4 Multiclass logistic regression

$p(1|\mathbf{x}; \mathbf{w}, b) = \sigma(\mathbf{x}; \mathbf{w}, b)$ and the probability of label 0 is: $p(0|\mathbf{x}; \mathbf{w}, b) = 1 - p(1|\mathbf{x}; \mathbf{w}, b) = 1 - \sigma(\mathbf{x}; \mathbf{w}, b)$. However, there are many text classification problems where two labels are not sufficient. For example, we might decide to implement a movie review classifier that produces five labels in order to capture ratings on a five-star scale. To accommodate this class of problems, we need to generalize the binary LR algorithm to multiclass scenarios, where the labels to be learned may take values from 1 to k, where k is the number of classes to be – for example, 5 in the previous example.

Figure 3.4 provides a graphical explanation of the multiclass LR. The key observation is that now, instead of maintaining a single weight vector $\mathbf{w}$ and bias b, we maintain one such vector and bias term *for each* class to be learned. Intuitively, this architecture is a merger of multiple "neurons," one for each class. This complicates our notations a bit: instead of using a single index to identify positions in an input vector $\mathbf{x}$ or in $\mathbf{w}$, we now have to maintain two. That is, we will use $\mathbf{w}_i$ to indicate the weight vector for class i, w_{ij} to point to the weight of the edge that connects the input x_j to the class i, and b_i to indicate the bias term for class i. The output of each "neuron" i in the figure produces a score for label i, defined as the sum between the bias term of class i and the dot product of the weight vector for class i and the input vector. More formally, if we use z_i to indicate the score for label i, then $z_i = \mathbf{w}_i \cdot \mathbf{x} + b_i$.

Note that these scores are not probabilities: they are not bounded between 0 and 1, and they will not sum up to 1. To turn them into probabilities, we are introducing a new function, called softmax, which produces probability values for the k classes. For each class i, softmax defines the corresponding probability as:

$$p(y = i|\mathbf{x}; \mathbf{W}, \mathbf{b}) = p(i|\mathbf{x}; \mathbf{W}, \mathbf{b}) = \frac{e^{z_i}}{\sum_{j=1}^{k} e^{z_j}} = \frac{e^{\mathbf{w}_i \cdot \mathbf{x} + b_i}}{\sum_{j=1}^{k} e^{\mathbf{w}_j \cdot \mathbf{x} + b_j}}, \qquad (3.16)$$

where the $\mathbf{W}$ matrix stores all $\mathbf{w}$ weight vectors – that is, row i in $\mathbf{W}$ stores the weight vector $\mathbf{w}_i$ for class i, and the $\mathbf{b}$ vector stores all bias values – that is, b_i is the bias term for class i.

Clearly, the softmax function produces probabilities: (a) the exponent function used guarantees that the softmax values are positives, and (b) the denominator, which sums over all the k classes, guarantees that the resulting values are between 0 and 1 and sum up to 1. Further, with just a bit of math, we can show that the softmax for two classes reduces to the logistic function. Using the softmax formula, the probability of class 1 in a two-class LR (using labels 1 and 0) is:

$$p(1|\mathbf{x}; \mathbf{W}, \mathbf{b}) = \frac{e^{\mathbf{w}_1 \cdot \mathbf{x} + b_1}}{e^{\mathbf{w}_0 \cdot \mathbf{x} + b_0} + e^{\mathbf{w}_1 \cdot \mathbf{x} + b_1}} = \frac{1}{\frac{e^{\mathbf{w}_0 \cdot \mathbf{x} + b_0}}{e^{\mathbf{w}_1 \cdot \mathbf{x} + b_1}} + 1}$$

$$= \frac{1}{e^{-((\mathbf{w}_1 - \mathbf{w}_0) \cdot \mathbf{x} + (b_1 - b_0))} + 1}. \tag{3.17}$$

Using a similar derivation, which we leave as an at-home exercise for the curious reader, the probability of class 0 is:

$$p(0|\mathbf{x}; \mathbf{W}, \mathbf{b}) = \frac{e^{\mathbf{w}_0 \cdot \mathbf{x} + b_0}}{e^{\mathbf{w}_0 \cdot \mathbf{x} + b_0} + e^{\mathbf{w}_1 \cdot \mathbf{x} + b_1}} = \frac{e^{-((\mathbf{w}_1 - \mathbf{w}_0) \cdot \mathbf{x} + (b_1 - b_0))}}{e^{-((\mathbf{w}_1 - \mathbf{w}_0) \cdot \mathbf{x} + (b_1 - b_0))} + 1}$$

$$= 1 - p(1|\mathbf{x}; \mathbf{W}, \mathbf{b}). \tag{3.18}$$

From these two equations, we can immediately see that the two formulations of binary LR – that is, logistic versus softmax – are equivalent when we set the parameters of the logistic to be equal to the difference between the parameters of class 1 and the parameters of class 0 in the softmax formulation, or $\mathbf{w} = \mathbf{w}_1 - \mathbf{w}_0$, and $b = b_1 - b_0$, where $\mathbf{w}$ and b are the logistic parameters in Equations 3.17 and 3.18.

The cost function for multiclass LR follows the same intuition and formalization as the one for binary LR. That is, during training, we want to maximize the probabilities of the correct labels assigned to training examples, or, equivalently, we want to minimize the negative log likelihood of the data. Similarly to Equation 3.7, the cost function for multiclass LR is defined as:

$$C(\mathbf{W}, \mathbf{b}) = -\log L(\mathbf{W}, \mathbf{b}) = -\sum_{i=1}^{m} \log p(y_i|\mathbf{x}_i; \mathbf{W}, \mathbf{b}), \tag{3.19}$$

or, for a single training example i:

$$C_i(\mathbf{W}, \mathbf{b}) = -\log p(y_i|\mathbf{x}_i; \mathbf{W}, \mathbf{b}), \tag{3.20}$$

where y_i is the correct label for training example i, and $\mathbf{x}_i$ is the feature vector for the same example. The probabilities in this cost function are computed using the softmax formula, as in Equation 3.16. This cost function, which generalizes the negative log likelihood cost function to multiclass classification, is

---

Algorithm 6 Learning algorithm for multiclass logistic regression

---

1   $\mathbf{W} = 0$

2   $\mathbf{b} = 0$

3   **while** *not converged* **do**

4   **for** *each training example x_i* **in X do**

5   **for** *each w_{jk} in* **W do**

6             $w_{jk} = w_{jk} - \alpha \frac{d}{dw_{jk}} C_i(\mathbf{W}, \mathbf{b})$

7   **end**

8   **for** *each b_j in* **b do**

9             $b_j = b_j - \alpha \frac{d}{db_j} C_i(\mathbf{W}, \mathbf{b})$

10   **end**

11   **end**

12 **end**

---

called *cross-entropy*. Its form for binary classification is called *binary cross-entropy*. Using Equations 3.17 and 3.18, it is easy to show that in the case of binary LR, Equation 3.19 is equivalent with our initial cost function from Equation 3.8. These are probably the most commonly used cost functions in NLP problems. We will see them a lot throughout the book.

The learning algorithm for multiclass LR stays almost the same as Algorithm 5, with small changes to account for the different cost function and the larger number of parameters – that is, we now update a matrix **W** instead of a single vector **w**, and a vector **b** instead of the scalar *b*. The adjusted algorithm is shown in Algorithm 6. We leave the computation of the derivatives used in Algorithm 6 as an at-home exercise for the interested reader. However, as we will see in the next chapter, we can now rely on automatic differentiation libraries such as PyTorch to compute these derivatives for us, so this exercise is not strictly needed to implement multiclass LR.

3.6   Evaluation Measures for Multiclass Text Classification

Now that we generalized our classifier to operate over an arbitrary number of classes, it is time to generalize the evaluation measures introduced in Section 2.3 to multiclass problems as well. Throughout this section, we will use as a walkthrough example a three-class version of the Medicaid application classification problem from Section 2.3. In this version, our classifier has to assign each application to one of three classes, where classes $C1$ and $C2$ indicate the high- and medium-priority applications and class $C3$ indicates regular applications that do not need to be rushed through the system. Same as before, most applications fall under class $C3$. Table 3.2 shows an example confusion matrix

Table 3.2 Example of a confusion matrix for three-class classification. The dataset contains 1,000 data points, with 2 data points in class $C1$, 100 in class $C2$, and 898 in class $C3$

	Classifier predicted $C1$	Classifier predicted $C2$	Classifier predicted $C3$
Gold label is $C1$	1	1	0
Gold label is $C2$	10	80	10
Gold label is $C3$	1	7	890

for this problem for a hypothetical three-class classifier that operates over an evaluation dataset that contains 1,000 applications.

The definition of accuracy remains essentially the same for multiclass classification – that is, accuracy is the ratio of data points classified correctly. In general, the number of correctly classified points can be computed by summing up the counts on the diagonal of the confusion matrix. For example, for the confusion matrix shown in Table 3.2, accuracy is $\frac{1+80+890}{1,000} = \frac{971}{1,000}$.

Similarly, the definitions of precision and recall for an individual class c, remain the same:

$$P_c = \frac{TP_c}{TP_c + FP_c} \qquad (3.21)$$

$$R_c = \frac{TP_c}{TP_c + FN_c}, \qquad (3.22)$$

where TP_c indicates the number of true positives for class c, FP_c indicates the number of positives for class c, and FN_c indicates the number of false negatives for the same class. However, because we now have more than two rows and two columns in the confusion matrix, we have to do a bit more math to compute the FP_c and FN_c counts. In general, the number of false positives for a class c is equal to the sum of the counts in the column corresponding to class c, excluding the element on the diagonal. The number of false negatives for a class c is equal to the sum of the counts in the corresponding row, excluding the element on the diagonal. For example, for class $C2$ in the table, the number of true positives is $TP_{C2} = 80$, the number of false positives is $FP_{C2} = 1+7 = 8$, and the number of false negatives is $FN_{C2} = 10+10 = 20$. Thus, the precision and recall for class $C2$ are: $P_{C2} = \frac{80}{80+8} = 0.91$ and $R_{C2} = \frac{80}{80+20} = 0.80$. We leave it as an at-home exercise to show that $P_{C1} = 0.08$, $R_{C1} = 0.5$, $P_{C3} = 0.99$, and $R_{C3} = 0.99$. From these values, one can trivially compute the respective F scores per class.

The important discussion for multiclass classification is how to average these sets of precision/recall scores into single values that will give us a quick understanding of the classifier's performance. There are two strategies to this end, both with advantages and disadvantages:

Macro averaging: Under this strategy, we simply average all the individual precision/recall scores into a single value. In our example, the macro precision score over all three classes is: macro $P = \frac{P_{C1}+P_{C2}+P_{C3}}{3} = \frac{0.08+0.91+0.99}{3} =$ 0.66. Similarly, the macro recall score is: macro $R = \frac{R_{C1}+R_{C2}+R_{C3}}{3} = \frac{0.50+0.80+0.99}{3} = 0.76$. The macro F_1 score is the harmonic mean of the macro precision and recall scores.

As discussed in Section 2.3, in many NLP tasks the labels are highly unbalanced, and we commonly care less about the most frequent label. For example, here, we may want to measure the performance of our classifier on classes $C1$ and $C2$, which requires rushed processing in the Medicaid system. In such scenarios, the macro precision and recall scores exclude the frequent class – for example, $C3$ in our case. Thus, the macro precision becomes: macro $P = \frac{P_{C1}+P_{C2}}{2} = \frac{0.08+0.91}{2} = 0.50$, which is more indicative of the fact that our classifier does not perform too well on the two important classes in this example.

The advantage of the macro scores is that they treat all the classes we are interested in as equal contributors to the overall score. But, depending on the task, this may also be a disadvantage. In our example, the latter macro precision score of 0.50 hides the fact that our classifier performs reasonably well on the $C2$ class ($P_{C2} = 0.91$), which is 100 times more frequent than $C1$ in the data!

Micro averaging: This strategy addresses the disadvantage of macro averaging by computing overall precision, recall, and F scores where each class contributes proportionally to its frequency in the data. In particular, rather than averaging the individual precision/recall scores, we compute them using the class counts directly. The micro precision and recall scores for the two classes of interest in the example, $C1$ and $C2$, are:

$$\text{micro } P = \frac{TP_{C1}+TP_{C2}}{TP_{C1}+TP_{C2}+FP_{C1}+FP_{C2}}$$
$$= \frac{1+80}{1+80+11+8} = 0.81 \tag{3.23}$$
$$\text{micro } R = \frac{TP_{C1}+TP_{C2}}{TP_{C1}+TP_{C2}+FN_{C1}+FN_{C2}}$$
$$= \frac{1+80}{1+80+1+20} = 0.79. \tag{3.24}$$

Similar to macro averaging, the micro F_1 score is computed as the harmonic mean of the micro precision and recall scores.

Note that in this example, the micro scores are considerably higher than the corresponding macro scores because: (a) the classifier's performance on the more frequent $C2$ class is higher than the performance on class $C1$, and (b) micro averaging assigns more importance to the frequent classes, which, in this case, raises the micro precision and recall scores. The decision of which averaging strategy to use is problem specific, and depends on the answer to the question: should all classes be treated equally during scoring, or should they be weighted by their frequency in the data? In the former case, the appropriate averaging is macro; in the latter, micro.

3.7   Drawbacks of Logistic Regression

The LR algorithm solves the lack of smooth updates in the perceptron algorithm through its improved update functions on its parameters. This seemingly small change has an important practical impact: in most NLP applications, LR tends to outperform the perceptron.

However, the other drawbacks observed with the perceptron still hold. Binary LR is also a linear classifier because its decision boundary remains a hyperplane. It is tempting to say this statement is not correct because the logistic is clearly a nonlinear function. However, the linearity of the binary LR classifier is easy to prove with just a bit of math. Remember that the decision function for the binary LR is: if $\frac{1}{1+e^{-(\mathbf{w}\cdot\mathbf{x}+b)}} \geq 0.5$ we assign one label, and if $\frac{1}{1+e^{-(\mathbf{w}\cdot\mathbf{x}+b)}} < 0.5$ we assign the other label. Thus, the decision boundary is defined by the equation $\frac{1}{1+e^{-(\mathbf{w}\cdot\mathbf{x}+b)}} = 0.5$. From this we can easily derive that $e^{-(\mathbf{w}\cdot\mathbf{x}+b)} = 1$, and $-(\mathbf{w} \cdot \mathbf{x} + b) = 0$, where the latter is a linear function on the parameters $\mathbf{w}$ and b. This observation generalizes to the multiclass LR introduced in Section 3.5. In the multiclass scenario, the decision boundary between all classes consists of multiple intersecting segments, each of which are fragments of a hyperplane. Figure 3.5 shows an example of such a decision boundary for a four-class problem, where each data point is described by two features: x_1 and x_2.[7] Clearly, forcing these segments to be linear reduces what the multiclass LR can learn.

Further, similar to the perceptron, the LR covered so far relies on hand-crafted features, which, as discussed in the previous chapter, may be cumbersome to generate and may generalize poorly. Last, LR also focuses on individual predictions rather than structured learning. We will address all these limitations in the following chapters. We will start by introducing nonlinear classifiers in Chapter 5.

3.8   Historical Background

The logistic function was discovered twice (see Cramer, 2002 for a fascinating history). In brief, Pierre François Verhulst introduced the logistic function

---

[7] This figure was generated by Clayton Morrison and is reproduced with permission.

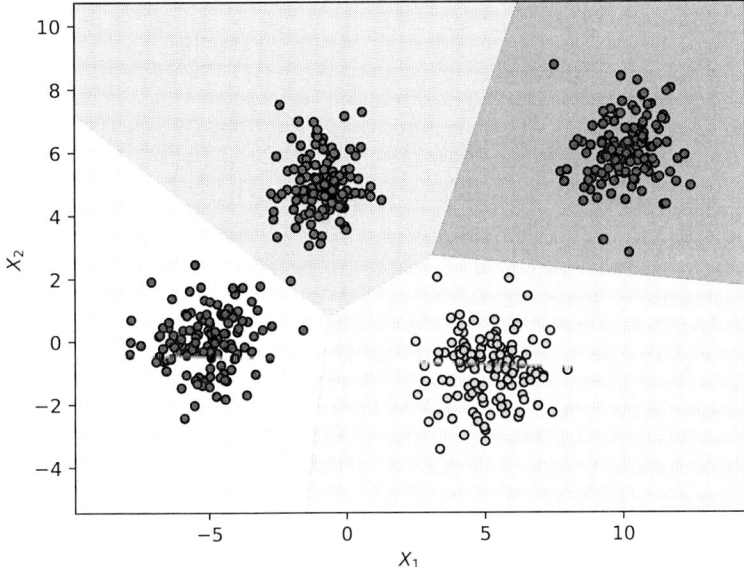

Figure 3.5 Example of a two-dimensional decision boundary for a four-class logistic regression classifier

(with this name) in the nineteenth century in the context of estimating population growth (Verhulst, 1838, 1845). Pearl and Reed rediscovered the logistic function in a 1920 study of population growth in the United States (Pearl and Reed, 1920). They were apparently unaware of Verhulst's precedent. In a follow-up publication, Reed, who was trained as a mathematician and biostatistician, applied the logistic function to autocatalytic reactions (Reed and Berkson, 1929). Once the repeated discovery was identified, Verhulst was rightfully credited for both the formula and the name (Yule, 1925).

   The gradient descent algorithm we used to train the LR classifier in this chapter was also discovered multiple times through history (Kelley, 1960; Dreyfus, 1990, 1962, inter alia). We will revisit this history in Chapter 5, where we will introduce a generalization of this algorithm to networks with an arbitrary number of neurons.

3.9   References and Further Readings

Because of its ubiquity, LR is described in many statistics and ML books and courses. The one that helped us the most is Andrew Ng's CS229 ML course at Stanford University (Ng, 2019).

3.10   Summary

This chapter introduced LR, which improves upon the perceptron by performing soft updates during training – that is, each parameter is updated based on its contribution to an incorrect decision. We also extended the vanilla LR, which was designed for binary classification, to handle multiclass classification.

Through LR, we introduced the concepts of cost function (i.e., the function we aim to minimize during training) and gradient descent (i.e., the algorithm that implements this minimization procedure).

4 Implementing Text Classification Using Perceptron and Logistic Regression

In the previous chapters, we have discussed the theory behind the perceptron and LR, including mathematical explanations of how and why they can learn from examples. In this chapter, we will transition from math to code. Specifically, we will discuss how to implement these models in the Python programming language. All the code that we will introduce throughout this book is available online as well: `http://clulab.github.io/gentlenlp`. The reader who is not familiar with the Python programming language is encouraged to read first Appendix A, for a brief introduction to the language, and Appendix B, for a discussion on how computers encode and preprocess text. Once done, please return here.

To get a better understanding of how these algorithms work under the hood, we will start by implementing them from scratch. However, as the book progresses, we will introduce some of the popular tools and libraries that make Python the language of choice for machine learning – for example, PyTorch[1] and Hugging Face's transformers.[2]

The code for all the examples in the book is provided in the form of Jupyter notebooks.[3] Important fragments of these notebooks will be presented in the implementation chapters so that the reader has the whole picture just by reading the book. However, we strongly encourage you to download the notebooks and execute them yourself. We also encourage you to modify them to conduct your own experiments!

4.1 Binary Classification

We begin this chapter with binary classification. That is, we aim to train classifiers that assign one of two labels to a given text. As the example for this task, we will train a review classifier using the Large Movie Review Dataset (Maas et al., 2011).[4] We tackle this task by implementing first a binary perceptron classifier, followed by a binary LR classifier. We will implement the latter from

---

[1] `https://pytorch.org`.

[2] `https://huggingface.co`.

[3] `https://jupyter.org`.

[4] `https://ai.stanford.edu/~amaas/data/sentiment`.

scratch as well as using PyTorch, so the reader has a clearer understanding of how PyTorch works.

4.1.1   Large Movie Review Dataset

This dataset contains movie reviews and their associated scores (between 1 and 10) as provided by IMDb.[5] Maas et al. converted these scores to binary labels by assigning each review a positive or negative label if the review score was above 6 or below 5, respectively. Reviews with scores 5 and 6 were considered too neutral and thus excluded. We follow the same protocol in this chapter.

The dataset is divided into two even partitions called *train* and *test*, each containing 25,000 reviews. The dataset also provides additional unlabeled reviews, but we will not use those here. Each partition contains two directories called pos and neg where the positive and negative examples are stored. Each review is stored in an independent text file, whose name is composed of an ID unique to the partition and the score associated with the review, separated by an underscore. An example of a positive and a negative review is shown in Table 4.1.

4.1.2   Bag-of-Words Model

As discussed in Section 2.2, we will encode the text to classify as a *bag of words*. That is, we encode each review as a list of numbers, with each position in the list corresponding to a word in our vocabulary, and the value stored in that position corresponding to the number of times the word appears in the review. For example, say we want to encode the following two reviews:

```
Review 1:  "I liked the movie. My friend liked it too."
Review 2:  "I hated it. Would not recommend."
```

First, we need to create a vocabulary that maps each word to a unique ID. Each of these numbers will be used as the index in a list, so they must start at zero and grow by one for each word in the vocabulary. For example, one possible vocabulary that encodes the previous reviews is:

```
{'would': 0,
 'hated': 1,
 'my': 2,
 'liked': 3,
 'not': 4,
 'it': 5,
```

[5] https://www.imdb.com.

Table 4.1 Two examples of movie reviews from IMDb. The first is a positive review of the movie *Puss in Boots (1988)*. The second is a negative review of the movie *Valentine (2001)*. These reviews can be found at www.imdb.com/review/rw0606396 and www.imdb.com/review/rw0721861, respectively

Filename	Score	Binary label	Review text
train/pos/24_8.txt	8/10	Positive	*Although this was obviously a low-budget production, the performances and the songs in this movie are worth seeing. One of Walken's few musical roles to date. (he is a marvelous dancer and singer and he demonstrates his acrobatic skills as well – watch for the cartwheel!) Also starring Jason Connery. A great children's story and very likable characters.*
train/neg/141_3.txt	3/10	Negative	*This stalk and slash turkey manages to bring nothing new to an increasingly stale genre. A masked killer stalks young, pert girls and slaughters them in a variety of gruesome ways, none of which are particularly inventive. It's not scary, it's not clever, and it's not funny. So what was the point of it?*

```
'movie': 6,
'recommend': 7,
'the': 8,
'I': 9,
'too': 10,
'friend': 11}
```

Using this mapping, we can encode the two reviews as follows:

Review 1: [0, 0, 1, 2, 0, 1, 1, 0, 1, 1, 1, 1]
Review 2: [1, 1, 0, 0, 1, 1, 0, 1, 0, 1, 0, 0]

Note that the word *liked* (fourth position) in the first review has a value of two. This is because this word appears twice in that review.

This is a small example with a vocabulary of only 12 terms. Of course, the same process needs to be implemented for our whole training dataset. For

this purpose, we will use scikit-learn's CountVectorizer class.[6] Using the
CountVectorizer class simplifies things, allowing us to get started quickly
with a bag-of-words approach. However, note that it makes several simplifying assumptions (e.g., text is lowercased and punctuation and single-character
tokens are removed). Some of these may not be adequate to other tasks.

First, we need to obtain the file names for the reviews in the training set:

```
[4]: from glob import glob

     pos_files = glob('data/aclImdb/train/pos/*.txt')
     neg_files = glob('data/aclImdb/train/neg/*.txt')

     print('number of positive reviews:', len(pos_files))
     print('number of negative reviews:', len(neg_files))
```

```
number of positive reviews: 12500
number of negative reviews: 12500
```

Once we have acquired the file names for the training reviews, we need to
read them using the CountVectorizer. In order for the CountVectorizer
to open and read the files for us, we make use of the input='filename' constructor parameter (otherwise it would expect the string content directly). The
CountVectorizer provides three methods that will be useful for us: a method
called fit() that is used to acquire the vocabulary, a method transform()
that converts the text into the bag-of-words representation, and a method
fit_transform() that conveniently acquires the vocabulary and transforms
the data in a single step. The resulting object is referred to as a *document-
term matrix*, where each row corresponds to a document and each column
corresponds to a term in the vocabulary.

```
[5]: from sklearn.feature_extraction.text import CountVectorizer

     cv = CountVectorizer(input='filename')

     doc_term_matrix = cv.fit_transform(pos_files + neg_files)
     doc_term_matrix
```

```
[5]: <25000x74849 sparse matrix of type '<class'numpy.int64'>'
          with 3445861 stored elements in Compressed Sparse Row format>
```

---

[6] https://scikit-learn.org/stable/modules/generated/sklearn.feature_
extraction.text.CountVectorizer.html.

As the output indicates, the resulting matrix has 25,000 rows (one for each review) and 74,849 columns (one for each term). Also, you may note that this matrix is *sparse*, with 3,445,861 stored elements. A regular matrix of shape $25,000 \times 74,849$ would have 1,871,225,000 elements. However, most of the elements in a document-term matrix are zeros because only a few words from the vocabulary appear in each document. A sparse matrix takes advantage of this fact by storing only the nonzero cells in order to reduce the memory required to store it. Thus, sparse matrices are convenient, especially when dealing with lots of data. Nevertheless, to simplify the downstream code in this example, we will convert it into a dense matrix – that is, a regular two-dimensional NumPy array.

```
[6]:  X_train = doc_term_matrix.toarray()
      X_train.shape
```

```
[6]:  (25000, 74849)
```

Finally, we also need the labels of the reviews. We assign a label of 1 to positive reviews and a label of 0 to negative ones. Note that the reviews in the first half are positive and those in the second half are negative. The label at the *i*th position of the y_train array corresponds to the review encoded in the *i*th row of the X_train matrix.

```
[7]:  # training labels
      y_pos = np.ones(len(pos_files))
      y_neg = np.zeros(len(neg_files))
      y_train = np.concatenate([y_pos, y_neg])
      y_train
```

```
[7]:  array([1., 1., 1., ..., 0., 0., 0.])
```

4.1.3   Perceptron

Now that we have defined our task and the data-processing pipeline, we will implement a perceptron classifier that classifies the movie reviews as positive or negative. The entire code discussed in this section is available in the chap4_perceptron notebook. Recall from Section 2.4 that the perceptron is composed of a weight vector **w** and a bias term *b*. These will be represented as a NumPy array w of the same length as our document vectors, and a variable b for the bias term. Both will be initialized with zeros.

```
[8]: n_examples, n_features = X_train.shape
     w = np.zeros(n_features)
     b = 0
```

The parameters **w** and *b* are learned through the following algorithm, which implements Algorithm 2 from Chapter 2:

```
[9]: n_epochs = 10

     indices = np.arange(n_examples)
     for epoch in range(10):
         n_errors = 0
         # shuffle training examples
         np.random.shuffle(indices)
         # traverse the training data
         for i in tqdm(indices, desc=f'epoch {epoch+1}'):
             x = X_train[i]
             y_true = y_train[i]
             # the perceptron decision based on the current model
             score = x @ w + b
             y_pred = 1 if score > 0 else 0
             # update the model is the prediction was incorrect
             if y_true == y_pred:
                 continue
             elif y_true == 1 and y_pred == 0:
                 w = w + x
                 b = b + 1
                 n_errors += 1
             elif y_true == 0 and y_pred == 1:
                 w = w - x
                 b = b - 1
                 n_errors += 1
         if n_errors == 0:
             break
```

There are a couple of details to point out. Line 3 of Algorithm 2 indicates that we need to repeat the training loop until convergence. Theoretically, convergence is defined as predicting all training examples correctly. This is an ambitious requirement, which is not always possible in practice, so in this code we also include a stop condition if we reach a maximum number of epochs. Another crucial difference between our implementation here and the theoretical Algorithm 2 is that we randomize the order in which the training examples are seen at the beginning of each epoch. This simple (but highly recommended!) change is necessary to avoid the introduction of spurious biases due to the arbitrary order of the examples in the original training partition.[7] We accomplish this by storing the indices corresponding to the X_train matrix rows in a NumPy array and shuffling these indices at the beginning of each epoch. We shuffle the indices instead of the examples so that we can preserve the mapping between examples and labels.

The training loop aligns closely with Algorithm 2. We start by iterating over each example in our training data, storing the current example in the variable x, and its corresponding label in the variable y_true.[8] Next, we compute the perceptron decision function shown in Algorithm 1. Note that NumPy (as well as PyTorch) uses Python's @ operator to indicate vector or matrix multiplication, depending on its operand types. Here we use it to calculate the dot product of the example x and the weights w. To this we add the bias b to obtain the predicted score, whose sign is used to assign a positive or negative predicted label. If the prediction is correct, then no update is needed and we can move on to the next training example. However, if the prediction is incorrect, then we need to adjust w and b, as described in Algorithm 2.

Sidebar 4.1   The tqdm function

This is our first exposure to the tqdm function. tqdm is a progress bar that "make your loops show a smart progress meter."[9] The name tqdm comes from the Arabic word *taqaddum*, which can mean progress. Using tqdm is as simple as wrapping it around the collection to be traversed.

After training, we evaluate the model's performance on the held-out test partition. The test data are loaded similarly to the training partition, but with one notable difference; we use CountVectorizer's transform() method instead of the fit_transform() method so that the vocabulary is not adjusted for the test data. We won't show here the loading of the test partition since it is so similar to the code already shown, but it is available in the Jupyter notebook that accompanies this section.

Using the model to assign labels to all the test data is easily done in one step – we simply multiply the entire test data document-term matrix by the previously learned weights and add the bias. Scores greater than zero indicate a positive review and those less than zero are negative.

---

[7] As an extreme example, consider a dataset where all the positive examples appear first in the training partition. This would cause the perceptron to artificially inflate the weights of the features that occur in these examples, a situation from which the learning algorithm may struggle to recover.

[8] We use typewriter font when we discuss variables in the code, to distinguish code from the theoretical discussion in the other chapters.

[9] https://github.com/tqdm/tqdm.

```
[11]:  y_pred = (X_test @ w + b) > 0
```

At this point, we can evaluate the classifier's performance, which we will do using precision, recall, and F_1 scores for binary classification (described in Section 2.3). For this purpose, we implement a function called `binary_classification_report` that computes these metrics and returns them as a dictionary:

```
[12]:  def binary_classification_report(y_true, y_pred):
           # count true positives, false positives,
           # true negatives, and false negatives
           tp = fp = tn = fn = 0
           for gold, pred in zip(y_true, y_pred):
               if pred == True:
                   if gold == True:
                       tp += 1
                   else:
                       fp += 1
               else:
                   if gold == False:
                       tn += 1
                   else:
                       fn += 1
           # calculate precision and recall
           precision = tp / (tp + fp)
           recall = tp / (tp + fn)
           # calculate f1 score
           fscore = 2 * precision * recall / (precision + recall)
           # calculate accuracy
           accuracy = (tp + tn) / len(y_true)
           # number of positive labels in y_true
           support = sum(y_true)
           return {
               "precision": precision,
               "recall": recall,
               "f1-score": fscore,
               "support": support,
               "accuracy": accuracy,
           }
```

We call this function to compare the predicted labels to the true labels and obtain the evaluation scores.

```
[13]:  binary_classification_report(y_test, y_pred)
```

```
[13]:  {'precision': 0.8288497891522466,
        'recall': 0.912,
        'f1-score': 0.8684390949950485,
        'support': 12500.0,
        'accuracy': 0.86184}
```

Our F_1 score here is 86.8%, which is much higher than the baseline that assigns labels randomly, which yields an F_1 score of about 50%. This is a good result, especially considering the simplicity of the perceptron! In the next sections and chapters, we will discuss a battery of strategies to considerably improve this performance.

4.1.4 Binary Logistic Regression from Scratch

Using the same task, dataset, and evaluation, we will now implement a logistic regression classifier, as described in Algorithm 5 from Chapter 3. To give the reader hands-on experience with the implementation of the gradient calculations for logistic regression, we start by implementing it from scratch using NumPy. All the code shown in this section is available in the chap4_logistic_regression_numpy notebook.

In the perceptron implementation, we represented the weights and the bias as two different variables. Here, however, we will use a different approach that will allow us to unify them into a single vector variable. Specifically, we take advantage of the similarity between the derivative of the cost function with respect to the weights (Equation 3.14) and the derivative of the cost with respect to the bias (Equation 3.15).

$$\frac{d}{dw_j} C_i(\mathbf{w}, b) = (\sigma_i - y_i)x_{ij} \qquad \text{(3.14 revisited)}$$

$$\frac{d}{db} C_i(\mathbf{w}, b) = \sigma_i - y_i. \qquad \text{(3.15 revisited)}$$

Note that the two derivative formulas are identical except that the former has a multiplication by x_{ij}, while the latter does not. However, since

$$\sigma_i - y_i = (\sigma_i - y_i)1,$$

we can multiply the derivative of the cost with respect to the bias by one without changing the semantics. This gives an opportunity for combining the computations, doing them both in a single pass. The idea is that we can treat the bias as a weight corresponding to a feature that always has a value of one.

```
[7]:  # Make an array with a one for each row/data point
       ones = np.ones(X_train.shape[0])
```

```
# Concatenate these ones to existing feature vectors
X_train = np.column_stack((X_train, ones))
X_train.shape
```

[7]: (25000, 74850)

As can be seen, we created a NumPy array of ones of the same length as the number of examples in our training set (i.e., the number of rows in the data matrix). Then we add this array as a new column to the data matrix, using NumPy's column_stack function.

Next, we need to initialize our model. This time we will use a single NumPy array w of the same length as the number of columns in the data matrix. The weight vector w is initialized randomly with values between 0 and 1:

[9]:
```
n_examples, n_features = X_train.shape
w = np.random.random(n_features)
```

Before implementing the learning algorithm, we need an implementation of the *logistic* function. Recall that the logistic function is

$$\sigma(x) = \frac{1}{1 + e^{-x}}. \hspace{2cm} \text{(3.1 revisited)}$$

This function can be easily implemented in NumPy as follows:

[10]:
```
def logistic(x):
    return 1 / (1 + np.exp(-x))
```

However, this naive implementation may produce the following warning during training:

```
RuntimeWarning: overflow encountered in exp
    return 1 / (1 + np.exp(-x))
```

The term *overflow* indicates that the result of evaluating exp(-x) is a number so large that it can't be represented by a float (specifically, we're using float64 numbers). We will avoid this issue by not calling exp with values that will overflow. NumPy provides the function finfo that can be consulted to find the limits of floating point numbers:

[11]: np.finfo(np.float64)

```
[11]: finfo(resolution=1e-15, min=-1.7976931348623157e+308,
      max=1.7976931348623157e+308, dtype=float64)
```

The `log` of the largest floating point number is the largest number for which `exp()` will not overflow, so we will use it as a threshold to filter out problematic values:

```
[10]: max_float = np.finfo(np.float64).max

      def logistic(x):
          if x > np.log(max_float):
              return 0.0
          return 1 / (1 + np.exp(-x))
```

We now have everything we need to implement Algorithm 4. The steps to follow for each example are: (1) use the model to make a prediction, (2) calculate the gradient of the loss function with respect to the model parameters, and (3) update the model parameters using the gradient. The size of the update is controlled by the learning rate.

```
[12]: learning_rate = 1e-1
      n_epochs = 10

      indices = np.arange(n_examples)
      for epoch in range(10):
          # randomize training examples
          np.random.shuffle(indices)
          # for each training example
          for i in tqdm(indices, desc=f'epoch {epoch+1}'):
              x = X_train[i]
              y = y_train[i]
              # make decision
              decision = x @ w
              # calculate derivative
              deriv_cost = (logistic(decision) - y) * x
              # update weights
              w = w - deriv_cost * learning_rate
```

Once the model has been trained, we evaluate it on the test dataset using our `binary_classification_report` function from the previous section. Loading and preprocessing the test dataset follows the same steps as with the previous classifier. We omit the code for brevity. These are the results:

```
[14]:  y_pred = X_test @ w > 0
       binary_classification_report(y_test, y_pred)
```

```
[14]:  {'precision': 0.8946762335016387,
        'recall': 0.808,
        'f1-score': 0.849131951742402,
        'support': 12500.0,
        'accuracy': 0.85644}
```

The performance is comparable with that of the perceptron. The difference in F_1 scores between the two classifiers (84.9% here vs. 86.8% for the perceptron) is not significant. Classifier parity is probably attributable to the fact that the signal distinguishing the two classes is easy to learn and the simpler perceptron training algorithm is sufficient in this case. Nevertheless, this task is useful in showing how to implement the logistic regression model from scratch – that is, by implementing the gradient calculation and parameter updates manually. Next, we will implement the same model again using PyTorch, highlighting how this ML library simplifies the process.

4.1.5   Binary Logistic Regression Utilizing PyTorch

While it is fairly straightforward to compute the derivatives for LR and implement them directly in NumPy, this will not scale well to arbitrary neural architectures. Fortunately, there are libraries that automate the computation of the derivatives of the cost function (assuming it is differentiable!) for any neural network, and use the resulting gradients to perform gradient descent or other more sophisticated optimization procedures. To this end, we will use the PyTorch deep learning library.[10] The corresponding notebook for this section is chap4_logistic_regression_pytorch_bce.

Our model for LR corresponds to PyTorch's Linear layer. When we instantiate this layer, we specify the size of the inputs (the size of our vocabulary) and the size of the output – that is, the number of output neurons (which is one because we're doing binary classification). The loss function we use is the binary cross-entropy loss (see Chapter 3), which is implemented as BCEWithLogitsLoss in PyTorch. In PyTorch, the gradients obtained from the loss function are applied to the model by an optimizer object, which implements and applies an optimization algorithm. Here, we will use the vanilla SGD optimizer; we set its learning rate to 0.1. This is equivalent to the discussion in Section 3.2.

Similarly to the manual implementation, the steps required to train the model for a given training example are: (1) ensure the gradients are set to zeros, (2)

---

[10] https://pytorch.org.

apply the model to obtain a prediction, (3) calculate the loss, (4) compute the gradient of the loss by backpropagation, and (5) update the model parameters.

```
[9]:  import torch
      from torch import nn
      from torch import optim

      lr = 1e-1
      n_epochs = 10

      model = nn.Linear(n_features, 1)
      loss_func = nn.BCEWithLogitsLoss()
      optimizer = optim.SGD(model.parameters(), lr=lr)

      X_train = torch.tensor(X_train, dtype=torch.float32)
      y_train = torch.tensor(y_train, dtype=torch.float32)

      indices = np.arange(n_examples)
      for epoch in range(10):
          n_errors = 0
          # randomize training examples
          np.random.shuffle(indices)
          # for each training example
          for i in tqdm(indices, desc=f'epoch {epoch+1}'):
              x = X_train[i]
              y_true = y_train[i]
              # ensure gradients are set to zero
              model.zero_grad()
              # make predictions
              y_pred = model(x)
              # calculate loss
              loss = loss_func(y_pred[0], y_true)
              # calculate gradients through back-propagation
              loss.backward()
              # optimise model parameters
              optimizer.step()
```

Recall that in our previous implementation everything was hardcoded: applying the model, computing the gradients, and optimizing the model parameters. Here, however, the implementation of the LR is expressed at a higher level of abstraction. This means that we are describing the logical steps without specifying a particular implementation. Instead, implementation details are the responsibility of the chosen model, loss function, and optimizer. Thus, we could even choose a different model, loss function, and/or optimizer, and use the same training steps with little or no modification. This decoupling of the training logic from the implementation details is one of the main advantages of libraries such as PyTorch.

As shown in the code, calling the model as a function, with the feature vectors as inputs, produces the predicted scores. Once again, a positive score corresponds to a positive label. When we evaluate this implementation on the test dataset, we obtain results that are in line with our previous models:

```
[13]:   y_pred = model(X_test) > 0
        binary_classification_report(y_test, y_pred)
```

```
[13]:   {'precision': 0.8908308222126561,
         'recall': 0.82776,
         'f1-score': 0.8581380883267676,
         'support': 12500.0,
         'accuracy': 0.86316}
```

Writing the perceptron and the LR from scratch is a good exercise, as it exposes us to the fundamentals of implementing ML algorithms. However, this becomes cumbersome for more complex neural architectures. For this reason, from this point on, we will use PyTorch for all our coding examples.

4.2   Multiclass Classification

So far in this chapter, we have discussed implementing binary classifiers. Next, we will modify these binary classifiers to perform multiclass classification, following the discussion in Section 3.5.

4.2.1   AG News Dataset

Before explaining the actual training/testing code, we have to choose a new dataset that is suitable for multiclass classification. To this end, we will use the AG News Dataset, a subset of the larger AG corpus of news articles collected from thousands of news sources (Zhang et al., 2015).[11] The classification dataset consists of four classes, and the data are equally balanced across all classes (30,000 articles per class for train and 1,900 articles per class for testing). The goal of the task is to classify each article as one of the four classes: World, Sports, Business, or Sci/Tech.

---

[11] http://groups.di.unipi.it/~gulli/AG_corpus_of_news_articles.html.

4.2.2   Preparing the Dataset

The AG News Dataset is distributed as two CSV files (one for training and one for testing), each containing three columns: the class index, the title, and the description. The dataset also provides a text file that maps these class indexes to more descriptive class labels.

Because of the tabular nature of the dataset, pandas, a Python library for tabular data analysis, is a natural choice for loading and transforming it.[12] To this end, our Jupyter notebook (chap4_multiclass_logistic_regress-ion) demonstrates the sequence of steps required to handle the data, as well as model training and evaluation. First, we show how to load the CSV, add column names, and inspect the result (the output for the code below is on page 64).

```
[2]: train_df = pd.read_csv('data/ag_news_csv/train.csv', header=None)
     train_df.columns = ['class index', 'title', 'description']
     train_df
```

Since the class labels themselves are in a separate file, we manually add them to the pandas data structure (called dataframe in pandas's terminology) to increase the interpretability of the data. We use the *class index* column as a starting point, and use its map method to create a new column with the corresponding labels (technically a new Series object) that is added to the dataframe using its insert method, which allows us to insert the column in a specific position. Note that the label indices are 1-based, so we subtract 1 to align them with their labels (the output for the code below is on page 65).

```
[3]: labels = open('data/ag_news_csv/classes.txt').read().splitlines()
     classes = train_df['class index'].map(lambda i: labels[i-1])
     train_df.insert(1, 'class', classes)
     train_df
```

Next, we will preprocess the text. First, we lowercase the title and descrip-tion, and then we concatenate them into a single string. Then we remove some spurious backslashes from the text. Once this is done, the preprocessed text is added to the dataframe as a new column. Note that pandas allows these steps to be applied to all rows simultaneously (the output for the code below is on page 66).

```
[6]: title = train_df['title'].str.lower()
     descr = train_df['description'].str.lower()
     text = title + " " + descr
     train_df['text'] = text.str.replace('\\', ' ', regex=False)
     train_df
```

---

[12] https://pandas.pydata.org.

class index	title	description
0	Wall St. Bears Claw Back Into the Black (Reuters)	Reuters - Short-sellers, Wall Street's dwindli...
1	Carlyle Looks Toward Commercial Aerospace (Reu...	Reuters - Private investment firm Carlyle Grou...
2	Oil and Economy Cloud Stocks' Outlook (Reuters)	Reuters - Soaring crude prices plus worrieslab...
3	Iraq Halts Oil Exports from Main Southern Pipe...	Reuters - Authorities have halted oil exportf...
4	Oil prices soar to all-time record, posing new...	AFP - Tearaway world oil prices, toppling reco...
...	...	...
119995	Pakistan's Musharraf Says Won't Quit as Army C...	KARACHI (Reuters) - Pakistani President Perve...
119996	Renteria signing a top-shelf deal	Red Sox general manager Theo Epstein acknowled...
119997	Saban not going to Dolphins yet	The Miami Dolphins will put their courtship of...
119998	Today's NFL games	PITTSBURGH at NY GIANTS Time: 1:30 p.m. Line: ...
119999	Nets get Carter from Raptors	INDIANAPOLIS -- All-Star Vince Carter was trad...

120000 rows × 3 columns

	class index	class	title	description
0	3	Business	Wall St. Bears Claw Back Into the Black (Reuters)	Reuters - Short-sellers, Wall Street's dwindli...
1	3	Business	Carlyle Looks Toward Commercial Aerospace (Reu...	Reuters - Private investment firm Carlyle Grou...
2	3	Business	Oil and Economy Cloud Stocks' Outlook (Reuters)	Reuters - Soaring crude prices plus worries\ab...
3	3	Business	Iraq Halts Oil Exports from Main Southern Pipe...	Reuters - Authorities have halted oil export\f...
4	3	Business	Oil prices soar to all-time record, posing new...	AFP - Tearaway world oil prices, toppling reco...
...	...	...	...	...
119995	1	World	Pakistan's Musharraf Says Won't Quit as Army C...	KARACHI (Reuters) - Pakistani President Perve...
119996	2	Sports	Renteria signing a top-shelf deal	Red Sox general manager Theo Epstein acknowled...
119997	2	Sports	Saban not going to Dolphins yet	The Miami Dolphins will put their courtship of...
119998	2	Sports	Today's NFL games	PITTSBURGH at NY GIANTS Time: 1:30 p.m. Line: ...
119999	2	Sports	Nets get Carter from Raptors	INDIANAPOLIS -- All-Star Vince Carter was trad...

120000 rows × 4 columns

	class index	class	title	description	text
0	3	Business	Wall St. Bears Claw Back Into the Black (Reuters)	Reuters - Short-sellers, Wall Street's dwindli...	wall st. bears claw back into the black (reute...
1	3	Business	Carlyle Looks Toward Commercial Aerospace (Reu...	Reuters - Private investment firm Carlyle Grou...	carlyle looks toward commercial aerospace (reu...
2	3	Business	Oil and Economy Cloud Stocks' Outlook (Reuters)	Reuters - Soaring crude prices plus worriesab...	oil and economy cloud stocks' outlook (reuters...
3	3	Business	Iraq Halts Oil Exports from Main Southern Pipe...	Reuters - Authorities have halted oil exportf...	iraq halts oil exports from main southern pipe...
4	3	Business	Oil prices soar to all-time record, posing new...	AFP - Tearaway world oil prices, toppling reco...	oil prices soar to all-time record, posing new...
...	...	...	...	...	...
119995	1	World	Pakistan's Musharraf Says Won't Quit as Army C...	KARACHI (Reuters) - Pakistani President Perve...	pakistan's musharraf says won't quit as army c...
119996	2	Sports	Renteria signing a top-shelf deal	Red Sox general manager Theo Epstein acknowled...	renteria signing a top-shelf deal red sox gene...
119997	2	Sports	Saban not going to Dolphins yet	The Miami Dolphins will put their courtship of...	saban not going to dolphins yet the miami dolp...
119998	2	Sports	Today's NFL games	PITTSBURGH at NY GIANTS Time: 1:30 p.m. Line: ...	today's nfl games pittsburgh at ny giants time...
119999	2	Sports	Nets get Carter from Raptors	INDIANAPOLIS -- All-Star Vince Carter was trad...	nets get carter from raptors indianapolis -- a...

120000 rows × 5 columns

At this point, the text is ready to be tokenized. For this purpose, we will use NLTK's word_tokenize function. This function can be applied to the whole column at once using the pandas map function, which returns a new column that we add to the dataframe. However, here, we actually use the progress_map function, which provides a visual progress bar. This visual feedback is especially helpful for tasks that take more time to complete (the output for the code below is on page 68).

[7]:
```
from nltk.tokenize import word_tokenize

train_df['tokens'] = train_df['text'].progress_map(word_tokenize)
train_df
```

From the tokens we just created, we then create a vocabulary for our corpus. Here, we keep only the words that occur at least 10 times, decreasing the memory needed and reducing the likelihood that our vocabulary contains noisy tokens. Note that each row in the tokens column contains a list of tokens. In order to create the vocabulary, we will need to convert the Series of lists of tokens into a Series of tokens using the explode() Pandas method. Then, we will use the value_counts() method to create a Series object in which the index are the tokens and the values are the number of times they appear in the corpus. The next step is removing the tokens with a count lower than our chosen threshold. Finally, we create a list with the remaining tokens, as well as a dictionary that maps tokens to token IDs (i.e., the index of the token in the list). We include in the vocabulary a special token [UNK] that will be used as a placeholder for tokens that do not appear in our vocabulary after the frequency pruning.

[8]:
```
threshold = 10
tokens = train_df['tokens'].explode().value_counts()
tokens = tokens[tokens > threshold]
id_to_token = ['[UNK]'] + tokens.index.tolist()
token_to_id = {w:i for i,w in enumerate(id_to_token)}
vocabulary_size = len(id_to_token)
print(f'vocabulary size: {vocabulary_size:,}')
```

```
vocabulary size: 19,671
```

Using this vocabulary, we construct a feature vector for each news article in the corpus. This feature vector will be encoded as a dictionary, with keys corresponding to token IDs, and values corresponding to the number of times the token appears in the article. As before, the feature vectors will be stored as a new column in the dataframe (the output for the code below is on page 70).

	class index	class	title	description	text	tokens
0	3	Business	Wall St. Bears Claw Back Into the Black (Reuters)	Reuters - Short-sellers, Wall Street's dwindli...	wall st. bears claw back into the black (reute...	[wall, st., bears, claw, back, into, the, blac...
1	3	Business	Carlyle Looks Toward Commercial Aerospace (Reu...	Reuters - Private investment firm Carlyle Grou...	carlyle looks toward commercial aerospace (reu...	[carlyle, looks, toward, commercial, aerospace...
2	3	Business	Oil and Economy Cloud Stocks' Outlook (Reuters)	Reuters - Soaring crude prices plus worries\lab...	oil and economy cloud stocks' outlook (reuters...	[oil, and, economy, cloud, stocks, ', outlook,...
3	3	Business	Iraq Halts Oil Exports from Main Southern Pipe...	Reuters - Authorities have halted oil exportf...	iraq halts oil exports from main southern pipe...	[iraq, halts, oil, exports, from, main, southe...
4	3	Business	Oil prices soar to all-time record, posing new...	AFP - Tearaway world oil prices, toppling reco...	oil prices soar to all-time record, posing new...	[oil, prices, soar, to, all-time, record, ,, p...
...	...	...	...	...	...	...
119995	1	World	Pakistan's Musharraf Says Won't Quit as Army C...	KARACHI (Reuters) - Pakistani President Perve...	pakistan's musharraf says won't quit as army c...	[pakistan, 's, musharraf, says, wo, n't, quit,...
119996	2	Sports	Renteria signing a top-shelf deal	Red Sox general manager Theo Epstein acknowled...	renteria signing a top-shelf deal red sox gene...	[renteria, signing, a, top-shelf, deal, red, s...
119997	2	Sports	Saban not going to Dolphins yet	The Miami Dolphins will put their courtship of...	saban not going to dolphins yet the miami dolp...	[saban, not, going, to, dolphins, yet, the, mi...
119998	2	Sports	Today's NFL games	PITTSBURGH at NY GIANTS Time: 1:30 p.m. Line: ...	today's nfl games pittsburgh at ny giants time...	[today, 's, nfl, games, pittsburgh, at, ny, gi...
119999	2	Sports	Nets get Carter from Raptors	INDIANAPOLIS -- All-Star Vince Carter was trad...	nets get carter from raptors indianapolis -- a...	[nets, get, carter, from, raptors, indianapoli...

120000 rows × 6 columns

```
[9]: from collections import defaultdict

     def make_features(tokens, unk_id=0):
         vector = defaultdict(int)
         for t in tokens:
             i = token_to_id.get(t, unk_id)
             vector[i] += 1
         return vector

     train_df['features'] = train_df['tokens'].progress_map(make_features)
     train_df
```

The final preprocessing step is converting the features and the class indices into PyTorch tensors. Recall that we need to subtract one from the class indices to make them zero-based.

```
[10]: def make_dense(feats):
          x = np.zeros(vocabulary_size)
          for k,v in feats.items():
              x[k] = v
          return x

      X_train = np.stack(train_df['features'].progress_map(make_dense))
      y_train = train_df['class index'].to_numpy() - 1

      X_train = torch.tensor(X_train, dtype=torch.float32)
      y_train = torch.tensor(y_train)
```

At this point, the data are fully processed and we are ready to begin training.

4.2.3 Multiclass Logistic Regression Using PyTorch

The model itself is a single linear layer whose input size corresponds to the size of our vocabulary, and its output size corresponds to the number of classes in our corpus. PyTorch's Linear layer includes a bias by default, so there is no need to handle that manually the way we did for our perceptron example.

```
[11]: from torch import nn
      from torch import optim

      # hyperparameters
      lr = 1.0
      n_epochs = 5
      n_examples = X_train.shape[0]
      n_feats = X_train.shape[1]
      n_classes = len(labels)

      # initialize the model, loss function, and optimizer
      model = nn.Linear(n_feats, n_classes).to(device)
```

	class index	class	title	description	text	tokens	features
0	3	Business	Wall St. Bears Claw Back Into the Black (Reuters)	Reuters - Short-sellers, Wall Street's dwindli...	wall st. bears claw back into the black (reute...	[wall, st., bears, claw, back, into, the, blac...	{427: 2, 563: 1, 1607: 1, 15062: 1, 120: 1, 73...
1	3	Business	Carlyle Looks Toward Commercial Aerospace (Reu...	Reuters - Private investment firm Carlyle Grou...	carlyle looks toward commercial aerospace (reu...	[carlyle, looks, toward, commercial, aerospace...	{15999: 2, 1076: 1, 855: 1, 1286: 1, 4251: 1, ...
2	3	Business	Oil and Economy Cloud Stocks' Outlook (Reuters)	Reuters - Soaring crude prices plus worries\ab...	oil and economy cloud stocks' outlook (reuters...	[oil, and, economy, cloud, stocks, ', outlook,...	{66: 1, 9: 2, 351: 2, 4565: 1, 158: 1, 116: 1...
3	3	Business	Iraq Halts Oil Exports from Main Southern Pipe...	Reuters - Authorities have halted oil export\f...	iraq halts oil exports from main southern pipe...	[iraq, halts, oil, exports, from, main, southe...	{77: 2, 7380: 1, 66: 3, 1787: 1, 32: 2, 900: 2...
4	3	Business	Oil prices soar to all-time record, posing new...	AFP - Tearaway world oil prices, toppling reco...	oil prices soar to all-time record, posing new...	[oil, prices, soar, to, all-time, record, ,, p...	{66: 2, 99: 2, 4390: 1, 4: 2, 3595: 1, 149: 1,...
...	...	...	...	...	...	...	...
119995	1	World	Pakistan's Musharraf Says Won't Quit as Army C...	KARACHI (Reuters) - Pakistani President Perve...	pakistan's musharraf says won't quit as army c...	[pakistan, 's, musharraf, says, wo, n't, quit,...	{383: 1, 23: 1, 1626: 2, 91: 1, 1809: 1, 285: ...
119996	2	Sports	Renteria signing a top-shelf deal	Red Sox general manager Theo Epstein acknowled...	renteria signing a top-shelf deal red sox gene...	[renteria, signing, a, top-shelf, deal, red, s...	{8428: 2, 2638: 1, 5: 4, 0: 3, 127: 1, 202: 3,...
119997	2	Sports	Saban not going to Dolphins yet	The Miami Dolphins will put their courtship of...	saban not going to dolphins yet the miami dolp...	[saban, not, going, to, dolphins, yet, the, mi...	{7762: 2, 68: 1, 661: 1, 4: 2, 1439: 2, 703: 1...
119998	2	Sports	Today's NFL games	PITTSBURGH at NY GIANTS Time: 1:30 p.m. Line: ...	today's nfl games pittsburgh at ny giants time...	[today, 's, nfl, games, pittsburgh, at, ny, gi...	{106: 1, 23: 1, 729: 1, 225: 1, 1586: 1, 22: 1...
119999	2	Sports	Nets get Carter from Raptors	INDIANAPOLIS -- All-Star Vince Carter was trad...	nets get carter from raptors indianapolis -- a...	[nets, get, carter, from, raptors, indianapoli...	{2170: 2, 226: 1, 2402: 2, 32: 1, 2995: 2, 219...

120000 rows × 7 columns

```python
loss_func = nn.CrossEntropyLoss()
optimizer = optim.SGD(model.parameters(), lr=lr)

# train the model
indices = np.arange(n_examples)
for epoch in range(n_epochs):
    np.random.shuffle(indices)
    for i in tqdm(indices, desc=f'epoch {epoch+1}'):
        # clear gradients
        model.zero_grad()
        # send datum to right device
        x = X_train[i].unsqueeze(0).to(device)
        y_true = y_train[i].unsqueeze(0).to(device)
        # predict label scores
        y_pred = model(x)
        # compute loss
        loss = loss_func(y_pred, y_true)
        # backpropagate
        loss.backward()
        # optimize model parameters
        optimizer.step()
```

The code for training this model (which implements Algorithm 6) is almost identical to that of the binary logistic repression. However, since we have to calculate a score for each of the four different classes, we need to replace the previous BCEWithLogitsLoss with CrossEntropyLoss, which applies a softmax over the scores to obtain probabilities for each class.

For each example, the model predicts four scores – one for each label. The label with the highest score is selected using the argmax function. We evaluate the predictions of our model for each class using scikit-learn's classification_report, which handles the results of multiclass classification.

[13]:
```python
from sklearn.metrics import classification_report

# set model to evaluation mode
model.eval()

# don't store gradients
with torch.no_grad():
    X_test = X_test.to(device)
    y_pred = torch.argmax(model(X_test), dim=1)
    y_pred = y_pred.cpu().numpy()
    print(classification_report(y_test, y_pred, target_names=labels))
```

	precision	recall	f1-score	support
World	0.94	0.82	0.88	1900
Sports	0.89	0.99	0.94	1900
Business	0.81	0.88	0.85	1900
Sci/Tech	0.89	0.83	0.86	1900
accuracy			0.88	7600
macro avg	0.88	0.88	0.88	7600
weighted avg	0.88	0.88	0.88	7600

4.3   Summary

In this chapter, we used movie review and news article classification to illustrate the implementation of the previously described algorithms for the binary perceptron, binary LR, and multiclass LR. For the binary LR, we made a direct comparison between the lower-level NumPy implementation and a higher-level version that made use of PyTorch.

We hope that through this series of exercises the reader has noted several key takeaways. First, data preparation is important and should be done thoughtfully. Certain tasks (e.g., text normalization or sentence splitting) are going to be frequently needed if you continue with NLP, so using or creating generic functions can be very helpful. However, what works for one dataset and one language may not be suitable for another scenario. For example, in our case, we selected different tokenizers for each of our tasks to account for the different registers of English, as well as removing diacritics during normalization.

Second, when it comes to implementing ML algorithms, it is often easier to use a higher-level library such as PyTorch instead of NumPy. For example, with the former, the gradients are calculated by the library, whereas in NumPy we have to code them ourselves. This becomes cumbersome quickly. Even the derivative of the softmax is nontrivial.

Third, PyTorch imposes a training structure that remains largely the same, regardless of what models are being trained. That is, at a high level, the same steps are always required: clearing the current gradients, predicting output scores for the provided inputs, calculating the loss, and optimizing. These features make PyTorch a very *powerful* and *convenient* deep learning library; we will continue to use it throughout the remainder of the book to implement more complex neural architectures.

5   Feed-Forward Neural Networks

So far we have explored classifiers with decision boundaries that are linear, or, in the case of the multiclass LR, a combination of linear segments. In this chapter, we will expand what we have learned so far to classifiers capable of learning nonlinear decision boundaries. The classifiers we will discuss here are called *feed-forward neural networks* (FFNNs), and are a generalization of both LR and the perceptron to systems of multiple "neurons." Without going into the theory behind it, it has been shown that, under certain conditions, these classifiers can approximate any function (Hornik, 1991; Leshno et al., 1993). That is, they can learn decision boundaries of any arbitrary shape. Intuitively, these complex decision boundaries are made possible through a nonlinear aggregation of several individual decision boundaries (one per neuron). Figure 5.1 shows a very simple example of a hypothetical situation where a nonlinear decision boundary is needed for a binary classifier.[1]

The good news is that we have already introduced the building blocks of FFNNs: the individual neuron, and the stochastic gradient descent algorithm. In this chapter, we are simply combining these building blocks in slightly more complicated ways, but without changing any of the fundamental operating principles.

5.1   Architecture of Feed-Forward Neural Networks

Figure 5.2 shows the general architecture of FFNNs. As seen in the figure, FFNNs combine multiple layers of individual neurons, where each neuron in a layer l is fully connected to all neurons in the next layer, $l + 1$. Because of this, architectures such as the one in the figure are often referred to as *fully connected FFNNs*. This is not the only possible architecture for FFNNs: any arbitrary connections between neurons are possible. However, because fully connected networks are the most common FFNN architecture seen in NLP, we will focus on these in this chapter and omit the fully connected modifier from now on, for simplicity.

---

[1]   See this blog post for a visual explanation of nonlinear decision boundaries in neural networks: https://towardsdatascience.com/a-visual-introduction-to-neural-networks-68586b0b733b.

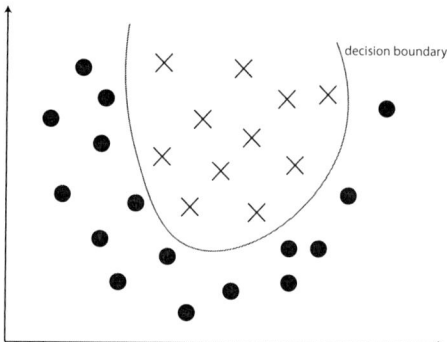

Figure 5.1 Decision boundary of a nonlinear classifier

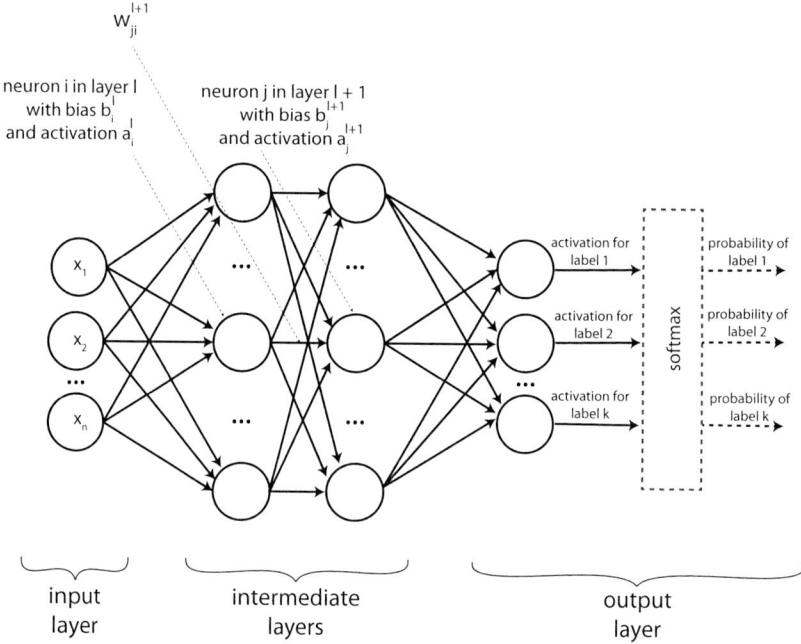

Figure 5.2 Fully connected feed-forward neural network architecture. The dashed lines indicate optional components.

Figure 5.2 shows that the neuron layers in an FFNN are grouped into three categories. These are worth explaining in detail:

Input layer: Similar to the perceptron or LR, the input layer contains a vector **x** that describes one individual data point. For example, for the review classification task, the input layer will be populated with features extracted from an individual review such as the presence (or count) of individual words. In Chapter 8, we will switch from such hand-crafted features to numerical representations of text that capture some of the underlying semantics of language and thus are better for learning. Importantly, the neural network is agnostic to the way the representation of an input data point is created. All that matters for now is that each input data point is summarized with a vector of real values, **x**.

Intermediate layers: Unlike the perceptron and LR, FFNNs have an arbitrary number of intermediate layers. Each neuron in an intermediate layer receives as inputs the outputs of the neurons in the previous layer, and produces an output (or activation) that is sent to all the neurons in the following layer. The activation of each neuron is constructed similarly to LR, as a nonlinear function that operates on the dot product of weights and inputs plus the bias term. More formally, the activation a_i^l of neuron i in layer l is calculated as:

$$a_i^l = f(\sum_{j=1}^{k} w_{ij}^l a_j^{l-1} + b_i^l) = f(\mathbf{w}_i^l \cdot \mathbf{a}^{l-1} + b_i^l) = f(z_i^l), \qquad (5.1)$$

where k is the total number of neurons in the previous layer $l-1$, w_{ij}^l are the weights learned by the current neuron (neuron i in layer l), a_j^{l-1} is the activation of neuron j in the previous layer, and b_i^l is the bias term of the current neuron. For simplicity, we group all the weights w_{ij}^l into the vector $\mathbf{w}_i^l$, and all activations a_j^{l-1} into the vector $\mathbf{a}^{l-1}$. Thus, the summation in the equation reduces to the dot product between the two vectors: $\mathbf{w}_i^l \cdot \mathbf{a}^{l-1}$. We further denote the sum between this dot product and the bias term b_i^l as z_i^l. Thus, z_i^l is the output of neuron i in layer l right before the activation function f is applied.

The function f is a nonlinear function that takes z_i^l as its input. For example, for the LR neuron, f is the logistic function, σ. Many other nonlinear functions are possible and commonly used in neural networks. We will discuss several such functions, together with their advantages and disadvantages, in Chapter 6. What is important to realize at this stage is that the aggregation of all these nonlinear functions gives neural networks the capability of learning nonlinear decision boundaries. A multilayer FFNN with identity activation functions – that is, where $a_i^l = z_i^l$, remains a linear classifier. As a simple example, consider the neural network in Figure 5.3, which has one intermediate layer with two neurons and a single neuron in the output layer. Let us consider that the

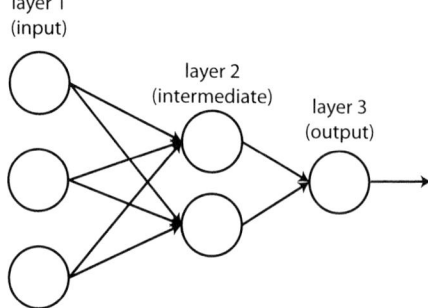

Figure 5.3 A feed-forward neural network with linear activation functions is a linear classifier

activation function in each neuron is a "pass through" (or identity) function $f(x) = x$. The activation of the output neuron is then computed as:

$$\begin{aligned}
a_1^3 &= w_{11}^3 a_1^2 + w_{12}^3 a_2^2 + b_1^3 \\
&= w_{11}^3(w_{11}^2 x_1 + w_{12}^2 x_2 + w_{13}^2 x_3 + b_1^2) + w_{12}^3(w_{21}^2 x_1 + w_{22}^2 x_2 \\
&\quad + w_{23}^2 x_3 + b_2^2) + b_1^3 \\
&= x_1(w_{11}^3 w_{11}^2 + w_{12}^3 w_{21}^2) + \\
&\quad x_2(w_{11}^3 w_{12}^2 + w_{12}^3 w_{22}^2) + \\
&\quad x_3(w_{11}^3 w_{13}^2 + w_{12}^3 w_{23}^2) + \\
&\quad w_{11}^3 b_1^2 + w_{12}^3 b_2^2 + b_1^3,
\end{aligned}$$
(5.2)

which is a linear function on the input variables x_1, x_2, and x_3. It is easy to show that this observation generalizes to any arbitrary FFNN, as long as the neuron activation functions are linear.

Output layer: Last, FFNNs have an output layer that produces scores for the classes to be learned. Similar to the multiclass LR, these scores can be aggregated into a probability distribution if the output layer includes a softmax function. However, the softmax function is optional (hence the dashed lines in the figure 5.2). If softmax is skipped, the class scores will not form a probability distribution, and they may or may not be bounded to the [0, 1] interval depending on the activation functions used in the final layer.

To distinguish between the final network layer and other intermediate layers, we use uppercase L to indicate the final layer, and lowercase l to identify an intermediate one. That is, z_i^L is the "raw" output of neuron i in the last layer – that is, before the activation function is applied – and a_i^L is the corresponding activation of the same neuron.

The values in the $\mathbf{z}^L$ vector are often referred to as *logits*. In the case of binary classification, the logits are usually normalized using sigmoid activations. In the case of multiclass classification, the logits become the input to a softmax function. We will see this terminology a lot in the coding chapters because PyTorch relies on it.

Sidebar 5.1   Tensor notation for feed-forward neural networks

Very often, you will see the equations discussed so far in this chapter summarized using tensor notation – that is, using vectors and matrices instead of the explicit math we introduced earlier. For example, Equation 5.1 is commonly summarized as: $\mathbf{a}^l = f(\mathbf{W}^l \cdot \mathbf{a}^{l-1} + \mathbf{b}^l)$, where the vectors $\mathbf{a}^l$ and $\mathbf{b}^l$ contain *all* activations and biases in layer l, and the matrix $\mathbf{W}^l$ contains *all* the weights that connect layer $l - 1$ to layer l. Thus, $\mathbf{W}^l$ has as many columns as the size of $\mathbf{a}^{l-1}$, and as many rows as the size of $\mathbf{a}^l$. Sometimes, the order of $\mathbf{W}^l$ and $\mathbf{a}^{l-1}$ is flipped in the equation: $\mathbf{a}^l = f(\mathbf{a}^{l-1} \cdot \mathbf{W}^l + \mathbf{b}^l)$. This does not really matter; one simply has to be careful about the dimensions of $\mathbf{W}^l$, which change in this case.

Operating with vectors and matrices, as shown in the above equation, is beneficial. Not only are the math and the resulting code (see Chapter 7) simpler, but we can take advantage of modern hardware – that is, graphics processing units (GPUs) – which have been designed for efficient tensor operations. We will discuss this more in Chapter 6. However, in this chapter, we will continue with the explicit notations used before this sidebar because they completely expose the underlying mathematical operations.

The architecture shown in Figure 5.2 can be reduced to most of the classifiers we introduced so far. For example:

Perceptron: The perceptron has no intermediate layers; has a single neuron in the output layer with a "pass through" activation function: $f(x) = x$; and no softmax.

Binary logistic regression: Binary LR is similar to the perceptron, with the only difference that the activation function of its output neuron is the logistic function: $f = \sigma$.

Multiclass logistic regression: Multiclass LR has multiple neurons in its output layer (one per class); their activation functions are the "pass through" function, $f(x) = x$; and it has a softmax.

5.2   Learning Algorithm for Neural Networks

At a very high level, one can view a neural network as a complex machinery with many knobs, one for each neuron in the architecture. In this analogy, the

---

Algorithm 7 Stochastic gradient descent algorithm for the training of neural networks

---

1 initialize parameters in Θ
2 **while** *not converged* **do**
3   **for** *each training example* $\mathbf{x}_i$ **in X do**
4   **for** *each θ in Θ* **do**
5             $\theta = \theta - \alpha \frac{d}{d\theta} C_i(\Theta)$
6   **end**
7   **end**
8 **end**

---

learning algorithm is the operating technician whose job is to turn all the knobs in order to minimize the machine's output – that is, the value of its cost function for each training example. If a neuron increases the probability of an incorrect prediction, its knob will be turned down. If a neuron increases the probability of a correct prediction, its knob will be turned up.

We will implement this learning algorithm that applies to any neural network with a generalization of the learning algorithm for multiclass LR (Algorithm 6). The first key difference is that the parameters we are learning are no longer a single weight vector and a single bias term per class as in the multiclass LR. Instead, the neural network parameters contain one weight vector and one bias term for *each* neuron in an intermediate or final layer (see Figure 5.2). Because the number of these neurons may potentially be large, let's use a single variable name, Θ, to indicate the totality of parameters to be learned – that is, all the weights and biases. We will also use θ to point to an individual parameter (i.e., one single bias term or a single weight) in Θ. Under these notations, we can generalize Algorithm 6 into Algorithm 7, which applies to any neural network we will encounter in this book.[2]

Note that the key functionality remains exactly the same between Algorithms 6 and 7: in each iteration, both algorithms update their parameters by subtracting the partial derivative of the cost from their current values. As discussed, this guarantees that the cost function incrementally decreases toward some local minimum. This observation is sufficient to understand how to implement the training algorithm for an FFNN using a modern ML library that includes auto-differentiation such as PyTorch. Thus, the impatient reader who wants to get to programming examples as quickly as possible may skip the remainder of this chapter and jump to Chapter 6 for code examples. However, we encourage the reader to stick around for the next sections in this chapter, where we will look "under the hood" of Algorithm 7 in order to understand better how it operates.

---

[2] We will revise this algorithm slightly in Chapter 6.

5.3 The Equations of Backpropagation

The key equation in Algorithm 7 is in row 5, which requires the computation of the partial derivative of the cost function for one training example $C_i(\Theta)$ with respect to *all* parameters in the network – that is, all edge weights and all bias terms. The intuition behind this algorithm is identical to what we discussed in Chapter 3: each parameter is updated proportionally to its contribution to the mistakes of the network.

While the formula in row 5 of Algorithm 7 looks mathematically simple, it is not intuitive: how are we to calculate the partial derivatives for parameters associated with neurons that are not in the final layer and thus do not contribute directly to the cost function computation? To achieve this, we will implement an algorithm that has two phases: a *forward* phase and a *backward* phase. In the forward phase, the algorithm runs the neural network with its current parameters to make a prediction on the given training example i. Using this prediction, we then compute the value of the cost function for this training example, $C_i(\Theta)$. Then, in the backward phase, we incrementally propagate this information backward – that is, from the final layer toward the first layer – to compute the updates to the parameters in each layer. Because of this, this algorithm is commonly called *backpropagation*, or, for people in a hurry, *backprop*.

Let us formalize this informal description. To do this, we need to introduce a couple of new notations. First, because in this section we will use only one training example i and refer to the same training parameters Θ throughout, we will simplify $C_i(\Theta)$ to C in all the equations. Second, and more importantly, we define the *error of neuron i*[3] in layer l as the partial derivative of the cost function C with respect to the neuron's output (z_i^l):

$$\delta_i^l = \frac{d}{dz_i^l} C, \tag{5.3}$$

where z_i^l is the output of neuron i in layer l before the activation function f is applied. Intuitively, the error of a neuron measures what impact a small change in its output z has on the cost C. Or, if we view z as a knob as in the previous analogy, the error indicates what impact turning the knob has. The error of a neuron is a critical component in backpropagation: we want to adjust the parameters of each neuron *proportionally* with the impact the neuron has on the cost function's value for this training example: the higher the impact, the bigger the adjustment. Last, we use the index L to indicate the *final* layer of the network – for example, the layer right before the softmax in Figure 5.2. Thus, δ_i^L indicates the error of neuron i in the final layer.

---

[3] Note that we are overloading the index i here. In Algorithm 7, we used it to indicate a specific training example x_i. Now we use it to indicate a specific neuron.

Algorithm 8 The back-propagation algorithm that computes parameter updates for a neural network

---

1 compute the errors in the final layer L, δ_i^L, using the cost function C (Equation 5.4)

2 backward propagate the computation of errors in all upstream layers (Equation 5.5)

3 compute the partial derivates of C for all parameters in a layer l, $\frac{d}{db_i^l}C$ and $\frac{d}{dw_{ij}^l}C$, using the errors in the same layer, δ_i^l (Equations 5.6 and 5.7)

Using these notations, we formalize the back-propagation algorithm with the three steps listed in Algorithm 8. Step 1 computes the error of neuron i in the final layer as the partial derivative of the cost function C with respect to neuron i's activation (a_i^L) multiplied by the partial derivative of the activation function f with respect to the neuron's output (z_i^L):

$$\delta_i^L = \frac{d}{da_i^L}C\frac{d}{dz_i^L}f(z_i^L). \tag{5.4}$$

This equation may appear daunting at first glance (two partial derivatives!), but it often reduces to an intuitive formula for given cost and activation functions. As a simple example, consider the case of binary classification – that is, a single neuron in the final layer with a logistic activation – coupled with the *mean squared error* (MSE) cost function. We will discuss the MSE cost in more detail in Chapter 6. For now, it is sufficient to known that MSE is a simple cost function commonly used for binary classification: $C = (y - a_1^L)^2$, where y is the gold label for the current training example – for example, 0 or 1 – assuming a logistic activation. That is, the MSE cost simply minimizes the difference between the prediction of the network (i.e., the activation of its final neuron) and the gold label. The derivative of the MSE cost with respect to the neuron's activation is: $\frac{d}{da_1^L}C = 2(a_1^L - y)$.[4] The derivative of the logistic with respect to the neuron's output is: $\frac{d}{dz_1^L}\sigma(z_1^L) = \sigma(z_1^L)(1 - \sigma(z_1^L))$ (see Table 3.1). Thus, δ_1^L in this simple example is computed as: $\delta_1^L = 2(a_1^L - y)\sigma(z_1^L)(1 - \sigma(z_1^L)) = 2(\sigma(z_1^L) - y)\sigma(z_1^L)(1 - \sigma(z_1^L))$. It is easy to see that this error formula follows our knob analogy: when the activation of the final neuron is close to the gold label y, which can take values of 0 or 1, the error approaches 0 because two of the terms in its product are close to 0. In contrast, the error value is largest when the classifier is "confused" between the two classes – that is, its activation is 0.5. The same can be observed for any (differentiable) cost and activation functions (see next chapter for more examples).

---

[4] This is trivially derived by applying the chain rule.

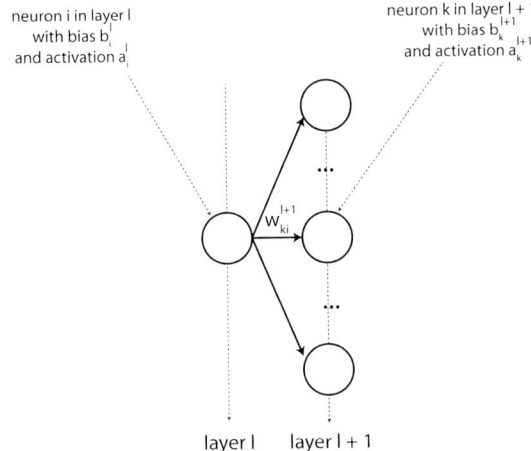

Figure 5.4 Visual helper for Equation 5.5

Equation 5.4 is easy to prove using a direct application of the chain rule:

$$\delta_i^L = \sum_k \frac{d}{da_k^L} C \frac{d}{dz_i^L} a_k^L$$

$$= \frac{d}{da_i^L} C \frac{d}{dz_i^L} a_i^L,$$

where k iterates over all neurons in the last layer. Note that we need to sum over all neurons in the final layer in the first line of the proof because C theoretically depends on all activations in the final layer. However, neuron i impacts only its own activation, and thus we can ignore all other activations (second line of the proof).

Equation 5.4 computes the errors in the *last* layer of the network. The next back-propagation equation incrementally propagates the computation of errors into the upstream layers – that is, the layers that are farther to the left in Figure 5.2. That is, this equation computes the errors in a layer l using the errors in the layer immediately downstream, $l + 1$, as follows:

$$\delta_i^l = \sum_k \delta_k^{l+1} w_{ki}^{l+1} \frac{d}{dz_i^l} f(z_i^l), \tag{5.5}$$

where k iterates over all neurons in layer $l + 1$.

We prove this equation by first applying the chain rule to introduce the outputs of the downstream layer, z_k^{l+1}, in the formula for the error of neuron i in layer l, and then taking advantage of the fact the outputs in the downstream layer $l + 1$ depend on the activations in the previous layer l. More formally:

$$\delta_i^l = \frac{d}{dz_i^l} C$$

$$= \sum_k \frac{d}{dz_k^{l+1}} C \frac{d}{dz_i^l} z_k^{l+1}$$

$$= \sum_k \delta_k^{l+1} \frac{d}{dz_i^l} z_k^{l+1}$$

$$= \sum_k \delta_k^{l+1} \frac{d}{dz_i^l} \left(\sum_j w_{kj}^{l+1} a_j^l + b_k^{l+1} \right)$$

$$= \sum_k \delta_k^{l+1} \frac{d}{dz_i^l} \left(w_{ki}^{l+1} a_i^l \right)$$

$$= \sum_k \delta_k^{l+1} w_{ki}^{l+1} \frac{d}{dz_i^l} a_i^l$$

$$= \sum_k \delta_k^{l+1} w_{ki}^{l+1} \frac{d}{dz_i^l} f(z_i^l),$$

where j iterates over all neurons in layer l. Similar to the previous proof, we need to sum over all the neurons in layer $l+1$ (second line of the proof) because the value of the cost function is impacted by all the neurons in this layer. The rest of the proof follows from the fact that $z_k^{l+1} = \sum_j w_{kj}^{l+1} a_j^l + b_k^{l+1}$. Figure 5.4 provides a quick visual helper to navigate the indices used in this proof.

Using Equations 5.4 and 5.5, we can compute the errors of all neurons in the network. Next we will use these errors to compute the partial derivatives for all weights and bias terms in the network, which we need for the SGD updates in Algorithm 7. First, we compute the partial derivative of the cost with respect to a bias term as:

$$\frac{d}{db_i^l} C = \delta_i^l. \tag{5.6}$$

The proof of this equation follows similar steps to the previous two proofs, but here we iterate over neurons in the same layer l (so we can access the error of neuron i). Thus, we can ignore all neurons other than neuron i, which depends on this bias term:

$$\frac{d}{db_i^l} C = \sum_k \frac{d}{dz_k^l} C \frac{d}{db_i^l} z_k^l$$

$$= \frac{d}{dz_i^l} C \frac{d}{db_i^l} z_i^l$$

$$= \delta_i^l \frac{d}{db_i^l} z_i^l$$

$$= \delta_i^l \frac{d}{db_i^l} (\sum_h w_{ih}^l a_h^{l-1} + b_i^l)$$

$$= \delta_i^l,$$

where k iterates over all neurons in layer l and h iterates over the neurons in layer $l - 1$.

Similarly, we compute the partial derivative of the cost with respect to the weight that connects neuron j in layer $l - 1$ with neuron i in layer l, $\frac{d}{dw_{ij}^l} C$, as:

$$\frac{d}{dw_{ij}^l} C = a_j^{l-1} \delta_i^l. \tag{5.7}$$

The proof of this equation follows the same structure as the proof just provided:

$$\frac{d}{dw_{ij}^l} C = \sum_k \frac{d}{dz_k^l} C \frac{d}{dw_{ij}^l} z_k^l$$

$$= \frac{d}{dz_i^l} C \frac{d}{dw_{ij}^l} z_i^l$$

$$= \delta_i^l \frac{d}{dw_{ij}^l} z_i^l$$

$$= \delta_i^l \frac{d}{dw_{ij}^l} (\sum_h w_{ih}^l a_h^{l-1} + b_i^l)$$

$$= \delta_i^l a_j^{l-1},$$

where k iterates over all neurons in layer l and h iterates over the neurons in layer $l - 1$.

Equations 5.4 to 5.7 provide a formal framework to update the parameters of any neural network (weights and biases). They also highlight several important observations:

(i) Implementing a basic FFNN is not that complicated. Equations 5.4 to 5.7 rely on only two derivatives: the derivative of the activation function f, and the derivative of the cost function. In theory, these could be hard-coded for the typical activation and cost functions to be supported. The rest of the mathematical operations needed to implement backpropagation are just additions and multiplications. However, in practice, additional issues need to be addressed for a successful neural network implementation. We will discuss these issues in Chapter 6.

(ii) Backpropagation is slow. As shown in the equations, updating the network parameters requires a considerable number of multiplications. For real-world neural networks that contain millions of parameters, this becomes

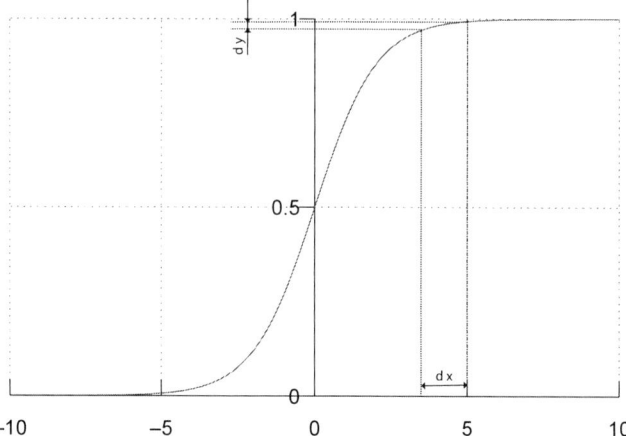

Figure 5.5 Visualization of the vanishing gradient problem for the logistic function: changes in x yield smaller and smaller changes in y at the two ends of the function, which means that $\frac{d}{dx}\sigma$ approaches zero in the two extremes

a significant part of the training runtime. In the next chapters, we will discuss multiple strategies for speeding up the training process such as batching multiple training examples and multiple operations together (e.g., updating all bias terms in a layer with a single vector operation rather than several scalar updates as in the equations). When these tensor operations are moved onto a GPU, which has hardware support for parallel tensor operations, they can be executed much faster.

(iii) Depending on the activation function, its partial derivative with respect to model parameters may be too small, which slows down the learning process. This happens because the equations to compute the errors in the network layers (Equations 5.4 and 5.5) both depend on this derivative. Multiplying this derivative repeatedly, as required by the recursive process described in the two equations, may have the unintended consequence of pushing some errors toward zero, which, in turn, means that the network parameters will not be updated in a meaningful way. This phenomenon is commonly called the "vanishing gradient problem." Figure 5.5 shows a visualization of this phenomenon for the logistic activation function. The softmax function, which generalizes the logistic function to multiclass classification, suffers from the same problem. For this reason, other activations that are more robust to this problem are commonly used in deep learning. We will discuss some of these in Chapter 6.

5.4   Drawbacks of Neural Networks (So Far)

In this chapter, we generalized LR into multilayered neural networks, which can learn nonlinear functions. This is a major advantage over LR, but it can be also be a drawback: because of their flexibility, neural networks can "hallucinate" classifiers that fit the training data well, but fail to generalize to previously unseen data (Domingos, 2015). This process is called *overfitting*. We will discuss multiple strategies to mitigate overfitting in Chapter 6.

In addition to overfitting, the training process of neural networks may suffer from other problems. We discussed the vanishing gradient problem in the previous section. Another problem commonly observed when training neural networks is the tendency to "Tony Hawk" the data, which slows down convergence or prevents it altogether. Chapter 6 discusses optimization algorithms that reduce this phenomenon.

Further, similar to the perceptron and LR, the neural networks covered so far continue to rely on handcrafted features. We will address this limitation in Chapter 8. Last, FFNNs focus on individual predictions rather than structured learning (i.e., where multiple predictions such as the part of speech in a sentence are jointly generated). We will introduce structured prediction using neural networks in Chapter 10. This will open the door to other important NLP applications such as part-of-speech tagging, named entity recognition, and syntactic parsing.

5.5   Historical Background

While the idea of neural networks was introduced by McCulloch and Pitts (1943), their network did not learn. The first general-purpose multilayer neural network that could learn from data was proposed by Ivakhnenko and Lapa (1966). However, they used a simpler (and more limited) method to train it rather than the back-propagation algorithm we discussed in this chapter. Backpropagation was codiscovered in the early 1960s by Kelley (1960) and Dreyfus (1962), and was formalized in the modern form we covered here by Linnainmaa (1970) (in a master's thesis!). However, he did not connect backpropagation to the training of neural networks. The first connection was made in the early 1980s by Werbos (1982) and soon after by Rumelhart et al. (1985).

Thus, we have had an assembled puzzle that connected "deep" neural networks to general-purpose training using backpropagation for almost 40 years now. So why did it take so long for neural networks to achieve the tremendous successes we see today? There are probably at least four reasons for the slow start. First, the interest in (and funding for) deep learning was negatively impacted by the "AI winter" we mentioned in Chapter 2 (despite the fact that the drawbacks observed by Minsky and Papert (1969) applied to the perceptron, not to multilayer neural networks). Second, it took a series of "tricks" (discussed in the next chapter) to bring stability to the process of training multilayer networks. Third, neural networks tend to require more data to learn than other ML algorithms. Until such datasets became available in image and language processing, other algorithms such as the support vector machines developed by Cortes and Vapnik (1995) dominated. One research direction that mitigated the lack of large annotated datasets was to pretrain neural networks using unsupervised algorithms. This idea was first proposed by Schmidhuber (1992) for recurrent neural networks (which we will introduce in Chapter 10). Pretraining is widely used today thanks to the advent of transformer networks, which are more amenable to expensive pretraining due to their architecture (see Chapters 12, 13, 14, 15, and 16). Last, deep learning was widely adopted when general-purpose GPUs became available. General-purpose GPUs, which were originally developed for graphics applications such as video games, provide hardware support for parallel matrix operations, which speed up the training and inference of neural networks considerably. This was first observed by Raina et al. (2009) and Cireşan et al. (2010), and then scaled up to a large network by Krizhevsky et al. (2012).

5.6   References and Further Readings

For a comprehensive description of deep learning, we recommend Goodfellow et al. (2016). We also found the online book of Nielsen (2019) to be an approachable introduction to deep learning, more in line with the scope of our book.

5.7   Summary

This chapter exposed us to FFNNs that assemble multiple "neurons" into arbitrary structures in which each neuron is itself a generalization of the perceptron and LR we saw in the previous chapters. Despite the more complicated structures presented, we showed that the key building blocks remain the same: the network is trained by minimizing a cost function. This minimization is implemented with backpropagation, which adapts the gradient descent algorithm introduced in the previous chapter to multilayer neural networks.

6 Best Practices in Deep Learning

The previous chapter introduced FFNNs and demonstrated that, theoretically, implementing the training procedure for an arbitrary FFNN is relatively simple: Algorithm 7 describes the learning algorithm that relies on SGD, and Algorithm 8 explains how the actual parameter updates are computed using backpropagation. Unfortunately, as described in Section 5.4, neural networks trained this way suffer from several problems including stability of the training process – that is, slow convergence due to parameters jumping around a good minimum – and overfitting.

In this chapter, we will describe several practical solutions that mitigate these problems. In particular, we will discuss minibatching and more modern optimization algorithms that improve the robustness of gradient descent; other activation functions that mitigate the vanishing gradient problem; a generalization of cost functions from binary to multiclass classification; techniques to reduce overfitting such as (a) regularization, (b) dropout, (c) temporal averaging, and, last, methods to initialize and normalize network parameters in order to increase our chances of finding a better minimum of the cost function during training. Note that most of these solutions are implemented in modern deep learning libraries such as PyTorch. We will see them in action in the next chapter.

6.1   Minibatching

Algorithm 7 updates the network parameters after each *individual* training example is seen. This means that the network changes its parameters at the fastest rate possible, with gradients that may have high variance (due to training examples that may be very different). These rapid changes may cause the resulting network to exhibit large differences in behavior (i.e., the network makes different predictions in response to the same inputs) in a short time. If you like, SGD is a training process that just had a triple espresso. Being highly caffeinated has several advantages and disadvantages. The pluses of this strategy are:

(i) In some cases, SGD converges to a good outcome more quickly due to the rapid parameter updates. This usually happens on easier problems, where the cost function has a minimum that is easy to find and yields a good solution.

(ii) Stochastic gradient descent has the capacity to "jump out" of local minima encountered during training due to the high variance in the gradients corresponding to different training examples. That is, similar to the function shown in Figure 3.3, the cost functions used by neural networks are not necessarily convex. At some point in the training process – that is, when only a subset of the training examples have been seen – the learning process may converge to a poor minimum – for example, similar to the one in the right part of the function shown in Figure 3.3. However, the following parameter updates, which can be drastically different from the previous ones that led to the suboptimal solution, increase the probability that the neural network leaves this local minimum and continues training.

(iii) Last but not least, SGD is easy to implement and has minimal memory requirements – that is, only one training example has to be kept in memory at a time.

The drawbacks of SGD are:

(i) The "jumping out" of suboptimal solutions advantage often translates into the disadvantage of slower convergence because the network "jumps around" good solutions rather than settling on one.

(ii) Stochastic gradient descent is computationally expensive due to the frequent updates of the network parameters. Further, these parameter updates are hard to parallelize due to the sequential traversal of the training examples.

The opposite of SGD is *batch gradient descent*, which updates the network parameters only after *all* the training examples have been seen. That is, batch gradient descent still computes the parameter gradients after each training example is processed, but updates them only at the end of each epoch with the average of all previously computed gradients. The process is summarized in Algorithm 9. In the algorithm, the variables $grad_\theta$ keep track of the sum of the partial derivatives of C with respect to θ for each training example i, and $|\mathbf{X}|$ indicates the size of the training dataset $\mathbf{X}$. If SGD is a highly caffeinated training process, batch gradient descent had a calming beverage such as chamomile tea. The advantages and disadvantages of this sedated training algorithm are opposite those of SGD. That is, its main advantages are:

(i) The average gradients used to update the network parameters tend to be more stable than the individual gradients used in SGD, and this often leads to convergence to better (local) minima on some problems.

Algorithm 9 Batch gradient descent algorithm

---

1 initialize parameters in Θ
2 **while** *not converged* **do**
3   **for** *each θ in Θ* **do**
4          $\text{grad}_\theta = 0$
5   **end**
6   **for** *each training example $\mathbf{x}_i$ in $\mathbf{X}$* **do**
7   **for** *each θ in Θ* **do**
8             $\text{grad}_\theta = \text{grad}_\theta + \frac{d}{d\theta} C_i(\Theta)$
9   **end**
10   **end**
11   **for** *each θ in Θ* **do**
12         $\theta = \theta - \alpha \frac{\text{grad}_\theta}{|\mathbf{X}|}$
13   **end**
14 **end**

(ii) Batch gradient descent is more computationally efficient because of the fewer updates of the network parameters. Further, batching is better suited for parallel implementations. That is, the **for** loop in line 6 of Algorithm 9 can theoretically be executed in parallel because the individual gradients are only used at the end of the loop.

Its disadvantages are:

(i) Batch gradient descent may prematurely converge to a less-than-ideal solution because its ability to "jump out" of an undesired local minimum is reduced.

(ii) Despite the computational efficiency within an individual epoch, batch gradient descent may take longer to train (i.e., more epochs) because the network parameters are updated only once per epoch.

(iii) The implementation of batch gradient descent is more complicated than that of SGD because it needs to keep track of the sum of all gradients for each network parameter throughout an epoch.

These two extreme strategies suggest that a middle ground may be the best practical solution. This middle ground is called *minibatch gradient descent*. Similar to batch gradient descent, the minibatch variant updates the network parameters only after a batch is completed, but its batches are smaller. For example, typical minibatch sizes for many NLP problems are 32 or 64 training examples. The minibatch gradient descent algorithm is described in Algorithm 10. Similar to Algorithm 9, the variables grad_θ keep track of the sum

---

Algorithm 10 Minibatch gradient descent algorithm

---

1 initialize parameters in Θ
2 **while** *not converged* **do**
3   **for** *each mini-batch* **M** *sampled from* **X do**
4   **for** *each θ in Θ* **do**
5              $\text{grad}_\theta = 0$
6   **end**
7   **for** *each training example* x_i **in M do**
8   **for** *each θ in Θ* **do**
9                  $\text{grad}_\theta = \text{grad}_\theta + \frac{d}{d\theta} C_i(\Theta)$
10   **end**
11   **end**
12   **for** *each θ in Θ* **do**
13             $\theta = \theta - \alpha \frac{\text{grad}_\theta}{|\mathbf{M}|}$
14   **end**
15   **end**
16 **end**

---

of the partial derivatives of C with respect to θ for each training example i in a given minibatch **M**. They are reset at the start of each minibatch (lines 4–6), and are used to update the parameter values after each minibatch completes (lines 12–14). $|\mathbf{M}|$ indicates the number of training examples in the minibatch **M**.

On the spectrum of caffeinated beverages, minibatch gradient descent consumed a green tea, a beverage that provides just enough energy without the jitters associated with large espressos. More formally, the advantages of minibatch gradient descent combine the best traits of the previous two training algorithms:

(i) Minibatch gradient descent tends to robustly identify good local minima because it reduces the "jumping around" disadvantage of SGD, while keeping some of its "jumping out" of undesired minima advantage.

(ii) Minibatch gradient descent allows for efficient, parallel implementation within an individual minibatch. We will show how this is done in PyTorch in the next chapter.

The main disadvantages of minibatch gradient descent are:

(i) Similar to batch gradient descent, its implementation is somewhat more complicated than that of SGD due to the additional bookkeeping necessary for each minibatch.

(ii) More importantly, minibatch gradient descent introduces a new *hyper parameter* – that is, a variable that needs to be tuned outside of the actual training process – the size of the minibatch. Unfortunately, the size of the minibatch tends to be specific to each task and dataset. Thus, the developer must search for the best minibatch size through an iterative trial-and-error *tuning process*, where various sizes are used during training, and the performance of the resulting model is evaluated on a separate tuning (or development) partition of the data.

In the next section, we describe other optimization algorithms that further increase the stability of the training process.

6.2   Other Optimization Algorithms

Beyond minibatching, several improvements have been proposed to increase the robustness of gradient descent algorithms. The first one we will discuss is *momentum* (Qian, 1999). Figure 6.1 provides a simple real-world analogy for it: imagine two sleds going down a hill, about to encounter a ravine. Sled 1 starts right before the ravine, whereas sled 2 starts further up the hill. Clearly, sled 1 is more likely to get stuck in the ravine than sled 2, which carries more speed (or momentum) as it enters the ravine, and is more likely to escape it. Sled 1 is the equivalent of the previous minibatch gradient algorithm, which is more likely to get stuck in a local minimum (the ravine). Algorithm 10 shows that for each minibatch – that is, step down the hill – we initialize each gradient – that is, the speed at this moment (line 5) – with zero. That is, we forget about the speed we had previously and compute the current speed simply based on the slope under our sled at this time. Gradient descent with momentum fixes this by initializing the gradients with a fraction of the final gradients computed for the previous minibatch. That is, at time t – that is, when the tth minibatch is processed – line 5 in Algorithm 10 changes to:

$$\text{grad}_\theta^t = \gamma \ \text{grad}_\theta^{t-1}, \tag{6.1}$$

where grad_θ^{t-1} is the gradient for θ computed for the previous minibatch, and γ is a hyper parameter with values between 0 and 1, which indicates how much of the previous momentum we want to preserve.[1]

A variant of momentum, called *Nesterov momentum* (Nesterov, 1983), builds upon this intuition by also changing line 9 of Algorithm 10. In particular, Nesterov momentum does not compute the partial derivative of the cost function, $\frac{d}{d\theta} C_i$, using the actual parameters in Θ. Instead, this algorithm subtracts the momentum – that is, a fraction of grad_θ^{t-1} – from each parameter θ when computing C_i. The intuition behind this operation is that this allows the

---

[1]   Common values for γ are around 0.9.

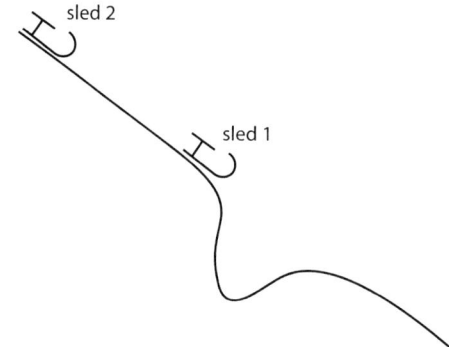

Figure 6.1 Illustration of momentum: sled 1 is more likely to get stuck in the ravine than sled 2, which starts farther up the hill, and carries momentum when it enters the ravine

algorithm to "peek into the future" by using values that estimate the parameter values at time $t+1$. This is possible because we know through the combination of the momentum initialization (discussed in the previous paragraph) and the actual update operation (line 13 in Algorithm 10) that the value of each parameter θ at the end of this minibatch will be computed by subtracting a fraction of its corresponding momentum from the old value of θ.[2] Thus, Nesterov momentum is informed by both the past (through the momentum initialization) and the future (through the modified parameter values when computing the cost function). Empirically, it has been shown that this brings more stability to the training process (Dean et al., 2012).

Another complication of gradient descent is identifying a good learning rate – that is, an appropriate value for the hyper parameter α in line 13 of Algorithm 10. Any deep learning practitioner will quickly learn that the performance of most deep learning models depends heavily on the learning rate used. In the opinion of the authors, the learning rate is the most important hyper parameter to be tuned. A value that is too big will yield faster training, but may cause the training process to "jump" over good minima. On the other hand, a value that is too small may cause training to be too slow and to risk getting stuck in a suboptimal minimum (i.e., the ravine in Figure 6.1). Further, the learning rate value should be adjusted based on the timeline of training. Earlier in the process, a larger training rate helps approaching a good solution more quickly from the randomly chosen starting point, but later in the course, a smaller value is generally preferable to avoid jumping out of good minima. Last, different features likely require different learning rates. That is, using a

---

[2]   We will, of course, also subtract the other $\frac{d}{d\theta} C_i$ computed for this minibatch, but these are unknown at this time.

single learning rate may cause the frequent features – that is, features commonly observed with nonzero values in training examples – to dominate in the learned model because their associated parameters – that is, the edges connecting them to the output neurons – will be updated more frequently. Thus, ideally, we would like to perform larger updates for parameters associated with less frequent features (which may still contain useful signal!) so they have a say in the final model.

The solution to all of these problems is to use *adaptive learning rates* – that is, have a distinct learning rate for each parameter θ in the network – and allow these values to change over time. A common strategy is to have each learning rate be inversely proportional to the square root of the sum of the squares of the gradients observed for this parameter in each minibatch up to the current one. That is, if we denote the sum of squares of the gradients for parameter θ as G_θ, then line 13 in Algorithm 10 becomes:

$$\theta = \theta - \frac{\alpha}{\sqrt{G_\theta + \epsilon}} \frac{\text{grad}_\theta}{|\mathbf{M}|}, \tag{6.2}$$

where ϵ is a small constant to avoid division by zero. There are several important observations about this seemingly simple change:

- Because G_θ is distinct for each θ, this formula yields different learning rates for different parameters.
- The summation of squares in G_θ guarantees that G_θ monotonically grows over time, regardless of the sign of the gradients. Thus, this formula captures our temporal intuition: learning rates will be larger in the beginning of the training process (when G_θ is small), and smaller later (when it is larger).
- Similarly, G_θ guarantees that parameters associated with frequent features will get smaller updates, while parameters associated with infrequent features will receive larger ones. This is because the gradients of parameters associated with frequent features will more often have nonzero values, which will lead to larger values for G_θ.

Most modern variants of gradient descent incorporate some form of momentum and adaptive learning rates. For example:

- The AdaGrad algorithm uses Nesterov momentum and the adaptive learning described earlier (Duchi et al., 2011).
- AdaDelta (Zeiler, 2012) and RMSProp address the fact that AdaGrad's learning rates are continuously diminishing due to the ever-increasing G_θ.[3] They both achieve this by changing the formula of G_θ to iterate only over

---

[3]   RMSProp was proposed by Geoff Hinton but never formally published.

the most recent few gradients. Further, instead of using a straight summation, these algorithms use a *decaying average*, where the contribution of older gradients decreases over time.[4]

- Adaptive Moment Estimation (Adam) (Kingma and Ba, 2015) builds on the previous two algorithms by also applying the same decaying average idea to the actual gradient values. That is, in Equation 6.2, Adam does not use the actual $\frac{grad_\theta}{|M|}$ value computed for this minibatch, but a decaying average of the past few gradients.

All these algorithms are supported by most deep learning libraries such as PyTorch. A superficial analysis of NLP publications performed by the authors suggests that Adam and RMSProp seem to be the ones most commonly used for NLP at the time this book was written.

6.3   Other Activation Functions

In the previous chapter, we mentioned that one important drawback of the logistic function (and its multiclass equivalent, the softmax) is the vanishing gradient problem. This is caused by the fact that the derivative of the logistic tends to be small. When several such derivatives are multiplied during backpropagation, the resulting value may be too close to zero to impact the network weights in a meaningful way. One solution to the vanishing gradient problem is to use other activation functions that have larger gradient values. One such activation function is the hyperbolic tangent function, or tanh:

$$tanh(x) = \frac{e^{2x} - 1}{e^{2x} + 1}. \tag{6.3}$$

Figure 6.2 shows a plot of the tanh function overlaid over the plot of the logistic function. The figure highlights the key advantage of the tanh function: the slope of the tanh is steeper than the logistic's and thus its derivative has larger values than the derivative of the logistic for most input values. This is the key reason tanh suffers less from the vanishing gradient problem, and why it is usually preferred over the logistic. However, as shown in the figure, for extreme input values (very large or very small), tanh also exhibits saturated gradients – that is, partial derivatives with very small values. One activation function that avoids this problem is the rectified linear unit, or ReLU:

$$ReLU(x) = \max(0, x) = \begin{cases} 0 & x < 0 \\ x & \text{otherwise.} \end{cases} \tag{6.4}$$

---

[4] For the exact math behind this change, and a more expanded discussion of optimization algorithms, we recommend Sebastian Ruder's excellent blog post: `https://ruder.io/optimizing-gradient-descent`.

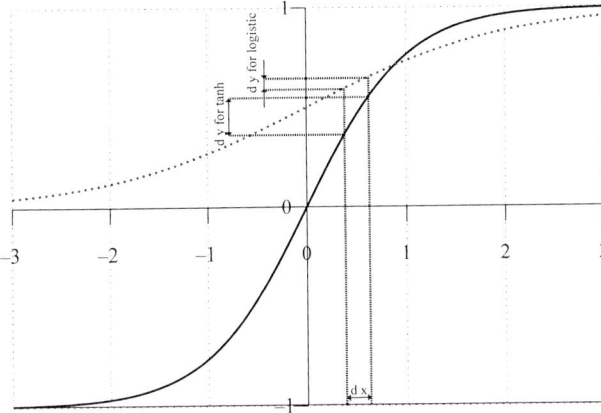

Figure 6.2 Comparison of the tanh (continuous line) and logistic (dashed line) functions. The derivative of the tanh is larger than the derivative of the logistic for input values around zero

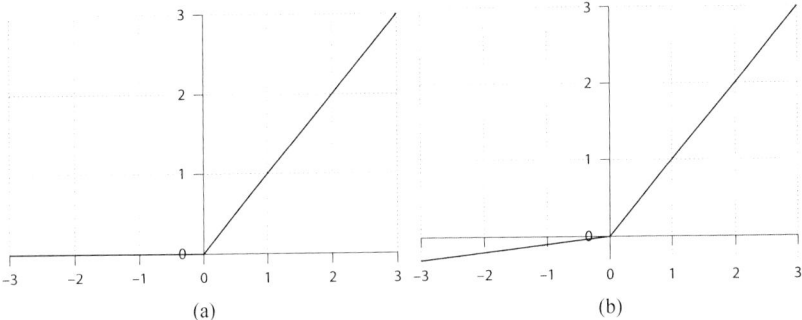

Figure 6.3 The ReLU (a) and Leaky ReLU (b) activation functions

Figure 6.3 (a) shows a plot of this function.[5] The figure shows that for positive input values, ReLU grows at a constant rate, without exhibiting the saturated gradients of the tanh of logistic functions, both of which taper at 1. In practice, this means that the training process for networks that rely on ReLU activation functions tends to converge faster than those that use tanh or logistic activations. As an empirical rule of thumb, ReLU tends to learn faster than tanh, which usually converges more quickly than logistic.

A second, more subtle advantage of ReLU is caused by the fact that, as the figure shows, the function's value is 0 for all negative input values. This means

---

[5] Figure 6.3 shows that ReLU and its variants are not differentiable for $x = 0$. For this input value, we typically set the derivative of ReLU to an arbitrary value – for example, 1 or 0.5.

that all neurons with ReLU activations whose dot product of input values and weights is negative become inactive – that is, their output is 0. Glorot et al. (2011) observed that "after uniform initialization of the weights, around 50% of the hidden units' continuous output values are real zeros" in such networks. This percentage increases when regularization is used (see Section 6.5). But why would sparse representations of neural networks be preferred? It turns out that sparsity has several advantages (Glorot et al., 2011):

- It allows a network to learn more flexible representations. That is, varying the number of neurons with nonzero activations in each layer allows better control of the actual dimensionality assigned to each layer (rather than relying on the hard-coded structure).
- It yields better explainability or better "information disentangling." That is, in the usual, dense networks where most activations are nonzero, it is hard to understand the underlying reason for a given output because all neurons contributed something to the final activation. Further, such networks tend to be more sensitive to small changes in inputs. Both these drawbacks are mitigated by sparse networks. They can explain their outputs easier because a smaller number of neurons contributes to the final activations. Further, some small input changes are dampened by the inactive neurons and thus do not affect the network's outputs.
- Sparse representations are more likely to be linearly separable, which means it is easier and cheaper to learn a good classifier.

A third advantage of ReLU is its computational simplicity. Unlike logistic and tanh, which require exponential operations, ReLU relies solely on the much simpler max function.

One drawback of ReLU's hard saturation at 0 is an extreme form of vanishing gradient commonly called the "dying ReLU." That is, not surprisingly, the derivative of an inactive ReLU neuron is always 0, which means that weights immediately downstream of it are not updated during backpropagation (see Equation 5.7). The solutions to this problem are variants of ReLU that keep the general ReLU behavior, but avoid the hard saturation at 0. One such function is Leaky ReLU:

$$LeakyReLU(x) = \begin{cases} \alpha x & x < 0 \\ x & \text{otherwise,} \end{cases} \tag{6.5}$$

where α typically takes values around 0.01. Figure 6.3 (b) shows a plot of Leaky ReLU.

All in all, this discussion suggests that, while some recommendations can be made, there is no universal answer to the question: which activation function is best? Many activation functions have been proposed (we recommend you check your favorite deep learning library's documentation for the list of

Table 6.1 Three cost functions commonly used in natural language processing tasks. m indicates the number of data points in the training dataset (or the minibatch, in the case of minibatch gradient descent). y_i is the correct label for example i

Name	Type of classification	Formula
Mean squared error	Binary	$\frac{1}{m}\sum_{i=1}^{m}(a_1^L - y_i)^2$
Binary crossentropy	Binary	$-\sum_{i=1}^{m}(y_i \log p(1|\mathbf{x}_i; \mathbf{w}, b) + (1 - y_i) \log p(0|\mathbf{x}_i; \mathbf{w}, b))$
		$= -\sum_{i=1}^{m}(y_i \log a_1^L + (1 - y_i) \log(1 - a_1^L))$
Crossentropy	Multiclass	$-\sum_{i=1}^{m} \log p(y_i|\mathbf{x}_i; \mathbf{W}, \mathbf{b})$

supported activation functions). All activation functions try to balance multiple desired, but sometimes conflicting properties such as mitigating vanishing gradient and producing sparse representations. Which one works best for you likely depends on the problem and data you work on at the moment. Be prepared to try several.

6.4 Cost Functions

So far, we have seen three cost functions in the book: cross-entropy and binary cross-entropy (Chapter 3) and, very briefly, the MSE cost (Chapter 5). These are probably the most common cost functions in NLP, so it is worth discussing them slightly more formally.

Recall that a cost function must have several properties: (a) it should return only positive values, (b) it should measure the distance between the classifier predictions and the corresponding correct (or gold) labels – that is, the higher the value of the cost function the more incorrect the underlying classifier is, and (c) it should be differentiable so we can "plug" it in some form of the gradient descent algorithm. All three cost functions, listed in Table 6.1, have these properties. Let us convince ourselves that this is indeed the case.

First, the MSE as shown here is designed for binary classification. That is, the underlying network has a single final neuron whose activation (a_1^L) is typically produced with either a logistic function or a hyperbolic tangent function. In the former case, the value of a_1^L should approach 1 for the positive class and 0 for the negative class. In the latter situation, the positive class has label 1 and the negative one -1. Regardless of the choice of activation function, MSE always returns positive values due to the square in its formula. Also, MSE explicitly measures the distance between the classifier prediction and the gold label, and this distance is differentiable. The MSE is easy to explain and trivial to implement, but it has one major disadvantage: it may lead to slow learning. To understand this disadvantage let's revisit Equation 5.4 in the context of MSE. For a binary classifier with a single final neuron, the error in the final layer is: $\delta_1^L = \frac{d}{da_1^L} C \frac{d}{dz_1^L} f(z_1^L)$. For a single training example, this error becomes: $\delta_1^L = 2(a_1^L - y)\frac{d}{dz_1^L}f(z_1^L)$, where y is the gold label for the corresponding example. Thus, δ_1^L depends on the derivative of the activation function, which becomes vanishingly small at the two ends of the function, for both the logistic and tanh functions. Because the weight and bias updates in any neural network (see Equations 5.6 and 5.7) depend on δ_1^L, a neural network trained using the MSE cost is likely to experience learning slowdown.

The binary cross-entropy addresses this limitation. Before we explain how, let us convince ourselves that the binary cross-entropy is a proper cost function. Because the logarithms in its formula take probabilities as parameters, we are restricted here to logistic activation functions for a_1^L. Thus, the summation in the formula is always smaller or equal to 0 (the natural logarithm of a number smaller than 1 is negative), and the resulting overall value is larger or equal to 0. Further, binary cross-entropy measures the quality of the classifier. For example, a good classifier that produces an a_1^L approaching 1 for a positive label ($y = 1$) will have a binary cross-entropy cost of $-\log a_1^L \approx -\log 1 = 0$. At the opposite extreme, a really bad classifier that produces a_1^L approaching 1 for a negative label ($y = 0$) will have a binary cross-entropy cost of $-\log(1 - a_1^L) \approx -\log 0 = \infty$.

To understand why the binary cross-entropy reduces the learning slowdown, let us derive δ_1^L in this context, for a single training example with gold label y:[6]

$$\delta_1^L = \frac{d}{da_1^L} C \frac{d}{dz_1^L} f\left(z_1^L\right)$$

$$= \frac{d}{da_1^L}(-y \log a_1^L - (1 - y) \log(1 - a_1^L))\frac{d}{dz_1^L}\sigma(z_1^L)$$

$$= (-\frac{y}{a_1^L} + \frac{1 - y}{1 - a_1^L})\frac{d}{dz_1^L}\sigma(z_1^L)$$

$$= \frac{a_1^L - y}{a_1^L(1 - a_1^L)}\frac{d}{dz_1^L}\sigma(z_1^L)$$

$$= \frac{a_1^L - y}{\sigma(z_1^L)(1 - \sigma(z_1^L))}\frac{d}{dz_1^L}\sigma(z_1^L)$$

$$= \frac{a_1^L - y}{\sigma(z_1^L)(1 - \sigma(z_1^L))}\sigma(z_1^L)(1 - \sigma(z_1^L))$$

$$= a_1^L - y.$$

[6] We recommend that the user verifies this derivation using the information in Table 3.1.

Thus, surprisingly, δ_1^L for binary cross-entropy does *not* depend on the derivative of the activation function! Because of this, the binary cross-entropy cost function is more resilient to learning slowdown, which makes it the most common choice for binary classification problems implemented with networks that have logistic activation in the output layer.

The cross-entropy cost (last row in Table 6.1) is simply a generalization of binary cross-entropy to multiclass classification.[7] Recall from the previous chapter that networks designed for multiclass classification have one neuron dedicated to each class in the output layer, and these neurons are usually followed by a softmax layer such that the final outputs form a probability distribution (see Figure 5.2). For such architectures, the cross-entropy cost maximizes the probability of the gold label for each training example i: $p(y_i|\mathbf{x}_i; \mathbf{W}, \mathbf{b})$. Because the denominator in the softmax formula iterates over the other activations (see Equation 3.16), maximizing the probability of the gold label for a given training example has the desired side effect of also minimizing the probabilities of all other (incorrect) labels for the same data point. Cross-entropy is the preferred cost function for multiclass classification in most NLP tasks for the same reason binary cross-entropy is the favored cost function for binary classification.

6.5 Regularization

In three of the four previous sections in this chapter, we discussed techniques to improve the stability of the training process for neural networks (e.g., mini-batching, improved optimizers, and better cost functions). Regularization is a fourth common technique used for this purpose. Recall from Chapter 2 that regularization is a family of techniques that control for the noise that is potentially present in the training data. Implementation-wise, regularization methods control for undesired fluctuations in parameter (i.e., weights or bias terms) values that may occur when the training process is not stable due to exposure to noise.

In Chapter 2, we have seen the averaged perceptron as one simple regularization method. Moving the same intuition into the space of cost functions, regularization is implemented for neural networks by adding an additional term to the cost:

$$C_{reg}(\mathbf{W}, \mathbf{b}) = C(\mathbf{W}, \mathbf{b}) + \lambda R(\mathbf{W}, \mathbf{b}), \qquad (6.6)$$

where $C(\mathbf{W}, \mathbf{b})$ is any cost function without regularization such as the ones listed in Table 6.1, R is the new regularization function, and λ is a positive

---

[7] Goodfellow et al. (2016) (Section 5.5) point out that calling this cost function "cross-entropy" is a misnomer, as the actual cross-entropy formula is more complex. However, minimizing the actual cross-entropy is equivalent to minimizing the formula in Table 6.1. For this reason, this abuse of terminology is widely spread. We will continue to use it throughout this book.

number (usually a small one) that indicates how much importance to impart on the regularization component of the cost. At a high level, the regularization function R aims to reduce the complexity of the network by pushing (some of) the network parameters toward zero.

Intuitively, there is a tug-of-war between the C and R functions in this equation when C_{reg} is minimized.[8] On one hand, C needs to be minimized, which has the effect of increasing the values of certain weights and biases – for example, to maximize the probabilities of the gold labels for the cross-entropy cost. On the other hand, minimizing R has, by definition, the effect of explicitly minimizing all weight and bias absolute values, which keeps the former component in check and has the desired effect of "squishing" unreliable parameter values.

There are many possible implementations for the regularization function R. *L2 regularization* is probably the most common one:

$$R(\mathbf{W}, \mathbf{b}) = \sum_{i=1}^{V} v_i^2, \tag{6.7}$$

where v_i iterates over all parameters in the network – that is, edge weights and bias terms – and V is the total number of weights and biases in the network.[9] In plain language, L2 regularization is simply the sum of the squared values of the weights and biases in the given network.

So, why would adding such a regularization term to the cost function have the effect of mitigating the parameter value fluctuation? To understand this, recall that gradient descent updates each network parameter v_i (again, these parameters include all weights and biases) by subtracting the partial derivative of the cost with respect to v_i, $\frac{d}{dv_i}C_{reg}(\mathbf{W}, \mathbf{b})$, from the current value of v_i. For each v_i, the partial derivative of the L2 regularization, which is part of C_{reg}, is: $\frac{d}{dv_i}R(\mathbf{W}, \mathbf{b}) = 2v_i$. Thus, during each back-propagation step, in addition to subtracting the partial derivative of the original C, we also subtract $2\lambda v_i$ from each parameter v_i. This has the effect of pushing parameter values toward zero. Importantly, this effect is more aggressive for values that are farther from zero because the corresponding derivative $(2v_i)$ has a larger absolute value.

Another common regularization function is L1 regularization, which is simply the sum of the parameter absolute values:

$$R(\mathbf{W}, \mathbf{b}) = \sum_{i=1}^{V} |v_i|. \tag{6.8}$$

Because the partial derivative of L1 regularization with respect to v_i is constant (i.e., $+1$ for positive values of v_i and -1 for negative values), the additional

[8] https://en.wikipedia.org/wiki/Tug_of_war.
[9] The more mathematically inclined reader may have realized that the name L2 regularization comes from the fact that this function is the square of the L2 norm of the vector containing all network parameters v_i.

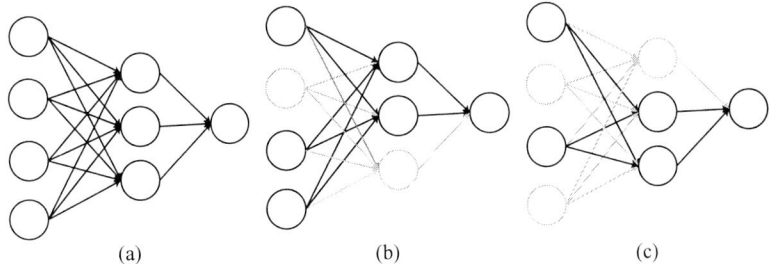

Figure 6.4 A simple neural network (a), and two views of it after dropout is applied (b and c). Greyed-out nodes and edges are dropped out and thus ignored during the corresponding forward pass and backpropagation in (b) and (c)

term introduced in $\frac{d}{dv_i}C_{reg}(\mathbf{W}, \mathbf{b})$ by L1 regularization is simply the constant $+\lambda$ or $-\lambda$. In practice, this means that L1 regularization "squishes" the parameter values by a constant value in each back-propagation step, regardless of the original values of these parameters. The consequence of this is that L1 regularization impacts parameter values that are closer to zero more aggressively than L2, which produces sparser networks (i.e., with more edge weights reduced to 0).

Empirically, L2 regularization tends to perform better than L1 for NLP tasks. But, as mentioned, L1 produces sparser networks, which can be represented in memory more efficiently. All in all, most forms of regularization (L1, L2, and others) are trivial to implement (they are just an additional, simple term in the cost function), and they tend to be beneficial.

6.6   Dropout

Dropout can be seen as another simple regularization strategy. However, instead of changing the cost function used to train the neural network to encourage weight "squishing," dropout changes the *structure* of the network during training. That is, for each training example, dropout ignores (or "drops") network neurons, as well as their incoming and outgoing connections, with probability p. For example, if $p = 0.3$, the network will remove 30% of its neurons on average, for each training example. Figure 6.4 visualizes this process for a simple feed-forward network (Figure 6.4 (a)), and two views obtained in different dropout iterations (Figure 6.4 (b) and (c)).

An important implementation detail about dropout is that dropout is applied only during training. That is, at testing time, the entire network – for example, the one in Figure 6.4 (a), is used. This difference has the effect that the output values generated by the network during training are smaller – that is, only $1 - p$ of the corresponding values seen during testing. For example, while the full network shown in Figure 6.4 (a) has eight neurons, the one in Figure 6.4 (b) has only six. Thus, on average, we expect the output of the final neuron in Figure 6.4 (b) to be only $\frac{6}{8}$ of the output of the original network. This introduces a problem, as it is best if the networks used in training and testing are similar, otherwise the weights learned during training will not be effective. This issue is solved in deep learning libraries with one of the two following strategies. The first strategy *scales down* by p the output of each neuron during testing. The second strategy *scales up* by $\frac{1}{p}$ the output of each active neuron (i.e., not dropped out) during training. The first strategy has the advantage of a simpler training procedure (no need to worry about scaling during training), but it complicates testing. That is, the one must remember the dropout probability used during training, and scale down neuron outputs accordingly during testing. The second strategy has the opposite benefits: more complicated training, but simpler testing that requires no knowledge about how dropout was applied during training.

The hyper parameter p is chosen based on the performance of the trained network on a development dataset. In many NLP tasks, typical values for p range between 20% and 30%.

There are two explanations for why dropout is useful. The first is that dropout forces the remaining nodes in the network to take more responsibility for transforming the received inputs into the correct prediction. The other side of the coin is that because dropout uses sparser networks, it encourages the learning of sparser representations – that is, learning smaller weights for neurons that the network learns can be safely ignored. This is exactly what regularization does as well! The second explanation is that dropout can be seen as an average of many different networks, one for each training example. For example, the training procedure may see the network in Figure 6.4 (b) for one training example, and the one in Figure 6.4 (c) for another example. But because these networks sample from the same group of neurons, the training procedure ends up aggregating all these updates into the same overall network, effectively learning an average of all the dropout views. Intuitively, this is very similar to the average perceptron we have seen in Chapter 2, with exactly the same benefits as discussed there.

6.7   Temporal Averaging

Temporal averaging is yet another simple strategy for constructing an ensemble model that builds upon the intuition behind the average perceptron. Temporal averaging works by averaging the network weights at the end of the best training epochs. More formally, temporal averaging is implemented through the following steps:

(i) At the end of each training epoch – for example, after each iteration in the outermost `while` loop in Algorithm 10 – the performance of the current network is evaluated on a development dataset, and the current weights are all saved.

(ii) After training completes, all epochs are sorted in descending order of their development performance.

(iii) The weights from the top k best epochs are averaged into the final model. The hyper parameter k is empirically chosen based on the performance of the final model on the development partition. Typical values range between 3 and 5.

In other words, temporal averaging is similar to the average perceptron, but the model snapshots that are averaged are not created every time the model is updated (which would be after each training example for gradient descent!) but less frequently – for example, at the end of an epoch. While less popular than dropout, in the experience of the authors temporal averaging is just as effective.

In the past few sections, we have discussed three different regularization strategies: "traditional" regularization through the cost function (Section 6.5), dropout (Section 6.6), and the temporal averaging discussed in this section. These three techniques are largely complementary to each other, and, because of this, often combined.

6.8   Parameter Initialization and Normalization

As we discussed before, gradient descent does not find the global minimum of the cost function but rather its nearest minimum. Thus, where we start matters. In other words, it is important to initialize the parameters of the network to be trained (i.e., edge weights and bias terms) to values that increase our chances of finding a good solution. There are several rules of thumb on what one should and should not do when initializing the network parameters:

(i) The parameters should be initialized randomly. It is trivial to observe that if all parameters were initialized with the same value, the network will learn the same feature for all its neurons. In contrast, initializing the parameters to different (random) values forces the network to assign different meaning to each of its neurons, which is the desired behavior.

(ii) The initial parameter values should not be too small. Values that are exceedingly small lead to the vanishing gradient problem, which slows down learning or stops it all together.

(iii) On the other hand, parameter values should not be too large either. Large parameter values yield large parameter updates during backpropagation, which causes unstable learning, or "Tony Hawking" the data, as we put it in Chapter 2. This problem is called the "exploding gradient problem." We will discuss this issue and practical solutions to mitigate it in Chapter 10.

(iv) The distribution of parameter values should be centered around zero because this is where the interesting things happen with most activation functions (see, e.g., the tanh and ReLU activations introduced earlier in this chapter).

One very common parameter initialization strategy that follows these rules is the Glorot method (Glorot and Bengio, 2010).[10] This method initializes a neuron's edge weights uniformly from the range $[-\frac{1}{\sqrt{n}}, \frac{1}{\sqrt{n}}]$, where n is the number of input edges to this neuron. For example, if layer $l-1$ in a network has 100 neurons, then the initial edge weights for neurons in layer l will be in the range $[-\frac{1}{10}, \frac{1}{10}]$. Without getting into the math, making the range of the initial weight values depend on the number of input edges makes the fluctuation (or variation) of the neuron's output activation similar to that of its inputs, which leads to more stable learning.

For the same reasons, the last three rules of thumb also apply to neuron activations throughout the entire training process – that is, the activation a_i^l of neuron i in layer l (see Equation 5.1) should not be too small or too large, and be centered around zero. To make sure that this holds, one common strategy normalizes each neuron's activation across every individual minibatch $\mathbf{M}$ (see Algorithm 10) such that the distribution of values for each activation fits in a small range centered around zero.[11] For obvious reasons, this method is called *batch normalization* or *batch norm*. In particular, for each minibatch $\mathbf{M}$, batch normalization computes the mean μ and standard deviation σ (or, informally, how dispersed the set of values is) for all values observed for an activation a_i^l in the given minibatch. Then, it *recenters* a_i^l around zero by subtracting μ from it, and *rescales* it to a standard deviation of 1 by dividing the resulting difference by σ. What is important here is that the recentering and rescaling operations are differentiable. In practice, batch normalization is implemented as another network component that follows each layer to be normalized in the network, and is trained using the same gradient descent algorithm used for the rest of the network.

As we will see in the next few chapters, in some situations, minibatches are not available, yet we still desire to normalize neuron activations. The normalization strategy used in such situations is called *layer normalization* (Ba et al., 2016). As its name implies, this method normalizes all activations *in the same network layer*. That is, for each layer l, layer normalization computes the mean and standard deviation for the set of values corresponding to *all* activations a_i^l

---

[10] This method is also commonly called Xavier initialization, based on the first author's first name.

[11] For the more mathematically inclined reader, batch normalization uses a mean of 0, and a variation of 1 for each activation.

in layer *l*, and then recenters and rescales them using the same strategy as batch normalization. However, unlike batch normalization, which is applied only at training time when minibatches are available, layer normalization applies the same procedure during both training and testing.

Both batch and layer normalization have been shown to lead to more stable and faster training, and are generally recommended. Batch normalization tends to perform better when larger minibatches are available (due to the more robust statistics), whereas layer normalization is recommended for small minibatches, or when batching is not possible.

6.9   References and Further Readings

It is unclear who invented minibatching. However, one of the first analyses of its impact in training neural networks comes from Wilson and Martinez (2003), who showed that small batches can act as regularizers during training.

The idea of momentum was proposed by Polyak (1964), who found that accumulating a velocity vector in "directions of persistent reduction in the objective across iterations" accelerates gradient descent (Sutskever et al., 2013). Nesterov (1983) generalized this idea by also "peeking into the future" using parameter values. Momentum and other observations such as adaptive learning rates motivated a series of gradient descent variants – for example, AdaGrad (Duchi et al., 2011), AdaDelta (Zeiler, 2012), and Adaptive Moment Estimation (Adam) (Kingma and Ba, 2015).

The rectified linear unit function was first used as an activation function by Jarrett et al. (2009), but it was widely adopted after Nair and Hinton (2010) popularized it. Since then, multiple ReLU variants were introduced – for example, Maas et al. (2013) proposed leaky ReLU, and He et al. (2015) introduced parametric ReLU, which includes a learnable parameter. Many other activation functions have been proposed in the deep learning literature. A comprehensive survey of activation functions is available in Dubey et al. (2021).

Shannon (1948) introduced information entropy, which is the foundation for the cross entropy loss function we discussed in this chapter. One of the first references to using the cross-entropy loss comes from Cox (1958), who used it to train an LR (or a single-neuron neural network). However, most early multilayer neural networks used the simpler MSE loss. Today, the majority of neural networks are trained using the cross-entropy loss, but it is unclear when this transition took place.

Dropout is a relatively recent addition to the deep learning quiver. Srivastava et al. (2014) proposed it as a technique for mitigating overfitting. In contrast, L1 and L2 regularization are adopted from older, well-known mathematical techniques. For example, Tikhonov (1943) proposed L2 regularization (also known as Tikhonov regularization) as a method of regularization of ill-posed problems.

Glorot and Bengio (2010) proposed the widely used parameter initialization we discussed in this chapter. Another similar initialization strategy was introduced by Mikolov et al. (2013b). Batch normalization was first proposed by Ioffe and Szegedy (2015), who showed that it yields a better model for image classification. Layer normalization was introduced by Ba et al. (2016).

6.10   Summary

In this chapter, we discussed several practical solutions for problems that affect neural networks, such as stability of the training process and overfitting. In particular, we introduced minibatching, multiple optimization algorithms, other activation and cost functions, regularization, dropout, temporal averaging, and parameter initialization and normalization.

7   Implementing Text Classification
with Feed-Forward Networks

In this chapter, we provide an implementation of the multilayer neural network described in Chapter 5, along with several of the best practices discussed in Chapter 6.

Remaining fairly simple, our network will consist of three neuron layers that are fully connected: an input layer that stores the input features, a hidden intermediate layer, and an output layer that produces the scores for each class to be learned. In between these layers we will include dropout and a nonlinearity (ReLU).

Sidebar 7.1   The PyTorch Linear layer implements the *connections* between layers of neurons

Before discussing the implementation of more complex neural architectures in PyTorch, it is important to address one potential source of confusion. In PyTorch, the Linear layer implements the connections between two layers of neurons rather than an actual neuron layer. That is, a Linear object contains the weights $\mathbf{W}^{l+1}$ that connect the neurons in layer l with the neurons in layer $l + 1$ in Figure 5.2. This is why the Linear constructor includes two dimensions: one for the input neuron layer (in_features) and one for the output neuron layer (out_features). If the parameter bias is assigned the value of True, which is its default setting, the corresponding Linear object also contains the bias weights for the output neurons – that is, $\mathbf{b}^{l+1}$ in Figure 5.2. Thus, in our Model with three neuron layers, we will have two Linear objects.

To stay close to the code, from this point forward when we mention the term *layer* in the implementation chapters, we refer to a PyTorch Linear layer, unless stated otherwise.

Further, we make use of two PyTorch classes: a Dataset and a DataLoader. The advantage of using these classes is that they make several things easy, including data shuffling and batching. Last, since the classifier's architecture has become more complex, for optimization, we transition from SGD to the Adam optimizer in order to take advantage of its additional features such as momentum and L2 regularization. As before, the code from this chapter is available in a Jupyter notebook: chap7_ffnn.

7.1   Data

In this chapter, we continue to use the AG News Dataset (Section 4.2.1), including the same loading and preprocessing steps. Also, we continue using the same train and test sets to be able to compare results to the ones obtained in Section 4.2. However, in this chapter, we will make use of a development set to tune the model's hyper parameters. For this purpose, we split the training set in two: 80% of the examples become a new training set, while the other 20% are the development set:

```
[8]:  from sklearn.model_selection import train_test_split

      train_df, dev_df = train_test_split(train_df, train_size=0.8)
      train_df.reset_index(inplace=True)
      dev_df.reset_index(inplace=True)

      print(f'train rows: {len(train_df.index):,}')
      print(f'dev rows: {len(dev_df.index):,}')
```

```
train rows: 96,000
dev rows: 24,000
```

In this code, we used scikit-learn's train_test_split function to split the training set into a development partition and a new training partition. Note that this function can split Python lists, NumPy arrays, and even Pandas dataframes. The returned dataframes preserve the index of the original training dataframe, which can be useful to keep the connection to the original data, but is not what we currently need, as we are trying to create two independent datasets. Therefore, we reset the index of the two new dataframes.

A second difference to what was done in Section 4.2 is the introduction of minibatches. PyTorch provides the DataLoader class which can be used for shuffling the data and splitting it into minibatches.[1] In order to create a DataLoader, we need the data to be in the form of a PyTorch Dataset.[2] There are two main types of PyTorch datasets: map-style and iterable-style. We will use the former, as it is simpler and meets our needs, but it is good to know that the other option is available for situations when, for example, you need to stream data from a remote source or random access is expensive.

---

[1] https://pytorch.org/docs/stable/data.html#torch.utils.data.DataLoader.
[2] https://pytorch.org/docs/stable/data.html#torch.utils.data.Dataset.

To create a map-style dataset we need to subclass `torch.utils.data.Dataset` and override its `__getitem__()` method (to return an example given a key), as well as its `__len__()` method (to return the number of examples in the dataset). Our dataset implementation stores two sequences: one for holding the features, and another for storing the corresponding labels. In our implementation, we store two Pandas `Series`, but Python lists or NumPy arrays would also work. The implementation `__len__()` is trivial: we simply return the length of the feature sequence – that is, the number of feature vectors. The implementation of `__getitem__()` is slightly more involved. Recall that each of our feature vectors is represented as a dictionary with word IDs as keys and word counts as values, and any word ID not in the dictionary has a count of zero. Our `__getitem__()` method transforms this representation into one that PyTorch can use. We first create two PyTorch tensors, one for the label and one for the features, which is initially populated with zeros. Then we retrieve the feature dictionary corresponding to the provided index and, for each key-value pair in the feature dictionary, we update the corresponding element of the tensor. Once this is complete, we return the feature and label tensors for the datum:

```
[11]:   from torch.utils.data import Dataset

        class MyDataset(Dataset):
            def __init__(self, x, y):
                self.x = x
                self.y = y

            def __len__(self):
                return len(self.x)

            def __getitem__(self, index):
                x = torch.zeros(vocabulary_size, dtype=torch.float32)
                y = torch.tensor(self.y[index])
                for k,v in self.x[index].items():
                    x[k] = v
                return x, y
```

7.2 Fully Connected Neural Network

Having completed the `Dataset` implementation, we next implement the model – that is, a fully connected neural network with two layers.[3] In Section 4.2, we used a `Linear` module directly to implement the simpler models discussed

---

[3] Recall that *layer* here refers to the PyTorch `Linear` layer that contains the connections between two neuron layers. See Sidebar 7.1 for more details.

there. This time, we will demonstrate how to implement a model as a new module, by subclassing `torch.nn.Module`. Although this is not necessary for this model, as it can be represented by a `Sequential` module, as models get more complex, it becomes helpful to encapsulate their behavior. To implement a `Module`, we need to implement the constructor and override the `forward()` method.

Note that, in our constructor, before initializing the object fields, we invoke the constructor of the parent class (i.e., `Module`) with the line `super().__init__()`. This allows PyTorch to set up the mechanisms through which any layers defined as attributes in the constructor are properly registered as model parameters. In our example, a `Sequential` instance is assigned to `self.layers`; this is enough for our model instance to know about it during backpropagation and parameter updating.

Here, our model consists of two linear layers, each one preceded by a dropout layer (which drops out input neurons from the corresponding linear layer). The input of the first linear layer has the same size as our vocabulary, and its output has the dimension of the hidden neuron layer (please see Section 5.1 for a refresher on the architecture of the FFNN). Consequently, the input size of the second linear layer is equal to the size of the hidden layer, and its output size is the number of classes. Additionally, between the two linear layers we add a ReLU nonlinearity.[4] All of the model layers are wrapped in a `Sequential` module, which simply connects the output of one layer to the input of the next.

The second method we need to implement is the `forward()` method, which defines how the model applies its layers to a given input during the forward pass. Our `forward()` method simply calls the sequential layer and returns its output. Note that while this method implements the model's forward pass, in general, this method should not be called directly by the user. Instead, the user should use the model as though it were a function (technically, invoking the `__call__()` method), and let PyTorch call the `forward()` method internally. This allows PyTorch to activate necessary features such as module hooks correctly.

```
[12]:  from torch import nn

       class Model(nn.Module):
           def __init__(self, input_dim, hidden_dim, output_dim, dropout):
               super().__init__()
               self.layers = nn.Sequential(
```

---

[4] Note that nonlinearities such as the ReLU function here are necessary to guarantee that the neural network can learn nonlinear decision boundaries. See Chapter 5 for an extended discussion on this topic. Further, nonlinearities can be added after each network layer, but, typically, the output layer is omitted. This is because a softmax or sigmoid function usually follows it. In PyTorch, the `nn.CrossEntropyFunction`, which we also use in this chapter, includes such a softmax function.

```
        nn.Dropout(dropout),
        nn.Linear(input_dim, hidden_dim),
        nn.ReLU(),
        nn.Dropout(dropout),
        nn.Linear(hidden_dim, output_dim),
    )

def forward(self, x):
    return self.layers(x)
```

7.3   Training

In order to train our model, we will first initialize the hyper parameters and the different components we need: model, loss function, optimizer, dataset, and data-loader. Notable differences with respect to Section 4.2 are the use of the Adam optimizer with a *weight decay* (this is just what PyTorch calls L2 regularization – see Chapter 6), and the use of a data-loader with shuffling and batches of 500 examples. We encourage you to take the time to examine the values we use for the hyper parameters, and to experiment with modifying them in the Jupyter notebook.

[24]:
```
from torch import optim
from torch.utils.data import DataLoader
from sklearn.metrics import accuracy_score

# hyperparameters
lr = 1e-3
weight_decay = 1e-5
batch_size = 500
shuffle = True
n_epochs = 5
input_dim = vocabulary_size
hidden_dim = 50
output_dim = len(labels)
dropout = 0.3

# initialize the model, loss function, optimizer, and data-loader
model = Model(input_dim, hidden_dim, output_dim, dropout).to(device)
loss_func = nn.CrossEntropyLoss()
optimizer = optim.Adam(
    model.parameters(),
    lr=lr,
    weight_decay=weight_decay)
train_ds = MyDataset(
    train_df['features'],
    train_df['class index'] - 1)
train_dl = DataLoader(
    train_ds,
    batch_size=batch_size,
    shuffle=shuffle)
dev_ds = MyDataset(
```

```
        dev_df['features'],
        dev_df['class index'] - 1)
    dev_dl = DataLoader(
        dev_ds,
        batch_size=batch_size,
        shuffle=shuffle)

    # lists used to store plotting data
    train_loss, train_acc = [], []
    dev_loss, dev_acc = [], []
```

The basic steps of the learning loop are the same as those in Section 4.2, except that we are now using a development set to keep track of the performance of the current model after each training epoch.

One important difference between using our model during training and evaluation is that, prior to each training session, we need to set the model to training mode using the `train()` method, and before evaluating on the development set, we need to set the model to evaluation mode using the `eval()` method. This is important, because some layers have different behavior depending on whether the model is in training or evaluation mode. In our model, this is the case for the `Dropout` layer, which randomly zeroes some of its input elements during training and scales its outputs accordingly (see Section 6.6), but during evaluation does nothing.

In order to plot some relevant statistics acquired from the training data, we collect the current loss and accuracy for each minibatch. Note that we call `detach()` on the tensors corresponding to the loss and the predicted/gold labels so they are no longer considered when computing gradients. Calling `cpu()` copies the tensors from the GPU to the CPU if we are using the GPU; otherwise it does nothing. Calling `numpy()` converts the PyTorch tensor into a NumPy array. Unlike the prediction sequence, which is represented as a vector of label scores, the loss is a scalar. For this reason, we retrieve it as a Python number using the `item()` method.

When evaluating on the development set, since we do not need to compute the gradients, we save computation by wrapping the steps in a `torch.no_grad()` context-manager. Since we are not learning, we do not perform backpropagation or invoke the optimizer.

```
[25]:  # train the model
       for epoch in range(n_epochs):
           losses, acc = [], []
           # set model to training mode
           model.train()
           for X, y_true in tqdm(train_dl, desc=f'epoch {epoch+1} (train)'):
               # clear gradients
               model.zero_grad()
               # send batch to right device
               X = X.to(device)
```

```
        y_true = y_true.to(device)
        # predict label scores
        y_pred = model(X)
        # compute loss
        loss = loss_func(y_pred, y_true)
        # compute accuracy
        gold = y_true.detach().cpu().numpy()
        pred = np.argmax(y_pred.detach().cpu().numpy(), axis=1)
        # accumulate for plotting
        losses.append(loss.detach().cpu().item())
        acc.append(accuracy_score(gold, pred))
        # backpropagate
        loss.backward()
        # optimize model parameters
        optimizer.step()
    # save epoch stats
    train_loss.append(np.mean(losses))
    train_acc.append(np.mean(acc))

    # set model to evaluation mode
    model.eval()
    # disable gradient calculation
    with torch.no_grad():
        losses, acc = [], []
        for X, y_true in tqdm(dev_dl, desc=f'epoch {epoch+1} (dev)'):
            # send batch to right device
            X = X.to(device)
            y_true = y_true.to(device)
            # predict label scores
            y_pred = model(X)
            # compute loss
            loss = loss_func(y_pred, y_true)
            # compute accuracy
            gold = y_true.cpu().numpy()
            pred = np.argmax(y_pred.cpu().numpy(), axis=1)
            # accumulate for plotting
            losses.append(loss.cpu().item())
            acc.append(accuracy_score(gold, pred))
        # save epoch stats
        dev_loss.append(np.mean(losses))
        dev_acc.append(np.mean(acc))
```

After completing training, we have gathered the loss and accuracy values after each epoch for both the training and development partitions. Next, we plot these values in order to visualize the classifier's progress over time. Plots such as these are important to determine how well our model is learning, which informs decisions regarding adjusting hyper parameters or modifying the model's architecture. Next, we only show the plot for the loss. Plotting the accuracy is very similar; the corresponding code as well as the plot itself is available in the Jupyter notebook.

[15]:
```
import matplotlib.pyplot as plt

x = np.arange(n_epochs) + 1

plt.plot(x, train_loss)
plt.plot(x, dev_loss)
plt.legend(['train', 'dev'])
plt.xlabel('epoch')
plt.ylabel('loss')
plt.grid(True)
```

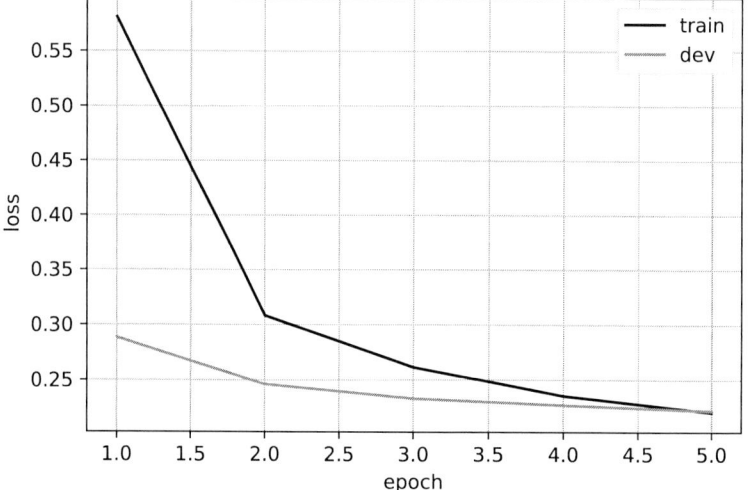

The plot indicates that both the training and development losses decrease over time. This is good! It indicates that our classifier is neither *overfitting* nor *underfitting*. Recall from Chapter 2 that overfitting happens when a classifier performs well in training, but poorly on unseen data. In the plot just presented, this would be indicated by a training loss that continues to decrease, but is associated with a development loss that does not. Underfitting happens when a classifier is unable to learn meaningful associations between the input features and the output labels. In this plot, this would be shown as loss curves that do not decrease over time.

This analysis means we are ready to evaluate our trained model on the test set, which must be a truly unseen dataset that was not used for training or to tune hyper parameters. In other words, this experiment will indicate how well our model performs "in the wild." Because we would like these results to be as close as possible to real-world results, the test set should be used sparingly, only after the entire architecture, its trained parameters, and its hyper parameters have been frozen.

[18]:
```
from sklearn.metrics import classification_report

# set model to evaluation mode
model.eval()

dataset = MyDataset(test_df['features'], test_df['class index'] - 1)
data_loader = DataLoader(dataset, batch_size=batch_size)
y_pred = []

# disable gradient calculation
with torch.no_grad():
    for X, _ in tqdm(data_loader):
        X = X.to(device)
        # predict one class per example
        y = torch.argmax(model(X), dim=1)
        # convert tensor to numpy array
        y_pred.append(y.cpu().numpy())

# print results
y_pred = np.concatenate(y_pred)
print(classification_report(dataset.y, y_pred, target_names=labels))
```

	precision	recall	f1-score	support
World	0.94	0.90	0.92	1900
Sports	0.96	0.99	0.97	1900
Business	0.89	0.88	0.89	1900
Sci/Tech	0.89	0.90	0.89	1900
accuracy			0.92	7600
macro avg	0.92	0.92	0.92	7600
weighted avg	0.92	0.92	0.92	7600

With our feed-forward neural architecture, we have achieved an accuracy of 92%, which is a substantial improvement over the 88% accuracy we obtained in Section 4.2. We strongly suggest that you experiment not only with the different hyper parameters, but also with different model architectures in the Jupyter notebook. Such exercises will help you develop an intuition about the different effects each design choice has, as well as how these decisions interact with each other.

7.4   Summary

In this chapter, we have shown how to implement an FFNN in PyTorch. We have also introduced several PyTorch features that encourage and simplify deep learning best practices. In particular, the built-in Dataset and DataLoader classes make minibatching straightforward while still allowing for customization such as sampling. The ability to create a custom `Dataset` object allows us to handle complex data and still have access to the features of a `DataLoader`. By convention, all the components provided by PyTorch are batch-aware and assume that the first dimension refers to the batch size, simplifying model implementation and improving readability.

In building the model itself, we also saw that PyTorch uses layer modularization – that is, both the network layers themselves and operations on them (such as dropout and activation functions) are modeled as layers in a pipeline. This makes it easy to interweave network layers, add various operations between them, and swap activation functions as desired. The weight initialization is also handled automatically when the layers are created, but can be customized as needed.

Further, one can tailor the training process in PyTorch by adding momentum, adaptive learning rates, and regularization through optimizer selection and configuration. In this chapter, we used the Adam optimizer, which, in the authors' experience, is a good default choice, but there are many other optimizers to choose from. We recommend that the reader read the PyTorch documentation on optimizers for more details: https://pytorch.org/docs/stable/optim.html.

8 Distributional Hypothesis and Representation Learning

As mentioned in the previous chapters, all the algorithms we covered so far rely on handcrafted features that must be designed and implemented by the ML developer. This is problematic for two reasons. First, designing such features can be a complicated endeavor. For example, even for the apparently simple task of designing features for text classification, questions arise quickly: How should we handle syntax? How do we model negation? Second, most words in any language tend to be very infrequent. This was formalized by Zipf (1932), who observed that if one ranks the words in a language in descending order of their frequency, then the frequency of the word at rank i is $\frac{1}{i}$ times the frequency of the most frequent word. For example, the most frequent word in English is *the*. The frequency of the second most frequent word according to Zipf's law is half the frequency of *the*; the frequency of the third most-frequent word is one third of the frequency of *the*, and so on.[1] In our context, this means that most words are very sparse, and our text classification algorithm trained on word-occurrence features may generalize poorly. For example, if the training data for a review classification dataset contain the word *great* but not the word *fantastic*, a learning algorithm trained on these data will not be able to properly handle reviews containing the latter word, even though there is a clear semantic similarity between the two. In the wider field of ML, this problem is called the "curse of dimensionality" (Bellman, 1957).

In this chapter, we will begin to address this limitation. In particular, we will discuss methods that learn numerical representations of words that capture some semantic knowledge. Under these representations, similar words such as *great* and *fantastic* will have similar forms, which will improve the generalization capability of our ML algorithms.

8.1 Traditional Distributional Representations

The methods in this section are driven by the distributional hypothesis of Harris (1954), who observed that words that occur in similar contexts tend to have

---

[1] Interestingly, this law was observed to hold even for nonhuman languages such as dolphin whistles (Ferrer-i Cancho and McCowan, 2009).

similar meanings. The same idea was popularized a few years later by Firth (1957) who, perhaps more elegantly, stated that "a word is characterized by the company it keeps." It is easy to intuitively demonstrate the distributional hypothesis. For example, when presented with the phrases *bread and ...* and *bagels with ...*, many people will immediately guess from the provided context that the missing words are *butter* and *cream cheese*, respectively.

In this section, we will formalize this observation. In particular, we will associate each word in a given vocabulary with a vector, which represents the context in which the word occurs. According to the distributional hypothesis, these vectors should capture the semantic meaning of words, and thus words that are similar should have similar vectors.

Traditionally, these vectors were built simply as co-occurrence vectors. That is, for each word w in the vocabulary, its vector counts the co-occurrence with other words in its surrounding context, where this context is defined as a window of size $[-c, +c]$ words around all instances of w in text. Here, we use negative values to indicate number of words to the left of w, and positive values to indicate number of words to the right. For example, consider the following text:

A bagel and cream cheese (also known as bagel with cream cheese) is a common food pairing in American cuisine. The bagel is typically sliced into two pieces, and can be served as-is or toasted.[2]

In this text, *bagel* occurs three times. Thus, we will have three context windows, one for each mention of the word. While common values for c range from 10 to 20, let us set $c = 3$ for this simple example. Under this configuration, the three context windows for *bagel* in this text are:

- *A* bagel *and cream cheese*
- *also known as* bagel *with cream cheese*
- *American cuisine The* bagel *is typically sliced*

Note that we skipped over punctuation signs when creating these windows.[3] If we aggregate the counts of words that appear in these context windows, we obtain the following co-occurrence vector for *bagel*:

---

[2] Text adapted from the *Bagel and cream cheese* Wikipedia page: https://en.wikipedia.org/wiki/Bagel_and_cream_cheese.

[3] Different ways of creating these context windows are possible. For example, one may skip over words deemed to contain minimal semantic meaning such as determiners, pronouns, and prepositions. Further, these windows may be restricted to content within the same sentence. Last, words may be normalized in some form – for example, through lemmatization. We did not use any of these heuristics in our example for simplicity.

A	1
also	1
American	1
and	1
as	1
cheese	2
cream	2
cuisine	1
is	1
known	1
sliced	1
The	1
typically	1
with	1

This example shows that the co-occurrence vector indeed captures meaningful contextual information: *bagel* is most strongly associated with *cream* and *cheese*, but also with other relevant context words such as *cuisine* and *sliced*. The larger the text used to compute these co-occurrence vectors is, the more meaningful these vectors become.

In practice, these co-occurrence vectors are generated from large document collections such as Wikipedia,[4] and are constructed to have size M, where M is the size of entire word vocabulary – that is, the totality of the words observed in the underlying document collection. Note that these vectors will be sparse – that is, they will contain many zero values, for all the words in the vocabulary that do not appear in the context of the given word. Having all co-occurrence vectors be of similar size allows us to formalize the output of this whole process into a single co-occurrence matrix $\mathbf{C}$ of dimension $M \times M$, where row i corresponds to the co-occurrence vector for word i in the vocabulary. A further important advantage of standardizing vector sizes is that we can easily perform vector operations (e.g., addition, dot product) between different co-occurrence vectors, which will become important soon.

Once we have this co-occurrence matrix, we can use it to improve our text classification algorithm. That is, instead of relying on an explicit feature matrix (see, e.g., the feature matrix in Table 2.4), we can build our classifier on top of the co-occurrence vectors. A robust and effective strategy to this end is to simply average the co-occurrence vectors for the words contained in a given training example (Iyyer et al., 2015). Take, for example, the first training example in Table 2.4: instead of training on the sparse feature vector listed in the first row in the table, we would train on a new vector that is the average of the context vectors for the three words present in the training example: *good, excellent,*

---

[4] www.wikipedia.org.

and *bad*. This vector should be considerably less sparse than the original feature vector, which contains only three nonzero entries. That is, the parameter vector **w** will contain many more nonzero elements because it is updated using training examples that are themselves less sparse. This means that our new classifier should generalize better to other, previously unseen words. For example, we expect other words that carry positive sentiment to occur in similar contexts with *good* and *excellent*, which means that the dot product of their co-occurrence vectors with the parameter **w** is less likely to be zero.

8.2   Matrix Decompositions and Low-Rank Approximations

But have we really solved the "curse of dimensionality" by using these co-occurrence vectors instead of the original lexical features? One may reasonably argue that we have essentially "passed the buck" from the explicit lexical features, which are indeed sparse, to the co-occurrence vectors, which are probably less sparse, but most likely have not eliminated the sparsity curse. This is intuitively true: consider the co-occurrence vector for the word *bagel* from our previous example. Regardless of how large the underlying document collection used to compute these vectors is and how incredible bagels are, it is very likely that the context vector for *bagel* will capture information about breakfast foods, possibly foods in general and other meal-related activities, but will not contain information about the myriad other topics that occur in these documents in bagel-free contexts. Thus, the context constructed for *bagel* contains some of the same lexical sparsity.

To further mitigate the curse of dimensionality, we will have to rely on a little bit of linear algebra. Without going into mathematical details, it is possible to decompose the co-occurrence matrix **C** into a product of three matrices:

$$\mathbf{C} = \mathbf{U}\Sigma\mathbf{V}^T, \tag{8.1}$$

where **U** has dimension $M \times r$, Σ is a squared matrix of dimension $r \times r$, and $\mathbf{V}^T$ has dimension $r \times M$.[5] Each of these three matrices has important properties. First, Σ is a diagonal matrix. That is, all its elements are zero with the exception of the elements on the diagonal: $\sigma_{ij} = 0$ for $i \neq j$.[6] The nonzero diagonal values, σ_{ii}, are referred to as the *singular values* of **C**, and, for this reason, this decomposition of **C** is called *singular value decomposition* or

---

[5]   The superscript T indicates the transpose operation. It is used here to indicate that $\mathbf{V}^T$ is computed as the transpose of another matrix **V**, which has certain mathematical properties. This is less important for our discussion. But we keep the same notation as the original algorithm for consistency.

[6]   For those of us not familiar with the Greek alphabet, σ and Σ are lowercase/uppercase forms of the Greek letter sigma. We use the former to indicate elements in the latter matrix.

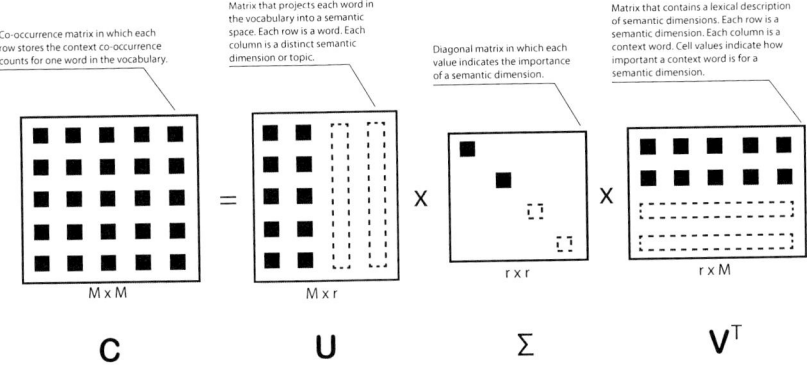

Figure 8.1 Summary of the four matrices in the singular value decomposition equation: $\mathbf{C} = \mathbf{U}\Sigma\mathbf{V}^T$. The empty rectangles with dashed lines indicate which elements are zeroed out under the low-rank approximation

SVD.[7] The dimension of Σ, r, is called the *rank* of the matrix $\mathbf{C}$.[8] Importantly, as we will see in a minute, the values σ_{ii} are listed in descending order in Σ. That is, $\sigma_{ii} > \sigma_{jj}$ for $i < j$. Further, the rows in $\mathbf{U}$ are orthogonal – that is, the dot product of any two rows in $\mathbf{U}$ is zero. Similarly, the rows in $\mathbf{V}$ (or columns in $\mathbf{V}^T$) are also orthogonal.

So, where does all this math leave us? It turns out that the output of the SVD process has important linguistic interpretations (see Figure 8.1 for a summary):

(i) Each row in the matrix $\mathbf{U}$ contains the numerical representation of a single word in the vocabulary, and each column in $\mathbf{U}$ is one semantic dimension, or topic, used to describe the underlying documents that were used to construct $\mathbf{C}$. For example, if row i contains the co-occurrence vector for the word *bagel* and column j contains a topic describing foods, we would expect c_{ij} to have a high value because the food topic is an important part of the semantic description of the word *bagel*. Importantly however, the SVD algorithm does not guarantee that the semantic dimensions encoded as columns in $\mathbf{U}$ are actually interpretable to human eyes. Assigning meaning to these dimensions is a post hoc, manual process that requires the inspection of the $\mathbf{V}^T$ matrix (see third item).

---

[7] The general form of singular value decomposition does not require the matrix $\mathbf{C}$ to be square. For this reason, the SVD form we discuss here, which relies on a square matrix $\mathbf{C}$, is referred to as *truncated* SVD. In this book, we will omit the *truncated* modifier for simplicity.

[8] In general, the rank of a matrix $\mathbf{C}$ is equal to the number of rows in $\mathbf{C}$ that are linearly independent of each other – that is, they cannot be computed as a linear combination of other rows. This is not critical to our discussion.

(ii) The singular values in Σ indicate the importance of topics captured in $\mathbf{U}$. That is, if $\sigma_{ii} > \sigma_{jj}$ then topic i (i.e., the column i in $\mathbf{U}$) is more important than column j. And, since the values in Σ are listed in descending order, we can state that topic i is more important than topic j, if $i < j$. This will become important in a minute.

(iii) Each row i in $\mathbf{V}^T$ contains a bag-of-words description of topic i, where the value at position j in row i indicates the importance of word j to topic i. For example, if the three highest values in a given row point to the words *bagel*, *bread*, and *croissant*, one can (subjectively) interpret this topic to be about bakery products. As mentioned before, such interpretations are not always easy to make. Because the SVD algorithm is completely agnostic to linguistic interpretations, it is possible that some of the produced topics will resist an immediate interpretation. This is an unfortunate drawback we will have to live with for the sake of mitigating the curse of dimensionality.

While the SVD process produces a new vector representation for each word in the vocabulary – that is, row i in the matrix $\mathbf{U}$ corresponds to the new representation of word i – we are not quite done. The rank of the matrix $\mathbf{C}$, r, which also indicates the number of columns in $\mathbf{U}$, is guaranteed to be smaller than M, but it is not necessarily much smaller. We would like to produce vector representations of dimension k, where k is much smaller than M, $k \ll M$. To generate these representations, we will take advantage of the fact that, as discussed, the diagonal matrix Σ contains the topic importance values listed from largest to smallest. Thus, intuitively, if one were to remove the *last $r - k$* topics we would not lose that much information because the top k topics that are most important to describe the content of $\mathbf{C}$ are still present. Formally, this can be done by zeroing out the last $r - k$ elements of Σ, which has the effect of ignoring the last $r - k$ columns in $\mathbf{U}$ and the last $r - k$ rows in $\mathbf{V}^T$ in the SVD multiplication. Figure 8.1 visualizes this process using empty squares and rectangles for the elements in Σ and rows/columns in $\mathbf{U}/\mathbf{V}^T$ that are zeroed out. The resulting matrix $\mathbf{C}$ that is generated when only the first k topics are used is called a *low-rank approximation* of the original matrix $\mathbf{C}$. To distinguish between the two matrices, we will use the notation $\mathbf{C}_k$ to denote the low-rank approximation matrix. There is a theory that demonstrates that $\mathbf{C}_k$ is the best approximation of $\mathbf{C}$ for rank k. What this means for us is that we can use the first k columns in $\mathbf{U}$ to generate numerical representations for the words in the vocabulary that approximate as well as possible the co-occurrence counts encoded in $\mathbf{C}$. In empirical experiments, k is typically set to values in the low hundreds – for example, 200. This means that, once this process is complete, we have associated each word in the vocabulary with a vector of dimension $k = 200$ that is its numerical representation according to the distributional hypothesis.

8.3 Drawbacks of Representation Learning Using Low-Rank Approximation

Although this approach has been demonstrated empirically to be useful for several NLP applications including text classification and search, it has two major problems. The first is that this method, in particular the SVD component, is expensive. Without going into mathematical details, we will mention that the cost of the SVD algorithm is cubic in the dimension of C. Since in our case, the dimension of C is the size of the vocabulary, M, our runtime cost is proportional to M^3. In many NLP tasks, the vocabulary size is in the hundreds of thousands of words (or more!), so this is clearly a very expensive process.

The second drawback is that this approach conflates all word senses into a single numerical representation. For example, the word *bank* may mean a financial institution or sloping land – for example, as in *bank of the river*. But because the algorithm that generates the co-occurrence counts is not aware of the various senses of a given word, all these different semantics are conflated into a single vector. We will address the first drawback in the remaining part of this chapter and the second in Chapter 12.

8.4 The Word2vec Algorithm

The runtime cost of learning word numerical representations has been addressed by Mikolov et al. (2013a), who proposed the word2vec algorithm.[9] Similar to our previous discussion, the goal of this algorithm is to learn numerical representations that capture the distributional hypothesis. More formally, word2vec introduces a training objective that learns "word vector representations that are good at predicting the nearby words." In other words, this algorithm flips the distributional hypothesis on its head. While the original hypothesis stated that "a word is characterized by the company it keeps" – that is, a word is defined by its context – word2vec's training objective predicts the context in which a given word is likely to occur – that is, the context is defined by the word. Mikolov et al. (2013a) proposed two variants of word2vec. For simplicity, we will describe here the variant called "skip-gram," which implements the aforementioned training objective. From here on, we will refer to the skip-gram variant of word2vec simply as word2vec.

Figure 8.2 illustrates the intuition behind word2vec's training process. Visually, the algorithm matches the distribution hypothesis exactly: it makes sure that the vector representation of a given word (e.g., *bagel* in the example shown in the figure) is close to those of words that appear near the given word (e.g., *cream* and *cheese*), and far from the vector representations of words that do not appear in its neighborhood (e.g., *computer* and *cat*). Importantly, to distinguish between input words and context words, the algorithm actually learns

---

[9] The name of this algorithm is an abbreviation of "word to vector."

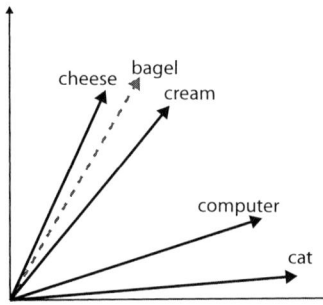

Figure 8.2 An illustration of the word2vec algorithm, the skip-gram variant, for the word *bagel* in the text: *A bagel and cream cheese (also known as bagel with cream cheese) is a common food pairing in American cuisine.* The dashed line indicates the "input" vector; continuous lines denote "output" vectors. The algorithm clusters together output vectors for the words in the given context window (e.g., *cream* and *cheese*) with the corresponding input vector (*bagel*), and pushes away output vectors for words that do not appear in its proximity (e.g., *computer* and *cat*)

two vectors for each word in the vocabulary: one for when it serves as an input word (e.g., *bagel* in the example), and one for when it serves as a context or output word (e.g., *cheese*).

More formally, the algorithm implements the distributional hypothesis as a prediction task. First, for each input word w_i in the vocabulary, the algorithm identifies the context windows of size $[-c, +c]$ around all instances of w_i in some large text.[10] This process is identical to the way we constructed the context windows at the beginning of this chapter. For example, the first context window for the word *bagel* and $c = 3$ is: *A* bagel *and cream cheese*. Second, all the context (or output) words that appear in these windows are added to the pool of words that should be predicted given w_i. Then, the training process maximizes the prediction probability for each word w_j in the context of w_i. That is, the theoretical cost function C for word2vec is:[11]

$$C = -\sum_{i=1}^{M} \sum_{w_j \text{ in the context of } w_i} \log(p(w_j|w_i)), \qquad (8.2)$$

where the probability $p(w_j|w_i)$ is computed using the input vector for w_i, the output vector for w_j and the softmax function introduced in Section 3.5:

[10] In practice, the algorithm uses only the most frequent k words in the vocabulary to reduce training runtimes.

[11] We call this cost function "theoretical" because, as we will see in a minute, this is not what is actually implemented.

$$p(w_j|w_i) = \frac{e^{\mathbf{v}_{w_j}^o \cdot \mathbf{v}_{w_i}^i}}{\sum_{k=1}^{M} e^{\mathbf{v}_{w_k}^o \cdot \mathbf{v}_{w_i}^i}}, \tag{8.3}$$

where $\mathbf{v}^i$ indicates an input vector (i.e., the dashed-line vector in Figure 8.2), $\mathbf{v}^o$ indicates a context (or output) vector (continuous-line vectors in the figure), and the denominator in the fraction iterates over all the words in the vocabulary of size M in order to normalize the resulting probability. All $\mathbf{v}^i$ and $\mathbf{v}^o$ vectors are updated using the standard SGD algorithm during training, similar to the procedure described in Chapter 3. That is, each weight u from a $\mathbf{v}^i$ and $\mathbf{v}^o$ vector is updated based on its partial derivative, $\frac{d}{du}C_i$, where C_i is the loss function for input word i in the vocabulary: $C_i = -\sum_{w_j \text{ in the context of } w_i} \log(p(w_j|w_i))$.

It is important to note at this stage that the last two equations provide a formalization of the intuition shown in Figure 8.2. That is, minimizing the cost function C has the effect of maximizing the probabilities $p(w_j|w_i)$ due to the negative sign in Equation 8.2. Further, maximizing these probabilities has the consequence of bringing the output vectors of context words ($\mathbf{v}_{w_j}^o$) and the input vector for word w_i ($\mathbf{v}_{w_i}^i$) closer together because that maximizes the dot product in the numerator in Equation 8.3. Similarly, maximizing these probabilities has the effect of minimizing the denominator of the fraction in Equation 8.3, which, in turn, means that the dot products with vectors of words *not* in the context of w_i will be minimized.

A second important observation is that there is a very close parallel between this algorithm and the multiclass LR algorithm introduced in Section 3.5. Similar to the multiclass LR algorithm, here we use data points described through a vector representation ($\mathbf{v}^i$ here vs. $\mathbf{x}$ in the standard LR algorithm) to predict output labels (context words vs. labels in $\mathbf{y}$ for LR). Both algorithms have the same cost function: the negative log likelihood of the training data. However, there are three critical differences between word2vec and multiclass LR:

Difference #1: while the formulas for the dot products in the two algorithms look similar, in LR, the $\mathbf{x}$ is static – that is, it doesn't change during training – whereas in word2vec both $\mathbf{v}^i$ and $\mathbf{v}^o$ vectors are dynamically adjusted through SGD. This is because the $\mathbf{x}$ vector in LR stores explicit features that describe the given training example (and thus does not change), whereas in word2vec, both $\mathbf{v}^i$ and $\mathbf{v}^o$ vectors are continuously moved around in their multidimensional space during training to match the distributional hypothesis in the training dataset. For this reason, the word2vec algorithm is also referred to as "dynamic logistic regression."

Difference #2: the $\mathbf{x}$ vector in LR stores explicit features whereas the weights u in the $\mathbf{v}^i$ and $\mathbf{v}^o$ vectors in word2vec are simply coordinates in a multidimensional space. For this reason, the output of the word2vec training process is considerably less interpretable than that of LR. For example, in multiclass LR, one can inspect the largest weights in the learned vector $\mathbf{w}_c$ for class c to understand which are the most important features for the classification of class c. This is not possible for word2vec. Further, word2vec is arguably even less interpretable than the SVD matrix $\mathbf{U}$ in Section 8.2. There we could use the $\mathbf{V}^T$ matrix to come up with a (subjective) interpretation of each column in $\mathbf{U}$. Again, this is not possible in word2vec, where no such descriptions exist.

Difference #3: Last, the number of classes in a multiclass LR problem is usually much smaller than the number of context words in word2vec, which is equal to the size of the vocabulary, M. Typically, the former is in the tens or hundreds, whereas M may be in the millions. Because of this, the denominator of the conditional probability in Equation 8.3 is prohibitively expensive to calculate.

Due to the latter issue, the actual word2vec algorithm does not implement the cost function in Equation 8.2 but an approximated form of it:

$$C = -\sum_{i=1}^{M} \left(\sum_{w_j \text{ in the context of } w_i} \log(\sigma(\mathbf{v}^o_{w_j} \cdot \mathbf{v}^i_{w_i})) \right.$$
$$\left. + \sum_{w_j \text{ not in the context of } w_i} \log(\sigma(-\mathbf{v}^o_{w_j} \cdot \mathbf{v}^i_{w_i})) \right), \qquad (8.4)$$

or, for a single input word w_i:

$$C_i = -\left(\sum_{w_j \in P_i} \log(\sigma(\mathbf{v}^o_{w_j} \cdot \mathbf{v}^i_{w_i})) + \sum_{w_j \in N_i} \log(\sigma(-\mathbf{v}^o_{w_j} \cdot \mathbf{v}^i_{w_i})) \right), \qquad (8.5)$$

where σ is the standard sigmoid function, $\sigma(x) = \frac{1}{1+e^{-x}}$, P_i is the set of context words for the input word w_i, and N_i is the set of words *not* in the context of w_i.

This new cost function captures the same distributional hypothesis: the first sigmoid maximizes the proximity of input vectors with the output vectors of words in context, whereas the second sigmoid minimizes the proximity of input vectors to output vectors of words not in context, due to the negative sign in the sigmoid parameter: $-\mathbf{v}^o_{w_j} \cdot \mathbf{v}^i_{w_i}$. However, this cost function is much easier to compute than the first cost function in Equation 8.2 for two reasons. First, we are no longer using conditional probabilities, which are expensive to normalize. Second, the right-most term of the cost function in Equation 8.5 does not operate over all the words in the vocabulary, but over a small sample of words that do not appear in the context of w_i. These words can be selected using various heuristics. For example, one can uniformly choose words from the training dataset such that they do not appear in the context of a given input word w_i. However, this has the drawback that it will oversample very frequent words (which are more common and thus more likely to be selected). To control for this, the word2vec algorithm selects a non-context word w proportional to the

---

Algorithm 11 word2vec training algorithm

---

1 **for** *each word w_i in the vocabulary* **do**
2 | initialize $\mathbf{v}_{w_i}^i$ and $\mathbf{v}_{w_i}^o$ randomly
3 **end**
4 **while** *not converged* **do**
5 | **for** *each word position i in the training dataset* **do**
6 | | w_i = word at position i
7 | | P_i = set of words in the window $[i - c, i + c]$ around w_i
8 | | N_i = sampled from the set of words not in P_i
9 | | compute cost function C_i using P_i, N_i and Equation 8.5
10 | | **for** *each dimension u in $\mathbf{v}_{w_i}^i$* **do**
11 | | | $u = u - \alpha \frac{d}{du} C_i$
12 | | **end**
13 | | **for** *each word $w_j \in P_i \cup N_i$* **do**
14 | | | **for** *each dimension u in $\mathbf{v}_{w_j}^o$* **do**
15 | | | | $u = u - \alpha \frac{d}{du} C_i$
16 | | | **end**
17 | | **end**
18 | **end**
19 **end**
20 **for** *each word w_i in the vocabulary* **do**
21 | **return** $(\mathbf{v}_{w_i}^i + \mathbf{v}_{w_i}^o)/2$
22 **end**

---

probability $p(w) = \frac{freq(w)^{3/4}}{Z}$, where $freq(w)$ indicates the frequency of word w in the training corpus, and Z is the total number of words in this corpus. The only difference between the probability $p(w)$ and the uniform probability is the 3/4 exponent. This exponent dampens the importance of the frequency term, which has the effect that very frequent words are less likely to be oversampled.

Algorithm 11 lists the pseudocode for the complete training procedure for word2vec that incorporates this discussion. This algorithm is another direct application of SGD, which is used to update both the input vectors (lines 10–12) and output vectors (lines 13–17) until convergence (or for a fixed number of epochs). In all update equations, α indicates the learning rate. At the end, the algorithm returns the average of the input and output vectors as the numeric representation of each word in the vocabulary (lines 20–22). Note that other ways of computing the final word numeric representations are possible, but the simple average has been observed to perform well in practice for downstream tasks (Levy et al., 2015).

In addition to the more efficient cost function, this algorithm has a second practical simplification over our initial discussion. The algorithm does not identify all context windows for each word in the vocabulary ahead of time, as we discussed when we introduced the cost function in Equation 8.2. This would require complex bookkeeping and, potentially, a considerable amount of memory. Instead, Algorithm 11 linearly scans the text (line 5) and constructs a *local* context P_i and a negative context N_i from the current context window at this position in the text (lines 7 and 8). This has several advantages. First, since only one pair of local P_i and N_i sets are kept in memory at a time, the memory requirements for this algorithm are much smaller. Second, the runtime cost of this algorithm is linear in the size of the training dataset because (a) all operations in the inner for loop depend on the size of the context window, which is constant (lines 6–17), and (b) the number of epochs used in the external while loop (line 4) is a small constant. This is a tremendous improvement over the runtime of the SVD procedure, which is cubic in the size of the vocabulary. One potential drawback of this strategy is that the local N_i used in the algorithm may not be accurate. That is, the words sampled to be added to N_i in line 8 may actually appear in another context window for another instance of the current word in the training dataset. However, in practice, this does not seem to be a major problem.

The vectors learned by word2vec have been shown to capture semantic information that has a similar impact on downstream applications as the vectors learned through the more expensive low-rank approximation strategy discussed earlier in this chapter (Levy et al., 2015). We will discuss some of these applications in the following chapters. The semantic information stored in the learned word vectors can also be directly analyzed. For example, Mikolov et al. (2013a) showed that a visualization of 1,000-dimensional vectors learned by word2vec surfaces interesting patterns. For example, the relation between countries and their capital cities (shown as the difference between the two respective vectors) tends to be same regardless of country and capital (Figure 8.3). That is, $\vec{China} - \vec{Beijing} \approx \vec{Portugal} - \vec{Lisbon}$, where the superscript arrow indicates the vector learned by word2vec for the corresponding word. Many other similar patterns have been observed. For example, the difference between the vectors of *king* and *man* is similar to the difference between the vectors of *queen* and *woman*: $\vec{king} - \vec{man} \approx \vec{queen} - \vec{woman}$, which suggests that this difference captures the semantic representation of a genderless monarch. In the following chapters, we will see how we use these vectors to replace the manually designed features in our NLP applications.

8.5   Drawbacks of the Word2vec Algorithm

Word2vec has overlapping drawbacks with the low-rank approximation algorithm previously discussed. Both approaches produce vectors that suffer from lack of interpretability, although one could argue that word2vec's vectors are

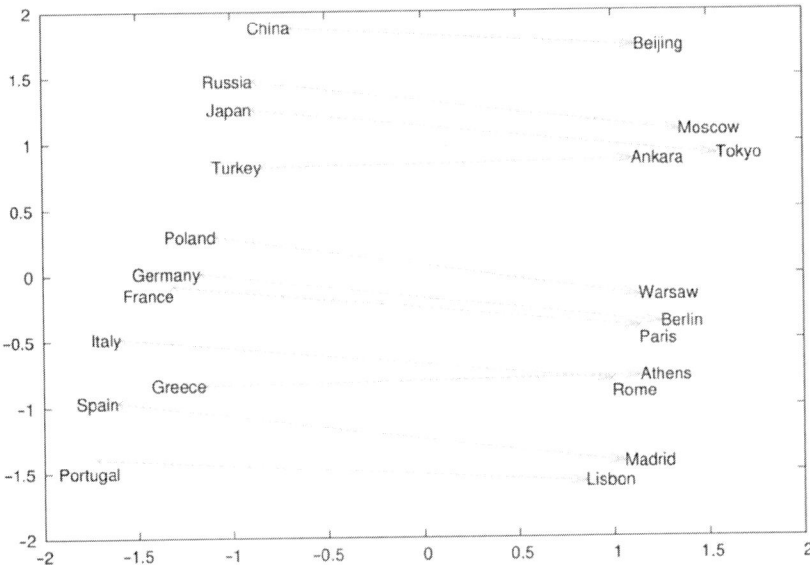

Figure 8.3 Two-dimensional projection of 1,000-dimensional vectors learned by word2vec for countries and their capitals (Mikolov et al., 2013a)

even less interpretable than the low-rank vectors in the $\mathbf{U}$ matrix, whose dimensions can be somewhat explained using the $\mathbf{V}^T$ matrix.

Further, similar to the SVD-based strategy, word2vec conflates all senses of a given word into a single numerical representation. That is, the word *bank* gets a single numerical representation regardless of whether its current context indicates a financial sense – for example, *Bank of America*, or a geological one – for example, *bank of the river*. In Chapter 12, we will discuss strategies to build word vector representations that are sensitive to the current context in which a word appears.

Last, Bolukbasi et al. (2016) showed that algorithms that learn numerical representations amplify the various biases present in the training data – for example, gender stereotypes – "to a disturbing extent." For example, they showed that for common representations, $\overrightarrow{father} - \overrightarrow{doctor} \approx \overrightarrow{mother} - \overrightarrow{nurse}$ (where $\vec{x}$ indicates the vector representation of the word x), or, in plain language, "doctor is to father what nurse is to mother." Or, similarly, $\overrightarrow{man} - \overrightarrow{computer\ programmer} \approx \overrightarrow{woman} - \overrightarrow{homemaker}$.

8.6 Historical Background

Representation learning is a beautiful interdisciplinary idea that spans a variety of fields ranging from linguistics and cognitive science to computer science (including computer architecture!). As expected, the distributional hypothesis idea comes from linguistics (Harris, 1954; Firth, 1957). Firth, the originator of the "London school of linguistics," elegantly summarized distributional hypothesis as "a word is characterized by the company it keeps."[12]

While the distributional hypothesis has been known since the 1950s, it took approximately three decades before distributional techniques became widespread. Some of the earliest distributional techniques included (Brown et al., 1992), which introduced a clustering algorithm that identified "semantically sticky" groups of words using distributional similarity. Schütze (1992) formalized a co-occurrence-based representation learning algorithm similar to what we described in Section 8.1. Yarowsky (1992) and Resnik (1993) used distributional techniques for word sense disambiguation. Deerwester et al. (1990) explored low-rank approximations of co-occurrence matrixes for document search.

Mikolov et al. (2013a) proposed two forms of the word2vec algorithm, one of which we discussed in Section 8.4. Pennington et al. (2014) introduced the GloVe (from *glo*bal *ve*ctors) algorithm, which includes a more general form of the word2vec's learning objective. Since then, these algorithms have been extended in various directions (see next section).

One can argue that these learned representations provide functionality similar to Kahneman's System 1 – that is, (mental) activities that "operate automatically and quickly, with little or no effort and no sense of voluntary control" (Kahneman, 2011). We will see some of these operations (e.g., word similarity, word analogies) in the next chapter.

Finally, algorithms such as word2vec motivated recent developments in deep learning. In Chapter 12, we will discuss the encoder part of transformer networks, which learn contextualized word representations – that is, representations that change depending on the context in which a word appears. Arguably, these representations are more faithful to the original distributional hypothesis. Importantly, learning such representations is computationally expensive, much more so than the word2vec algorithm, which learns a single (static) representation for each word in the vocabulary. As mentioned in Chapter 5, these algorithms were realistically possible only once hardware support for parallel operations, in the form of general-purpose GPUs, became available.

8.7   References and Further Readings

In this chapter, we contrasted distributional techniques based on low-rank approximations with methods that learn dense representations directly. While the latter unquestionably train faster, Levy et al. (2015) have shown that under similar training regimes, the representations learned by the two directions are empirically equivalent.

[12] www.britannica.com/biography/John-R-Firth.

The distributional techniques discussed in this chapter have been extended in several important directions. Lin (1997) and Levy and Goldberg (2014) proposed representation learning algorithms that use syntactic context rather than surface information, as we used in this chapter. Mrkšić et al. (2016) and Vulić and Mrkšić (2018) showed how to learn representations that capture not only the distributional hypothesis but also other linguistic constraints such as antonymy and hypernymy. Yu and Dredze (2015), Shwartz (2019), and Vacareanu et al. (2020b) proposed compositional algorithms that learn representations of multi-word expressions (rather than individual words).

Bolukbasi et al. (2016) showed that distributional techniques may amplify potential biases in the training data. Garg et al. (2018) exploited these learned biases in order to develop metrics to show how gender and minority stereotypes in the United States evolved during the twentieth and twenty-first centuries. Bolukbasi et al. (2016) proposed a methodology for modifying representations to remove gender stereotypes while maintaining desired associations. This problem is just one instance in the important field of bias and fairness in ML (Mehrabi et al., 2021).

8.8 Summary

This chapter discussed methods that learn numerical representations of words based on the distributional hypothesis of Harris (1954) and Firth (1957), who observed that words that occur in similar contexts tend to have similar meanings. In particular, we introduced traditional distributional representations, which rely on co-occurrence vectors and, optionally, low-rank approximation methods such as SVD, and the word2vec algorithm, which learns directly word numerical representations that are good at predicting the words in the neighborhood.

9   Implementing Text Classification Using Word Embeddings

In the previous chapter, we introduced word embeddings, which are real-valued vectors that encode semantic representation of words. We discussed how to learn them, and how they capture semantic information that makes them useful for downstream tasks.

In this chapter, we show how to *use* word embeddings that have been pretrained using a variant of the algorithm discussed in the previous chapter. We show how to load them, explore some of their characteristics, and show their application for a text classification task. As usual, the code for this chapter is available in our repository. It is organized into two notebooks: one corresponding to the explorations shown in the first half of this chapter (chap9_embeddings), and a second one in which we modify our previous classifier to use word embeddings (chap9_classification).

9.1   Pretrained Word Embeddings

There are several algorithms for training word embeddings, including the original word2vec algorithm (Mikolov et al., 2013a) (which we discussed in the previous chapter), GloVe (Pennington et al., 2014), and fastText (Bojanowski et al., 2017). They all provide the software for training the embeddings as well as pretrained word embeddings on their respective websites. In general, most open-domain word embeddings are trained on large corpora that cover a variety of topics such as Wikipedia[1] and Gigaword.[2] Commonly, these embeddings are freely distributed so that practitioners can use them in downstream tasks. We will use one such set of vectors in this chapter.

Pretrained embeddings are usually distributed as a text file in which each line represents a word vector. The first element in the line is the word itself, and the rest of the elements are the vector components. This is usually referred to as the word2vec format. For example, Figure 9.1 shows the line in the glove.6B.50d.txt file (from the GloVe website) corresponding to

---

[1]   https://en.wikipedia.org/wiki/Wikipedia:Database_download.
[2]   https://catalog.ldc.upenn.edu/LDC2011T07.

```
house 0.60137 0.28521 −0.032038 −0.43026 0.74806 0.26223
−0.97361 0.078581 −0.57588 −1.188 −1.8507 −0.24887 0.055549
 0.0086155 0.067951 0.40554 −0.073998 −0.21318 0.37167
−0.71791 1.2234 0.35546 −0.41537 −0.21931 −0.39661 −1.7831
−0.41507 0.29533 −0.41254 0.020096 2.7425 −0.9926 −0.71033
−0.46813 0.28265 −0.077639 0.3041 −0.06644 0.3951 −0.70747
−0.38894 0.23158−0.49508 0.14612−0.02314 0.56389 −0.86188
−1.0278 0.039922 0.20018
```

Figure 9.1 GloVe embedding corresponding to the word *house*, found in the GloVe file glove.6B.50d.txt. We have broken the vector in several lines for display purposes, but this is a single line in the text file

the word *house*. This vector is represented by the word itself, followed by 50 floating-point numbers corresponding to the 50-dimensional vector.

Note that some embeddings files have a header line composed of two numbers: the number of vectors (i.e., the number of lines in the file), and the vector dimensionality. However, this is not always the case. For example, the original word2vec implementation includes this header line, but the more recent GloVe does not (probably because this information can be inferred from the content of the file).

For the examples in the rest of the chapter, we will use the glove.6B.300d.txt embeddings that can be downloaded from the GloVe website.[3] This file provides 400,000 word embeddings of 300 dimensions trained on texts from Wikipedia 2014 and Gigaword 5.

We will begin our exploration of word embeddings using Gensim,[4] a Python library that provides excellent support for loading and using word embeddings, among other more advanced features.

```
[1]:  from gensim.models import KeyedVectors

      fname = "glove.6B.300d.txt"
      glove = KeyedVectors.load_word2vec_format(fname, no_header=True)
      glove.vectors.shape
```

```
[1]:  (400000, 300)
```

As we can see, the embeddings have been loaded and assigned to the glove variable. Note that we had to specify that this file doesn't contain the header

---

[3] https://nlp.stanford.edu/projects/glove.
[4] https://radimrehurek.com/gensim.

that is usually present in the word2vec format. The `glove.vectors` attribute contains a two-dimensional NumPy array with 400,000 rows and 300 columns, each row corresponding to a word embedding.

9.1.1   Word Similarity

Gensim's `KeyedVectors` class provides a method called `most_similar` that receives a word and computes its cosine similarity to all other embeddings, and returns the `topn` most-similar words. By default, `topn` is set to 10.

```
[2]:  # common noun
      glove.most_similar("cactus")
```

```
[2]:  [('cacti', 0.6634564399719238),
       ('saguaro', 0.6195855140686035),
       ('pear', 0.5233486890792847),
       ('cactuses', 0.5178282260894775),
       ('prickly', 0.515631914138794),
       ('mesquite', 0.4844854772090912),
       ('opuntia', 0.4540084898471832),
       ('shrubs', 0.4536206126213074),
       ('peyote', 0.45344963669776917),
       ('succulents', 0.45127877593040466)]
```

The example shows the top 10 most-similar words to the word *cactus* when using the 300-dimension GloVe embeddings trained on Wikipedia and Giga-word. All 10 most-similar words are related to *cactus* in different ways: *cacti* and *cactuses* are its plural forms; *saguaro, peyote, opuntia,* and *prickly pear* are types of cacti; and *mesquite, shrubs,* and *succulents* are other plants from arid climates.

As we discussed in Section 8.5, word2vec and other similar algorithms con-flate all senses of a given word into a single numerical representation. This can be observed when searching for the most-similar words for a word that has multiple meanings such as *fall*:

```
[3]:  # a word with multiple senses
      glove.most_similar("fall")
```

```
[3]:  [('falling', 0.6513392925262451),
       ('rise', 0.6301450133323669),
       ('drop', 0.6298139691352844),
       ('decline', 0.6145920753479004),
       ('beginning', 0.6086390614509583),
       ('spring', 0.5864908695220947),
```

```
('year', 0.5789673328399658),
('coming', 0.5778051018714905),
('fallen', 0.5676990747451782),
('fell', 0.5675972104072571)]
```

The example shows that the most similar words to *fall* capture its different meanings – for example, *falling*, *rise*, *drop* are related to the "sudden drop from an upright position" meaning, whereas *spring* and *year* are related to the season meaning.

You can find more examples of word similarity queries in the Jupyter notebook that accompanies this chapter. Also, as an exercise, try loading a different set of embeddings trained with a different corpus (e.g., Twitter) to see if you obtain different results!

9.1.2   Word Analogies

As we discussed in the previous chapter, the semantic information encoded by word embeddings captures much more than word similarity. To surface this additional information, we will use word analogies represented using additional vector operations. For example, a well-known analogy that highlights gender information is: $\vec{king} - \vec{man} \approx \vec{queen} - \vec{woman}$, or, in plain language: "man is to king what woman is to queen."[5] From this, it immediately follows that one can subtract the meaning of *man* and add the meaning of *woman* to obtain the definition of female royalty: $\vec{king} - \vec{man} + \vec{woman} \approx \vec{queen}$.

The same `most_similar` method we've been using can be repurposed to find word analogies such as this. To this end, two sets of words have to be provided to the `most_similar` method: a list of *positive* words that should be added, and a list of *negative* words that should be subtracted. For example, the code that follows implements the left-hand side of the previous analogy:

```
[8]:   # king - man + woman
       glove.most_similar(positive=["king", "woman"], negative=["man"])
```

```
[8]:   [('queen', 0.6713276505470276),
        ('princess', 0.5432624220848083),
        ('throne', 0.5386105179786682),
        ('monarch', 0.5347574353218079),
        ('daughter', 0.498025119304657),
        ('mother', 0.49564430117607117),
        ('elizabeth', 0.4832652509212494),
        ('kingdom', 0.47747090458869934),
```

5 A word with an arrow on top refers to the embedding vector corresponding to that word. Please see Section 1.4 for a summary of the notations used in this book.

```
('prince', 0.4668239951133728),
('wife', 0.46473267674446106)]
```

Another interesting analogy relation that shows how the embeddings have captured information about currencies is shown next. More examples are discussed in the Jupyter notebook.

```
[12]: # japan - yen + peso
      glove.most_similar(positive=["japan", "peso"], negative=["yen"])
```

```
[12]: [('mexico', 0.5726831555366516),
       ('philippines', 0.5445369482040405),
       ('peru', 0.4838225543498993),
       ('venezuela', 0.4816672205924988),
       ('brazil', 0.46643102169036865),
       ('argentina', 0.4549050033092499),
       ('philippine', 0.4417840242385864),
       ('chile', 0.4396097660064697),
       ('colombia', 0.43862593173980713),
       ('thailand', 0.4339679479598999)]
```

9.1.3   Looking under the Hood

Let us understand now how these queries are actually implemented. First, we need to know what components we need. Clearly, we need the embedding vectors themselves. They are stored in the vectors attribute of the KeyedVectors object.

```
[14]: glove.vectors.shape
```

```
[14]: (400000, 300)
```

As we mentioned previously, this is a two-dimensional NumPy array, each row corresponding to a word in the vocabulary. These embeddings are not normalized, but normalized embeddings can be obtained using the get_normed_vectors method.

```
[15]: normed_vectors = glove.get_normed_vectors()
      normed_vectors.shape
```

```
[15]: (400000, 300)
```

We also need to know the mapping between words and the matrix rows. The KeyedVectors object stores this mapping in a list of terms called index_to_key, and a term-to-index dictionary called key_to_index. Next, we show only the first five terms in order to save space, but you can inspect the whole vocabulary in the Jupyter notebook.

```
[16]: glove.index_to_key
```

```
[16]: ['the', ',', '.', 'of', 'to', ...]
```

```
[17]: glove.key_to_index
```

```
[17]: {'the': 0, ',': 1, '.': 2, 'of': 3, 'to': 4, ...}
```

9.1.4   Word Similarity from Scratch

Implementing the word similarity function ourselves is a good exercise to ensure that we understand how cosine similarity works, and to practice our NumPy skills.

We will write a function called most_similar_words that will take a word, the embeddings matrix, the vocabulary in the form of the index_to_key list and key_to_index dictionary, and the number of similar words to return (defaults to 10).

```
[18]: import numpy as np

      def most_similar_words(word, vectors, index_to_key,
      ↪key_to_index, topn=10):
          # retrieve word_id corresponding to given word
          word_id = key_to_index[word]
          # retrieve embedding for given word
          emb = vectors[word_id]
          # calculate similarities to all words in out vocabulary
          similarities = vectors @ emb
          # get word_ids in ascending order with respect to sim score
          ids_ascending = similarities.argsort()
          # reverse word_ids
          ids_descending = ids_ascending[::-1]
          # get bool array with word_id position set to false
          mask = ids_descending != word_id
          # obtain new array of indices that doesn't contain word_id
          ids_descending = ids_descending[mask]
          # get topn word_ids
```

```
    top_ids = ids_descending[:topn]
    # retrieve topn words with their similarity scores
    top_words = [(index_to_key[i], similarities[i]) for i in␣
↪top_ids]
    # return results
    return top_words
```

The implementation of `most_similar_words` is straightforward. First, we find the *word ID* for the given word, using the `key_to_index` dictionary. Then, we retrieve the row from the vectors matrix that corresponds to that word. The next step is computing the cosine similarity between the word of interest and the rest of the vocabulary. Recall that the cosine similarity is equivalent to a dot product if the vectors are normalized. We use this equivalence by performing a matrix-vector multiplication between the word embedding and the embedding matrix using Python's *at* operator (denoted as @ in code). This means that we must pass the normalized embeddings as an argument to this function. Next, we need to sort the similarities preserving the mapping to the words in the vocabulary. We achieve this using the `argsort` NumPy method, which returns the indices in sorted (ascending) order. Since we need them in descending order, the next step is to reverse this list of indices. Obviously, the most similar word to whichever word we're querying is the word itself, but that is not an interesting result, so we will remove it from the results. We do this by using NumPy's ability to index arrays using Booleans. We first create a new array in which the position corresponding to the query word is set to `False` and every other element is set to `True`, and we use this Boolean array to index the list of IDs. Lastly, we create a list of tuples of the form (word, similarity) for the `topn` words, and return the results.

Now we will test our implementation of word similarity using the word *cactus*. You can compare the results to the ones obtained by KeyedVectors's `most_similar` method.

```
[19]:  vectors = glove.get_normed_vectors()
       index_to_key = glove.index_to_key
       key_to_index = glove.key_to_index
       most_similar_words("cactus", vectors, index_to_key,␣
       ↪key_to_index)
```

```
[19]:  [('cacti', 0.66345644),
        ('saguaro', 0.6195855),
        ('pear', 0.5233487),
        ('cactuses', 0.5178282),
        ('prickly', 0.51563185),
        ('mesquite', 0.4844855),
```

```
('opuntia', 0.45400846),
('shrubs', 0.4536207),
('peyote', 0.4534496),
('succulents', 0.45127875)]
```

9.1.5   Word Analogies from Scratch

The implementation of the word analogy function is not that much different from our `most_similar_word` function. The main difference between this function and `most_similar_words` is that now we have two lists of words that we need to combine into a single embedding. We first add the positive words into a single vector, and we do the same for the negative words. Then we subtract the negative vector from the positive one and normalize the result. The similarity scores are computed the same way as before, but now we need to remove several words from the results, so this time we use NumPy's `isin` function, which checks for any of the words in `given_word_ids`. We then package the results the same way we did before, and return them.

```
[20]:   from numpy.linalg import norm

        def analogy(positive, negative, vectors, index_to_key,
         ⤷key_to_index, topn=10):
            # find ids for positive and negative words
            pos_ids = [key_to_index[w] for w in positive]
            neg_ids = [key_to_index[w] for w in negative]
            given_word_ids = pos_ids + neg_ids
            # get embeddings for positive and negative words
            pos_emb = vectors[pos_ids].sum(axis=0)
            neg_emb = vectors[neg_ids].sum(axis=0)
            # get embedding for analogy
            emb = pos_emb - neg_emb
            # normalize embedding
            emb = emb / norm(emb)
            # calculate similarities to all words in out vocabulary
            similarities = vectors @ emb
            # get word_ids in ascending order with respect to sim score
            ids_ascending = similarities.argsort()
            # reverse word_ids
            ids_descending = ids_ascending[::-1]
            # get bool array with given_word_ids set to false
            given_words_mask = np.isin(ids_descending, given_word_ids,
         ⤷invert=True)
            # obtain new array of indices excluding given_word_ids
            ids_descending = ids_descending[given_words_mask]
```

```
# get topn word_ids
top_ids = ids_descending[:topn]
# retrieve topn words with their similarity scores
top_words = [(index_to_key[i], similarities[i]) for i in
↪top_ids]
# return results
return top_words
```

Now let's try our implementation with the same $\vec{king} - \vec{man} + \vec{woman}$ query we discussed previously. Please compare the results to the ones obtained by Gensim.

[21]:
```
positive = ["king", "woman"]
negative = ["man"]
vectors = glove.get_normed_vectors()
index_to_key = glove.index_to_key
key_to_index = glove.key_to_index
analogy(positive, negative, vectors, index_to_key, key_to_index)
```

[21]:
```
[('queen', 0.67132765),
 ('princess', 0.5432625),
 ('throne', 0.53861046),
 ('monarch', 0.53475744),
 ('daughter', 0.49802512),
 ('mother', 0.49564427),
 ('elizabeth', 0.48326525),
 ('kingdom', 0.47747087),
 ('prince', 0.46682402),
 ('wife', 0.46473265)]
```

9.2   Text Classification with Pretrained Word Embeddings

In this section, we will continue using the AG News classification dataset introduced in previous chapters. Most of the data preparation is the same, up to tokenization. However, we need to remember that the embeddings were trained on a different corpus, so it would be a good idea to estimate how well they cover the words AG News dataset. To achieve this, we load the embeddings just like we did previously. Then we count the tokens in our corpus that do not appear in the embeddings vocabulary, as well as the total number of tokens. We use these numbers to print informative statistics such as the proportion of unknown tokens in the corpus. We also print the top 10 unknown tokens. You can use the Jupyter notebook to explore this task further.

```
[9]:  from collections import Counter

      def count_unknown_words(data, vocabulary):
          counter = Counter()
          for row in tqdm(data):
              counter.update(tok for tok in row if tok not in
      ↪vocabulary)
          return counter

      # compute how many times each unknown token occurs in the corpus
      c = count_unknown_words(train_df['tokens'], glove.key_to_index)

      # find the total number of tokens in the corpus
      total_tokens = train_df['tokens'].map(len).sum()

      # find some statistics about occurrences of unknown tokens
      unk_tokens = sum(c.values())
      percent_unk = unk_tokens / total_tokens
      distinct_tokens = len(list(c))

      print(f'total number of tokens: {total_tokens:,}')
      print(f'number of unknown tokens: {unk_tokens:,}')
      print(f'number of distinct unknown tokens: {distinct_tokens:,}')
      print(f'percentage of unkown tokens: {percent_unk:.2%}')
      print('top 10 unknown words:')
      for token, n in c.most_common(10):
          print(f'\t{n}\t{token}')
```

```
total number of tokens: 5,273,730
number of unknown tokens: 65,847
number of distinct unknown tokens: 24,631
percentage of unkown tokens: 1.25%
top 10 unknown words:
        2984    /b
        2119    href=
        2117    /a
        1813    //www.investor.reuters.com/fullquote.aspx
        1813    target=/stocks/quickinfo/fullquote
        537     /p
        510     newsfactor
        471     cbs.mw
        431     color=
        417     /font
```

Our analysis indicates that only 1.25% of the tokens are not accounted for in the embeddings vocabulary. Further, the most common unknown words seem to be URL fragments. This is encouraging. However, for more robustness, we will introduce a couple of special embeddings that are often needed when dealing with word embeddings. The first one is an embedding used to represent unknown words. A common strategy is to use the average of all the embeddings in the vocabulary for this purpose. The second embedding will be used for padding. Padding is required when we want to train with (mini)batches because the lengths of all the examples in a given batch have to match in order for the batch to be efficiently processed in parallel. The padding embedding consists only of zeros, which essentially excludes these virtual tokens from the forward/backward passes. None of these embeddings are included in the pretrained GloVe embeddings, but other pretrained embeddings may already include them, so it is a good idea to check if they are included with the embeddings we are using before adding them.

```
[10]:  # string values corresponding to the new embeddings
       unk_tok = '[UNK]'
       pad_tok = '[PAD]'

       # initialize the new embedding values
       unk_emb = glove.vectors.mean(axis=0)
       pad_emb = np.zeros(300)

       # add new embeddings to glove
       glove.add_vectors([unk_tok, pad_tok], [unk_emb, pad_emb])

       # get token ids corresponding to the new embeddings
       unk_id = glove.key_to_index[unk_tok]
       pad_id = glove.key_to_index[pad_tok]

       unk_id, pad_id
```

```
[10]:  (400000, 400001)
```

The new embeddings were added at the end of embedding collection, so their IDs are 400,000 and 400,001.

Before fine-tuning the word embeddings for our task, it is a good idea to remove infrequent words from our vocabulary as their embeddings are likely to be seldom updated during training. Here, we discard the words that appear 10 times or fewer. However, this is a hyper parameter that can be adjusted.

```
[12]:  threshold = 10
       tokens = train_df['tokens'].explode().value_counts()
       vocabulary = set(tokens[tokens > threshold].index.tolist())
       print(f'vocabulary size: {len(vocabulary):,}')
```

```
vocabulary size: 17,446
```

Now we need to generate a list of token IDs for each training example. Since we decided to ignore tokens that appear fewer than 10 times, we need to replace those with [UNK] too, even if they appear in the embedding vocabulary.

```
[13]:  # find the length of the longest list of tokens
       max_tokens = train_df['tokens'].map(len).max()

       # return unk_id for infrequent tokens too
       def get_id(tok):
           if tok in vocabulary:
               return glove.key_to_index.get(tok, unk_id)
           else:
               return unk_id

       # function that gets a list of tokens and returns a list
       # of token ids, with padding added accordingly
       def token_ids(tokens):
           tok_ids = [get_id(tok) for tok in tokens]
           pad_len = max_tokens - len(tok_ids)
           return tok_ids + [pad_id] * pad_len

       # add new column to the dataframe
       train_df['token ids'] = train_df['tokens'].
        ↪progress_map(token_ids)
       train_df
```

Next, we create a Dataset object from the padded lists of token IDs. This one is even easier since the lists of token IDs are ready. So all that is required is turning them into tensors.

```
[15]:  from torch.utils.data import Dataset

       class MyDataset(Dataset):
           def __init__(self, x, y):
               self.x = x
               self.y = y

           def __len__(self):
               return len(self.y)

           def __getitem__(self, index):
               x = torch.tensor(self.x[index])
               y = torch.tensor(self.y[index])
               return x, y
```

Last, we need to modify the model class in order to indicate that we now use embedding vectors. To this end, we will add an nn.Embedding layer that stores the embedding vectors for all words in the vocabulary. We will use this object to look up embeddings by their token IDs. This layer will be initialized from a tensor containing the pretrained embeddings for the entire vocabulary. Also, the pad_id is specified when creating the embedding layer. When a nn.Embedding layer gets initialized using the from_pretrained method with other arguments set to default values, the embeddings are not updated during training. We will keep it that way for this example, but that could be changed by setting the freeze parameter to False.

The rest of the layers are the same as in our previous example from Chapter 7 – that is, one intermediate layer and one output layer, with a nonlinearity (ReLU) between them. The only major difference is that now the input size of the intermediate layer is the size of one embedding (e.g., 300) instead of the size of the vocabulary like last time. This is because, as we explain next, the intermediate layer receives an average of the numerical representations of the words in the current text.

The forward function of the Model class changes significantly. This time, we are encoding the text as an average of the embeddings of all the words it contains. To compute the denominator of this average, we obtain the length of each text by counting how many of its words are not the virtual padding token. Then we sum all the embeddings and divide by the number of non-padding tokens. Adding all embeddings is safe because padding embeddings are comprised of zeros. This process leaves us with a single embedding for the whole text, which is then passed to the rest of the layers.

```
[16]:  from torch import nn
       import torch.nn.functional as F

       class Model(nn.Module):
           def __init__(self, vectors, pad_id, hidden_dim, output_dim,␣
       ↪dropout):
               super().__init__()
               # embeddings must be a tensor
               if not torch.is_tensor(vectors):
                   vectors = torch.tensor(vectors)
               # keep padding id
               self.padding_idx = pad_id
               # embedding layer
               self.embs = nn.Embedding.from_pretrained(vectors,␣
       ↪padding_idx=pad_id)
               # feedforward layers
               self.layers = nn.Sequential(
```

```
            nn.Dropout(dropout),
            nn.Linear(vectors.shape[1], hidden_dim),
            nn.ReLU(),
            nn.Dropout(dropout),
            nn.Linear(hidden_dim, output_dim),
        )

    def forward(self, x):
        # get boolean array with padding elements set to false
        not_padding = torch.isin(x, self.padding_idx,
↪invert=True)
        # get lengths of examples (excluding padding)
        lengths = torch.count_nonzero(not_padding, axis=1)
        # get embeddings
        x = self.embs(x)
        # calculate means
        x = x.sum(dim=1) / lengths.unsqueeze(dim=1)
        # pass to rest of the model
        output = self.layers(x)
        return output
```

The training and evaluation steps are the same as before. The results of this model on the AG News test partition are displayed next:

	precision	recall	f1-score	support
World	0.92	0.88	0.90	1900
Sports	0.95	0.97	0.96	1900
Business	0.85	0.86	0.85	1900
Sci/Tech	0.86	0.87	0.87	1900
accuracy			0.90	7600
macro avg	0.90	0.90	0.90	7600
weighted avg	0.90	0.90	0.90	7600

Comparing these results with the ones obtained by the multilayer percep-tron with explicit features in Chapter 7, we observe that on this particular task, utilizing embeddings as features does not yield a performance improvement. Notably, this is a small dataset and a rather simplistic task where the presence of certain words is sufficient to distinguish the category of an article (e.g., the word *basketball* is highly indicative of the label *Sports*). Nevertheless, in other tasks where distinctions are more nuanced, or in which there is less likely to be word overlap between texts of interest, word embeddings do provide neces-sary signal. Additionally, when there are class imbalances, word embeddings can supplement underrepresented classes by bringing the external knowledge gained during their pretraining.

9.3   Summary

In this chapter, we showed how to explore the semantic space encoded by word embeddings through word similarity and analogies, as well as one way to use them for text classification. At this point, we have not taken into consideration the order in which the words appear – that is, we averaged the embeddings for all the words in the text using a bag-of-words representation of text. In subsequent chapters, we will explore how to incorporate word order into the learned representations of text.

10   Recurrent Neural Networks

Up to this point, we have only discussed neural approaches for text classification (e.g., review and news classification) that handle the text as a *bag of words*. That is, we aggregate the words either by representing them as explicit features in a feature vector, or by averaging their numerical representations (i.e., embeddings). Although this strategy completely ignores the order in which words occur in a sentence, it has been repeatedly shown to be a good solution for many practical NLP applications that are driven by text classification, to the frustration of many of us who care about linguistic information (Baldridge, 2013).

Nevertheless, for many NLP tasks, we need to capture the word-order information more explicitly. We will discuss several of these applications in Chapter 16. For a simple example, in this chapter, we will use part-of-speech (POS) tagging, which is the task of assigning each word in a sequence its part of speech – that is, a category that captures its grammatical properties such as noun, verb, or adjective. The Penn Treebank corpus,[1] one of the most widely used English corpora that contains texts annotated with POS information (and other linguistic information) uses 36 such parts of speech.[2] Assigning POS tags to words clearly benefits from word order. For example, in English, once we see a determiner, it is much more likely that the *next* word will be a noun or an adjective, and not a verb. Thus, assigning a POS tag to a word depends not only on the word itself, but also on the context in which it appears. To continue the previous example, assume we are trying to tag the word *teaching*. From a large corpus such as the Penn Treebank, we may learn that this word can be a noun or a verb, and is more commonly seen as a verb. However, if the preceding word in the sentence is *the*, we can confidently assign it the noun POS tag. Sequence models capture exactly this scenario, where classification decisions must be made using not only the current information but also the context in which it appears.

---

[1]   https://catalog.ldc.upenn.edu/LDC99T42.
[2]   www.ling.upenn.edu/courses/Fall_2003/ling001/penn_treebank_pos.html.

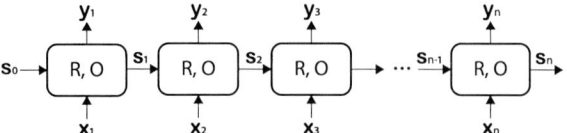

Figure 10.1 "Vanilla" recurrent neural network, where $\mathbf{s}_i$ are state vectors, $\mathbf{x}_i$ are input vectors, and $\mathbf{y}_i$ are output vectors. R and O are functions that compute the next state and the current output vector, respectively

10.1   Vanilla Recurrent Neural Networks

Figure 10.1 shows a generic neural architecture that models sequence information called a recurrent neural network (RNN). Each of the blocks in the architecture is called a *cell*. Each cell is assigned to one individual element in the sequence – for example, for POS tagging, there is one cell per word. The job of cell i is to produce an output vector $\mathbf{y}_i$ using as input an input vector $\mathbf{x}_i$ – for example, for POS tagging the vectors $\mathbf{x}_i$ are word embeddings (either static or contextualized), and a state vector $\mathbf{s}_{i-1}$, which captures the information that has "flown" through the sequence until this cell. The state and output vectors are computed using the R and O functions, respectively. That is, in general:

$$\mathbf{s}_i = R(\mathbf{s}_{i-1}, \mathbf{x}_i) \qquad (10.1)$$
$$\mathbf{y}_i = O(\mathbf{s}_i). \qquad (10.2)$$

Note that even though Equation 10.2 states that the output vector $\mathbf{y}_i$ depends only on the current state vector $\mathbf{s}_i$, this state vector encodes *both* the current input $(\mathbf{x}_i)$ as well as the information that has flown through the network thus far $(\mathbf{s}_{i-1})$.

The simplest RNN (Elman, 1990) uses the following implementations for R and O:

$$\mathbf{s}_i = R(\mathbf{s}_{i-1}, \mathbf{x}_i) = f(\mathbf{W}^\mathbf{s} \cdot \mathbf{s}_{i-1} + \mathbf{W}^\mathbf{x} \cdot \mathbf{x}_i + \mathbf{b}) \qquad (10.3)$$
$$\mathbf{y}_i = O(\mathbf{s}_i) = \mathbf{s}_i, \qquad (10.4)$$

where $\mathbf{W}^\mathbf{x}$, $\mathbf{W}^\mathbf{s}$, and $\mathbf{b}$ are the parameters that are shared between all cells in the network.[3] Thus, the R function is very similar to one layer of an FFNN (see Chapter 5), but it has two sets of weights: one that operates over the input vector $\mathbf{x}_i$ ($\mathbf{W}^\mathbf{x}$), and one that operates over the state vector $\mathbf{s}_{i-1}$ ($\mathbf{W}^\mathbf{s}$). Similar to an FFNN, $\mathbf{b}$ contains the bias weights, and f is a nonlinear function such

---

[3]   In this chapter, we will use the tensor notation briefly introduced in Chapter 5, which expresses the weights in the network as matrices. This is because we now must handle multiple weight matrices at the same time, and the explicit notation used in Chapter 5 becomes too cumbersome in this situation.

as a sigmoid or hyperbolic tangent. The dimensions of vectors and matrices in this architecture are: $\mathbf{x}_i \in \mathbb{R}^{d_{in}}$, $\mathbf{s}_i, \mathbf{y}_i, \mathbf{b} \in \mathbb{R}^{d_{out}}$, $\mathbf{W}^{\mathbf{x}} \in \mathbb{R}^{d_{out} \times d_{in}}$, and $\mathbf{W}^{\mathbf{s}} \in \mathbb{R}^{d_{out} \times d_{out}}$, where d_{in} indicates the dimension of the input vector, d_{out} is the dimension of the state and output vectors, $\mathbb{R}^d$ indicates a real-valued vector of dimension d, and $\mathbb{R}^{r \times c}$ indicates a real-valued matrix with r rows and c columns.

Thus, Equations 10.3 and 10.4 capture both input information, which flows in through the vertical arrows in Figure 10.1, and information about the sequence, which travels through the horizontal arrows in the figure. Importantly, though somewhat different from the FFNNs we have seen so far, an RNN is just another neural network architecture. Hence, we can train it, which in this case, means learning the cell parameters $\mathbf{W}^{\mathbf{x}}$, $\mathbf{W}^{\mathbf{s}}$, and $\mathbf{b}$ (and possibly adjusting the input embeddings $\mathbf{x}_i$) using essentially the same back-propagation algorithm we discussed in Chapter 5. This back-propagation variant for sequences is called *backpropagation through time* (Werbos, 1990) because these sequence architectures have also been applied to time series where each cell in the network corresponds to a step in time.

But where does the signal for backpropagation through time come from? In other words, what training data and loss functions do we need for sequence models? This depends on their application. There are three typical ways to use RNNs:

Acceptor RNN: This configuration adds a classification layer – for example, a simple FFNN, on the *last* output vector $\mathbf{y}_n$, where n is the length of the sequence – that is, the *right-most* $\mathbf{y}$ vector in Figure 10.1. Theoretically, $\mathbf{y}_n$ captures information about the entire sequence and thus should be a good summary for the whole text. (This is not entirely true. We will revisit this observation in Section 10.3.) For example, one could implement a binary review classifier by adding a single layer on top of $\mathbf{y}_n$ that projects the $\mathbf{y}_n$ vector into a single neuron with a sigmoid nonlinearity. The training data for this network would use values of 1 for positive reviews, and 0 for negative ones. The network would be trained using either the MSE loss or the binary cross-entropy (see Chapter 6). A multiclass review classifier would be implemented using a layer that projects $\mathbf{y}_n$ into an output layer with as many neurons as the number of classes, and be trained using the regular cross-entropy loss.

Transducer RNN: This network receives supervision during training for *each* cell. For example, when training a POS tagger, we want to make sure that the network learns to predict the correct POS tag for every word in the sequence. That is, a transducer adds one or more layers on top of *each* $\mathbf{y}_i$ vector to predict something specific for each cell such as the POS tag of word i. During training, the transducer calculates the individual loss of each cell

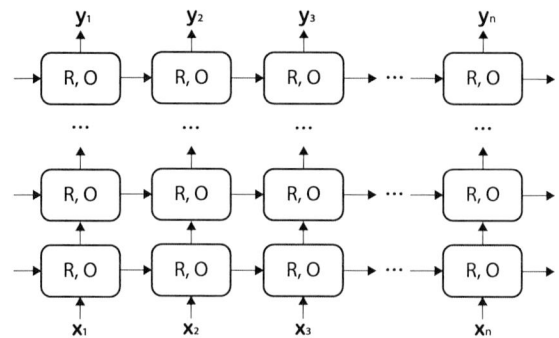

Figure 10.2 Stacked or "deep" recurrent neural network

(e.g., cross-entropy for POS tagging); these individual losses are summed up into a single value that is used for backpropagation.

Encoder-decoder RNN: This architecture combines one encoder RNN, which encodes an input sequence into a single vector (similar to the acceptor RNN), with a decoder that generates one element at a time from an output sequence. This architecture is typically used for machine translation, where the input sequence is text in the source language, and the output sequence is the translation in the target language. We will discuss this architecture in detail in Chapter 14.

We will introduce multiple applications of all these RNN architectures in detail in Chapter 16.

10.2   Deep Recurrent Neural Networks

Before we discuss the problems that plague RNNs and their corresponding solutions, it is worth mentioning that RNNs can be composed into more complex architectures in multiple ways. The two most common are shown in Figures 10.2 and 10.3. Stacked or deep RNNs (Figure 10.2) add multiple RNNs on top of each other. That is, the output vector $\mathbf{y}_i$ for the jth RNN becomes the input vector $\mathbf{x}_i$ for the next RNN ($j + 1$). In contrast, bidirectional RNNs (Figure 10.3) employ two RNNs in parallel. Both operate over the same input embeddings, but one traverses the text from left to right (or "forward," assuming that the writing system is left to right), and another scans the text right to left (or "backward"). Let's denote the output vectors of the former RNN with $\mathbf{y}^{\mathbf{f}}$, and the output vectors of the latter with $\mathbf{y}^{\mathbf{b}}$. Then, the output vector of the bidirectional RNN for word i is generated by concatenating $\mathbf{y}^{\mathbf{f}}_i$ and $\mathbf{y}^{\mathbf{b}}_{n+1-i}$:
$y_i = [\mathbf{y}^{\mathbf{f}}_i; \mathbf{y}^{\mathbf{b}}_{n+1-i}]$.

It has been empirically observed that these more complex RNNs perform better for harder NLP tasks such as syntactic parsing (Dyer et al., 2015; Dozat

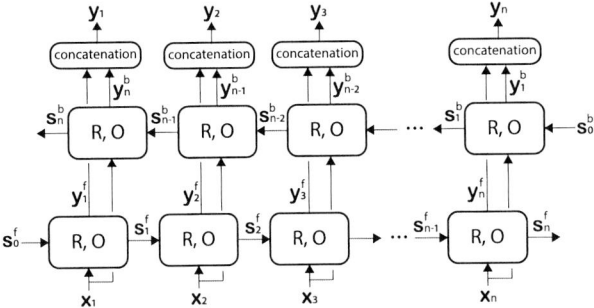

Figure 10.3 Bidirectional recurrent neural network

and Manning, 2016; Vacareanu et al., 2020a). This is not that surprising. For example, syntactic signal for the English language occurs both to the left and right of the verb – for example, the subject typically precedes the verb in a sentence while the object follows it. This information is handled much better by a bidirectional RNN.

10.3 The Problem with Simple Recurrent Neural Networks: Vanishing Gradient

While the previous section ends on a note of victory (deep recurrent networks capture language intricacies well!), we are not quite done. There are several problems that we need to address. The first one is that the simple RNN cell introduced in Section 10.1 suffers from the vanishing gradient problem – that is, when gradient values become too small to impact the parameter updates in a meaningful way. In other words, vanishing gradients cause learning to stop, which is clearly undesirable. This is easy to demonstrate: let's assume an RNN with two cells. In this case, the state vector after the second cell is computed using the equation:

$$s_2 = R(s_1, x_2) = R(R(s_0, x_1), x_2)$$
$$= f(W^s \cdot f(W^s \cdot s_0 + W^x \cdot x_1 + b) + W^x \cdot x_2 + b).$$

Thus, the computation of the state vector relies on chaining several multi-plications of the nonlinear function f. Recall that the activation functions f typically produce small values – for example, the hyperbolic tangent is bound to the interval $[-1, 1]$ (see Section 6.3). Even with unbounded activation functions such as ReLU, it is common that their outputs are normalized to be centered around zero and have a standard deviation of one (see Section 6.8). Multiplying several such values will quickly yield very small values, which,

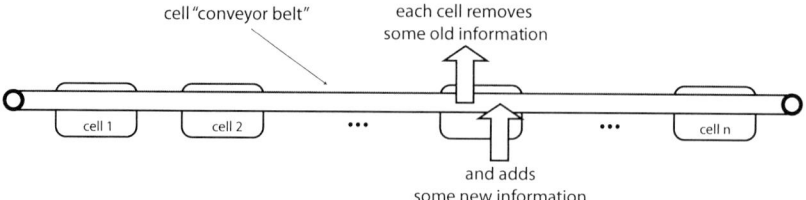

Figure 10.4 Intuition behind the long short-term memory architecture

in turn, will cause vanishingly small parameter updates (see the equations of backpropagation introduced in Section 5.3).

To mitigate the vanishing gradient problem in RNNs, we introduce next a recurrent network architecture that replaces the multiplicative approach in the aforementioned simple RNN cell with an additive method.

10.4   Long Short-Term Memory Networks

Long short-term memory networks, or LSTMs, are recurrent neural networks that address the vanishing gradient problem (Hochreiter and Schmidhuber, 1997). To this end, LSTMs replace the multiplicative architecture used by the vanilla RNNs with an *additive architecture* – that is, transitions between cells are handled (mostly) with additions and subtractions rather than multiplications. While the math behind this gets a little complicated, the intuition behind LSTMs is simple: imagine that cells in an RNN are connected by a conveyor belt that carries information throughout the whole sequence (see Figure 10.4). Each cell subtracts information that is no longer needed from the conveyor belt and adds new information from the current input. Then, the state vector s_i for cell i is computed using the information available at this moment on the conveyor belt.

Sidebar 10.1   Significance of the "long short-term memory" name

The mouthful of a name "long short-term memory" aims to encapsulate the fact that this RNN captures both local ("short-term") and global ("long-term") information. The local information comes from the input vector corresponding to the current cell; the global information comes from all previous cells in the network. This global information is better managed than in a vanilla RNN due to the underlying additive architecture.

Before we explain the mathematical operations that implement this intuition, we need to introduce neural *gates*, which are simple mechanisms that control access to multidimensional vectors. In the simplest scenario, these gates

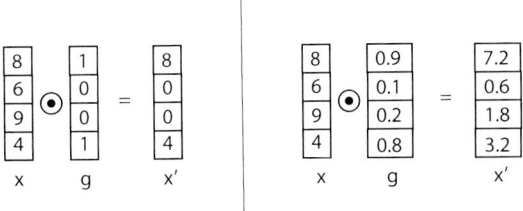

Figure 10.5 Example of a binary gate (left) and gate with real-valued elements (right)

are binary – that is, they either let an element pass through or not. The left part of Figure 10.5 shows an example of such a binary gate. These gates are implemented as element-wise products (indicated with the operator $\odot$) of the gate vector **g** with the original vector **x**.[4] In practice, we would like to learn *how much* of the original vector to let pass through the gate. Thus, most gate mechanisms (including the ones we will introduce later in this chapter) use gate vectors that contain real-valued numbers in the interval $[0, 1]$, which are learned with the rest of the network parameter during backpropagation. The right part of Figure 10.5 shows an example of such a gate.

The LSTMs use three types of gates, defined as follows:

Forget gate (f) – controls how much of the content on the conveyor belt to preserve in the current cell,

Input gate (i) – decides how much of the input local to the current cell to add to the conveyor belt, and

Output gate (o) – controls how much of the conveyor belt vector to include in the hidden state vector for each cell.

To formalize, let us denote with $\mathbf{c}_t$ the vector that stores the content of the conveyor belt for the cell at position t in the sequence, and with $\mathbf{f}_t, \mathbf{i}_t, \mathbf{o}_t$ the forget, input, and output gate vectors for cell t. Then, the $\mathbf{c}_t$ vector is computed as:

$$\mathbf{c}_t = \mathbf{f}_t \odot \mathbf{c}_{t-1} + \mathbf{i}_t \odot \tilde{\mathbf{c}}_t, \qquad (10.5)$$

where $\mathbf{c}_{t-1}$ is the conveyor belt vector for the previous cell at position $t-1$, and $\tilde{\mathbf{c}}_t$ is the input available to this cell (we will formally define it shortly). Further, the hidden state $\mathbf{h}_t$ for cell t is then computed as:

$$\mathbf{h}_t = \mathbf{o}_t \tanh(\mathbf{c}_t), \qquad (10.6)$$

[4]   The element-wise vector product operation is often called a Hadamart product in the literature.

where tanh is the hyperbolic tangent function we saw before. These two equations implement the intuition previously described. Equation 10.5 shows that the vector that stores the information on the conveyor belt is adjusted at position t to forget some of the previous information, while adding some new content. Equation 10.6 shows that the LSTM hidden states are computed by filtering the conveyor belt vector through a dedicated gate.

The cell's input vector $\tilde{c}_t$ is computed as a transformation of the vector that concatenates the input embedding vector to this cell $\mathbf{x}_t$ (e.g., the word embedding that corresponds to this cell) and the hidden state of the previous cell $\mathbf{h}_{t-1}$:

$$\tilde{c}_t = \tanh(\mathbf{W}^c[\mathbf{x}_t; \mathbf{h}_{t-1}] + b^c). \tag{10.7}$$

Thus, $\tilde{c}_t$ is the output of an FFNN defined by the parameters $\mathbf{W}^c$ and b^c, which receives as input information specific to cell t – that is, the concatenation of $\mathbf{x}_t$ and $\mathbf{h}_{t-1}$. Note that the parameters of this FFNN ($\mathbf{W}^c$ and b^c) are shared across all cells in the sequence. This reduces the number of the parameters that the LSTM has to learn while preserving cell-specific inputs.

The three LSTM gates are computed in a similar way:

$$\mathbf{f}_t = \tanh(\mathbf{W}^f[\mathbf{x}_t; \mathbf{h}_{t-1}] + b^f) \tag{10.8}$$
$$\mathbf{i}_t = \tanh(\mathbf{W}^i[\mathbf{x}_t; \mathbf{h}_{t-1}] + b^i) \tag{10.9}$$
$$\mathbf{o}_t = \tanh(\mathbf{W}^o[\mathbf{x}_t; \mathbf{h}_{t-1}] + b^o) \tag{10.10}$$

Thus, the parameters to be learned for a LSTM consist of four matrices, $\mathbf{W}^c$, $\mathbf{W}^f$, $\mathbf{W}^i$, and $\mathbf{W}^o$ (and the corresponding bias terms). In some cases, the input embedding vectors ($\mathbf{x}^t$) are also learned together with these matrices. That is, the $\mathbf{x}$ vectors can be initialized with embeddings produced by a distributional similarity algorithm (e.g., the word2vec algorithm from Chapter 8), and then adjusted to the task at hand through the regular back-propagation algorithm that learns the other LSTM parameters.

Since the introduction of LSTMs in 1997 (Hochreiter and Schmidhuber, 1997), several other RNN variants have been proposed. For example, the gated recurrent unit (GRU) is a popular LSTM alternative (Cho et al., 2014). The key change in the GRU is that the forget gate is connected to the input gate: $\mathbf{f}_t = 1 - \mathbf{i}_t$. That is, we introduce as much new information on the conveyor belt as what we forget from the previous cell. The key advantage of this simplification is that it reduces the number of parameters to be learned (from four matrices to three). While the many LSTM variants differ in their implementation, the intuition behind all of them is the same: they mitigate the vanishing gradient problem behind multiplicative architectures by switching to additive approaches. For more details as well as visualizations of the gating mechanism in LSTMs, we refer the reader to Christopher Olah's excellent blog (Olah, 2015).

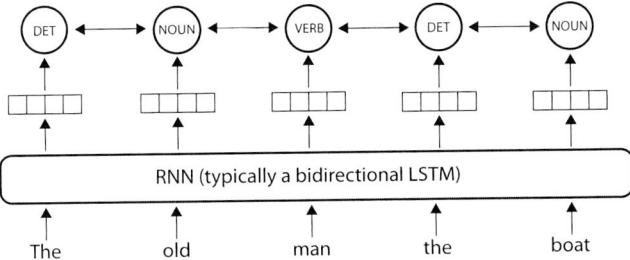

Figure 10.6 Conditional random fields architecture on top of a recurrent neural network

10.5 Conditional Random Fields

As discussed, our POS tagger can be implemented using a transducer RNN. That is, the transducer RNN predicts the POS tag for *every* word in the sequence using the hidden-state embedding produced by an RNN. The typical RNN used in this scenario is a bidirectional LSTM (biLSTM), which guarantees that information both to the left and to the right of each word is used in the prediction. However, during training, the transducer's loss maximizes the probability of the correct tag for each word in the sequence *independently of the other tags*. This is not always ideal: in many situations, the assignment of labels to words in a sequence needs to be performed by looking at the entire sequence *jointly* to understand what the most meaningful assignment overall is. Probably the most famous examples of such situations are garden-path sentences – that is, sentences whose understanding requires a double take because the first and most likely interpretation (from both people and machines!) tends to be incorrect (Fowler, 1994; Wikipedia, n.d.). Figures 10.6 and 10.7 show a simple example of such a garden-path sentence: *The old man the boat.* In this example, the reader scanning the sentence from left to right tends to assign the POS tag noun (NOUN) to the word *man*, which turns out to be incorrect. The correct POS tag for *man* is verb in the present tense (VERB), which can be inferred only after the entire text is scanned. Computationally, the fact that the second determiner *the* is highly unlikely to follow a noun (*man* in this case) is the hint that the initial assignment is incorrect.

Conditional random fields (CRF) were introduced to handle this joint assignment problem (Lafferty et al., 2001). While they were introduced in the context of a different probabilistic framework, here we will discuss them as they were adapted to operate on top of RNNs (Lample et al., 2016). Informally, CRFs model not only the individual tag probabilities (i.e., the vertical arrows that connect the hidden-state embeddings to the tags in the figure 10.6), but also the *transition probabilities* between two adjacent tags (the bidirectional horizontal arrows in the figure 10.6), which are not explicitly modeled by RNNs.

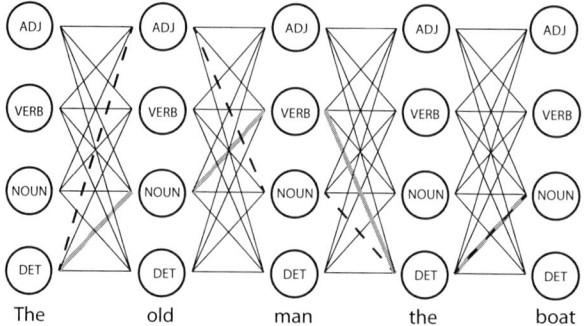

Figure 10.7 Lattice of possible tag assignments for the example sentence from Figure 10.6. For simplicity, we show only four of the possible POS tags: DET – determiner, NOUN – common noun (either singular or plural), VERB – verb (any tense), and ADJ – adjective. The thick lines indicate the correct path in the lattice; the dashed lines indicate the incorrect path suggested by the first interpretation of the garden-path sentence

A bit more formally, CRFs inspect all possible paths in the lattice that contains all tag assignments and the transitions between them (as shown in Figure 10.7), and select the path in this lattice that has the highest overall probability. The probability of a path includes the probabilities of individual tags as well as the probabilities of transitions between tags. Because of the latter, CRFs can discard paths with improbable transitions such as the transition from *man* as a noun to the determiner *the*. Next, we will discuss how to train CRFs, and then, how to efficiently apply a trained CRF model to new texts.

10.5.1   *Training a Conditional Random Field with the Forward Algorithm*

As mentioned, unlike the traditional transducer RNN that maximizes the probabilities of individual labels during training, the CRF optimizes the probability of the *complete* sequence of correct labels – that is, the probability of the correct path in the lattice of possible label assignments. Before we discuss how this happens, let's formalize the probability of an entire sequence of labels $\mathbf{y} = (y_1, y_2, \ldots, y_n)$, where y_i is the label assigned to word i in a sequence of n words. The overall score s of such a path is computed by aggregating two types of scores: the individual score produced by the RNN for each word in the sequence, and the transition score between any two consecutive labels:

$$s(\mathbf{y}) = \sum_{i=1}^{n} R[y_i, i] + \sum_{i=0}^{n} T[y_i, y_{i+1}], \qquad (10.11)$$

where **R** is the matrix of individual label scores produced by the RNN – that is, $R[y, i]$ is the score of label y for word i in the sequence – and **T** is a matrix that stores the transition scores between any two labels – that is, $T[y_i, y_j]$ is the transition score from label y_i to y_j.[5] Note that the second summation in Equation 10.11 includes two virtual labels y_0 and y_{n+1}, which are set to indicate the beginning and end of the sequence: $y_0 = start$ and $y_{n+1} = stop$. Thus, **R** has dimension $k \times n$; **T** has dimension $k \times k$, where n is the number of words in the sequence, and k is the total number of distinct labels including the virtual *start* and *stop* labels (e.g., the number of possible POS tags in our example, plus two for the virtual labels).

Importantly, the **R** scores are computed just like in any vanilla transducer. For example, the most common setting is to use a one-layer (sometimes two-layer) FFNN that takes as input the hidden state of the underlying RNN and outputs as many neurons as the number of labels to be predicted. The values in the transition matrix **T** are learned directly from the training data – that is, we initialize them randomly, and adjust them through backpropagation. After training, **T** should contain high values for likely transitions – for example, from the determiner POS tag (DET) to the common noun tag (NOUN) – and low values for unlikely ones – for example, from DET to the verb POS tag (VERB).

Next, we convert the score into an actual probability using the softmax formula:

$$p(\mathbf{y}) = \frac{e^{s(\mathbf{y})}}{\sum_{\tilde{\mathbf{y}} \in \mathbf{Y}} e^{s(\tilde{\mathbf{y}})}},   \tag{10.12}$$

where $\tilde{\mathbf{y}}$ iterates through *all* the possible paths in the complete lattice of words $\times$ labels **Y** (e.g., such as the one shown in Figure 10.7). Working with probabilities of label sequences rather than scores has multiple advantages. First, it keeps the values bounded to the interval $[0, 1]$, which mitigates the exploding gradient problem. Second, maximizing this probability has two desired effects: it maximizes the score of the correct sequence of labels (which appears in the numerator in the probability formula), and it minimizes the scores of all other incorrect paths in the lattice (which are part of the denominator). This is achieved with a cost function that is the negative log likelihood of the correct sequence of labels **y**: $-\log(p(\mathbf{y}))$. Thus, the cost function of the CRF for the correct prediction for the entire sequence, **y**, becomes:

$$C(\mathbf{y}) = -\log(p(\mathbf{y})) = \log\left(\sum_{\tilde{\mathbf{y}} \in \mathbf{Y}} e^{s(\tilde{\mathbf{y}})}\right) - s(\mathbf{y}) = \text{logSumExp}_{\tilde{\mathbf{y}} \in \mathbf{Y}}(s(\tilde{\mathbf{y}})) - s(\mathbf{y}),$$
$$\tag{10.13}$$

---

[5]   To avoid an abuse of subscript font, here we use square brackets to indicate indices in a matrix.

where logSumExp is a function that computes the logarithm of sum of exponentials. That is:

$$\mathrm{logSumExp}(\mathbf{x}) = \log\left(\sum_i e^{x_i}\right), \qquad (10.14)$$

Importantly, the implementation of logSumExp computes the formula *in a stable way*. To clarify what this means, consider a naive implementation where we first compute each exponential, then the sum, and only then the logarithm. Because the CRF scores (Equation 10.11) are unbounded, we may run into overflows (i.e., numbers that are too large to be represented with the given range of digits available in the computer) in the very first step of the procedure. For our purposes, it is sufficient to know that logSumExp avoids this problem through a mathematical trick.[6]

Going back to the cost function of the CRF (Equation 10.13), $s(\mathbf{y})$ can be trivially computed using Equation 10.11. The complicated part of the loss is the first component, which needs to apply logSumExp on *all* the possible paths $\tilde{y}$ in the lattice. A naive enumeration of all these paths has a cost of n^k, where n is the number of words in the sentence, and k is the number of possible labels (Jurafsky and Martin, 2009). The forward algorithm was introduced to address this problem; as we will see, it reduces the cost of computing the sum of probabilities for all possible label sequences in the lattice from n^k to nk^2, which is much more manageable.[7]

Algorithm 12 lists a version of the forward algorithm that is adapted to compute the logSumExp part of the loss function. The intuition behind this algorithm is fairly simple: for every node in the lattice, the algorithm computes the sum (or rather logSumExp in log space) of the $s(\tilde{y})$ values for all paths $\tilde{y}$ that end in that node. For example, for the word *old* with the POS NOUN in Figure 10.7, the algorithm computes the logSumExp for four paths: one that goes through the word *The* with POS tag DET, one that goes through *The* as NOUN, one that goes through *The* as VERB, and a last one that traverses *The* as ADJ. These values are stored in the variable *forward*, where *forward*[i,j] indicates this logSumExp for the ith label and the jth word in the sequence.

More specifically, lines 1–3 in the algorithm initialize the first column in the lattice with the cost of producing the corresponding label and transitioning to each of these nodes from *start*. Note that because all paths begin at the virtual *start* node, there is no **R** cost for this label. Further, we can simplify the

---

[6] For the mathematically inclined reader, Wikipedia as well as the PyTorch CRF tutorial provide the implementation for this function. See, for example, the implementation of the `log_sum_exp` here: https://pytorch.org/tutorials/beginner/nlp/advanced_tutorial.html.

[7] The terminology is unfortunately overloaded here: *forward* in the context of this algorithm refers to traversing the lattice of possible label assignments and computing the sum of all path probabilities, not to a feed-forward layer in a neural network.

Algorithm 12 Variant of the forward algorithm that computes logSumExp for all possible paths in the lattice

---

1 **for** *each label index l* **from** *1* **to** k **do**
2 $\quad$ $forward[l, 1] = T[start, y_l] + R[y_l, 1]$
3 **end**
4 **for** *each word position i* **from** *2* **to** n **do**
5 $\quad$ **for** *each label index l* **from** *1* **to** k **do**
6 $\quad\quad$ $forward[l, i] =$
   $\quad\quad\quad logSumExp_{l'=1}^{k}(forward[l', i-1] + T[y_{l'}, y_l] + R[y_l, i])$
7 $\quad$ **end**
8 **end**
9 $output = logSumExp_{l=1}^{k}(forward[l, n] + T[y_l, stop])$
10 **return** output

---

forward values in the first column to: $forward[l, 1] = logSumExp(T[start, y_l] + R[y_l, 1]) = \log(e^{T[start, y_l] + R[y_l, 1]}) = T[start, y_l] + R[y_l, 1]$ (line 2 in the algorithm). Lines 4–7 iteratively compute the *forward* values for the other columns in the lattice. That is, for the ith word in the sequence and the label index l, $forward[l, i]$ accumulates the path scores from *all* nodes in the previous lattice column to the current one. Each of these scores is computed as the logSumExp of the *forward* value for the corresponding previous node ($forward[l', i-1]$), which accumulates all scores up to that point, plus the score of transitioning from the previous node to the current one ($T[y_{l'}, y_l]$) and the score of producing the current label ($R[y_l, i]$). Last, line 9 computes the logSumExp for all paths ending in the *stop* node by accumulating the transition score to *stop* from all nodes in the last layer in the lattice ($T[y_l, stop]$).

Algorithm 12 looks deceptively simple. But it is actually not trivial to show that it is indeed computing what we need – that is, $logSumExp(s(\tilde{y}))$ for all possible paths $\tilde{y}$ – correctly. To convince ourselves of this, let's trace the execution of the algorithm on the simpler lattice shown in Figure 10.8. In this figure, s_{ij} indicates the node in the lattice corresponding to label y_i for word w_j. For example, s_{21} corresponds to the second label, y_2, assigned to the first word in the sentence, w_1. The first block in Algorithm 12 (lines 1–3) computes:

$$forward[1, 1] = T[start, y_1] + R[y_1, 1]$$
$$forward[2, 1] = T[start, y_2] + R[y_2, 1],$$

which capture the cost of transitioning from the *start* node to s_{11} and s_{21}, respectively. The next block in the algorithm (lines 4–8) computes the forward values for the second word in the sequence. For example, for the first label, this computation is:

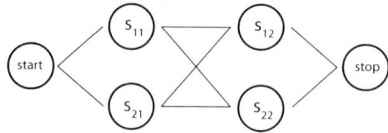

Figure 10.8 A simple lattice for the walkthrough example of the forward algorithm

$$forward[1,2] = \text{logSumExp}_{l'=1}^2(forward[l',1] + T[y_{l'},y_1] + R[y_1,2])$$

$$= \log\left(e^{forward[1,1]+T[y_1,y_1]+R[y_1,2]} + e^{forward[2,1]+T[y_2,y_1]+R[y_1,2]}\right)$$

$$= \log\left(e^{T[start,y_1]+R[y_1,1]+T[y_1,y_1]+R[y_1,2]}\right.$$

$$\left. + e^{T[start,y_2]+R[y_2,1]+T[y_2,y_1]+R[y_1,2]}\right)$$

$$= \log\left(e^{\{start,y_1,y_1\}} + e^{\{start,y_2,y_1\}}\right)$$

$$= \text{logSumExp}(\{start,y_1,y_1\}, \{start,y_2,y_1\}),$$

where we used $\{\dots\}$ as syntactic sugar to indicate the sum of all T and R scores along the included path. Similarly, the forward score for the s_{22} node is:

$$forward[2,2] = \text{logSumExp}(\{start,y_1,y_2\}, \{start,y_2,y_2\}).$$

Line 9 in the algorithm computes the logSumExp for all paths ending in the *stop* node:

$$output = \text{logSumExp}_{l=1}^2(forward[l,2] + T[y_l,stop])$$

$$= \log\left(e^{forward[1,2]+T[y_1,stop]} + e^{forward[2,2]+T[y_2,stop]}\right)$$

$$= \log\left(e^{forward[1,2]}e^{T[y_1,stop]} + e^{forward[2,2]}e^{T[y_2,stop]}\right)$$

$$= \log($$
$$(e^{\{start,y_1,y_1\}} + e^{\{start,y_2,y_1\}})e^{T[y_1,stop]} +$$
$$(e^{\{start,y_1,y_2\}} + e^{\{start,y_2,y_2\}})e^{T[y_2,stop]}$$
$$)$$

$$= \log($$
$$e^{\{start,y_1,y_1\}+T[y_1,stop]} +$$
$$e^{\{start,y_2,y_1\}+T[y_1,stop]} +$$
$$e^{\{start,y_1,y_2\}+T[y_2,stop]} +$$
$$e^{\{start,y_2,y_2\}+T[y_2,stop]}$$
$$)$$

$$= \log(\,$$
$$e^{\{start, y_1, y_1, stop\}} +$$
$$e^{\{start, y_2, y_1, stop\}} +$$
$$e^{\{start, y_1, y_2, stop\}} +$$
$$e^{\{start, y_2, y_2, stop\}}$$
$$)$$

$$= \text{logSumExp}(\,$$
$$\{start, y_1, y_1, stop\},$$
$$\{start, y_2, y_1, stop\},$$
$$\{start, y_1, y_2, stop\},$$
$$\{start, y_2, y_2, stop\}$$
$$).$$

Thus, for every node in the lattice, the algorithm indeed computes what we need – that is, the logSumExp of all paths ending in that node. For the *stop* node, this value is the logSumExp for all paths in the lattice (since all of them end in *stop*), which is the first term of the cost function in Equation 10.13.

The runtime cost of Algorithm 12 is driven by the three loops in lines 4–7 (there is a hidden loop in the logSumExp in line 6). Because the algorithm traverses the sequence of n words once (line 4), and the set of labels twice (lines 5 and 6), this cost is nk^2, which is much better than the n^k runtime cost of the naive implementation. This is possible because the *forward* values *cache* the intermediate results for every node in the lattice, which are then reused in the downstream computations (i.e., for the following words in the lattice).

Sidebar 10.2   Dynamic programming

The forward algorithm described here belongs to a class of programs in computer science called *dynamic programming*. In general, dynamic programming reduces complex programs into a series of iterative or recursive steps, where the solution at each step is computed once and stored for future use (Bellman, 1954).

10.5.2   *Applying the Conditional Random Field Using the Viterbi Algorithm*

Once the CRF has been trained – that is, we have trained parameters for the RNN that produces the **R** scores and meaningful transition scores in **T** – we can apply it to arbitrary sequences to find the path **y** with the maximum score $s(\mathbf{y})$ in the corresponding lattice. The algorithm that finds this path efficiently is another dynamic programming algorithm known as the Viterbi algorithm,

---

Algorithm 13 Viterbi algorithm that finds the path with the maximum score in the lattice

---

1 **for** *each label index l* **from** *1* **to** *k* **do**
2      $viterbi[l, 1] = T[start, y_l] + R[y_l, 1]$
3      $backPointer[l, 1] = 0$
4 **end**
5 **for** *each word position i* **from** *2* **to** *n* **do**
6   **for** *each label index l* **from** *1* **to** *k* **do**
7          $viterbi[l, i] = \max_{l'=1}^{k}(viterbi[l', i-1] + T[y_{l'}, y_l] + R[y_l, i])$
8          $backPointer[l, i] =$
            $\operatorname{argmax}_{l'=1}^{k}(viterbi[l', i-1] + T[y_{l'}, y_l] + R[y_l, i])$
9   **end**
10 **end**
11 $bestPathScore = \max_{l=1}^{k}(viterbi[l, n] + T[y_l, stop])$
12 $bestPathPointer = \operatorname{argmax}_{l=1}^{k}(viterbi[l, n] + T[y_l, stop])$
13 **return** $viterbi, bestPathScore, bestPathPointer$

---

after its author (Viterbi, 2006). It is very similar to the forward algorithm we discussed before, with the key difference being that in each node, we store the score of the *best path that ends in that node* (rather than the logSumExp of all path scores ending there as in the forward algorithm). In particular, $viterbi[i, j]$ stores the score of the best path ending on the ith label for the jth word in the sequence. The additional variable $backPointer$ stores the index in the previous lattice column that produced the last component in the best path for this node. For example, the back pointer for the node *man* with the POS tag VERB in Figure 10.7 points back to the POS tag NOUN for *old*.

The value *bestPathScore* stores the overall score for the best path in the lattice. *bestPathPointer* points back from the *stop* node to the node in the last lattice column that is part of the best path found. Thus, traversing these pointers back from *bestPathPointer* all the way to *start* recovers the sequence of nodes that form the best path.

The Viterbi algorithm has the same runtime cost as the forward algorithm: nk^2. Thus, the combination of the forward and Viterbi algorithms provide an efficient framework to train and apply a CRF that is linear in the number of words in the sequence. In several NLP applications (see Chapter 16), using a CRF layer on top of an RNN yields considerable improvements in performance. For example, the authors have observed an increase in four F_1 percentage points in named entity recognition performance when a bidirectional LSTM is coupled with a CRF over the standalone bidirectional LSTM.

10.6   Drawbacks of Recurrent Neural Networks

As we will detail in Chapter 16, sequence models are fundamental for many applications that rely on sequences such as language models, POS tagging, named entity recognition, or syntactic parsing. However, as the tweet at the beginning of this chapter hints, many NLP applications – for example, text classification or question answering – do not need sequence modeling (despite one's linguistic intuitions). For such applications, simpler bag-of-words techniques perform just as well or better.

Further, RNNs are hard to parallelize. They must process the input words sequentially because each RNN cell depends on the output of the previous cell. In contrast, other newer architectures such as transformer networks (see Chapter 12) are designed to facilitate parallel processing, which makes them considerably faster when executed on GPUs, which have hardware support for parallel tensor operations.

Last, despite the fact that RNN architectures such as LSTMs are designed to capture arbitrarily long sequences, in practice, they become "fuzzy far away" – that is, they tend to forget about word order for long-range contexts such as beyond 50 words (Khandelwal et al., 2018).

10.7   Historical Background

Recurrent neural networks were invented by Rumelhart et al. (1986). Interestingly, Rumelhart was yet another psychologist who made a fundamental contribution to the field of deep learning (see the history of the perceptron in Chapter 2). As we discussed in this chapter, these original RNNs suffered form the vanishing gradient problem. This issue was addressed by LSTM networks, which were invented by Hochreiter and Schmidhuber (1997) approximately 10 years later. Until the late 2010s when the transformer networks were introduced (see Chapter 12), LSTMs were ubiquitous in the NLP field. They were used to implement virtually all applications where modeling sequences are important, from automated speech recognition to machine translation.

Conditional random fields were invented by Lafferty et al. (2001), but they were initially applied to a non-neural sequence modeling algorithm. The adaptation to RNNs that we discussed in this chapter was proposed by Lample et al. (2016). Conditional random fields incorporate other, older algorithms. For example, the Viterbi algorithm, which CRFs use for inference, was discovered by Andrew Viterbi in 1967 (Viterbi, 2006).[8] The forward algorithm, which is used during the training of CRFs, was initially proposed in the context of training hidden Markov models, a different sequence modeling framework (Baum and Petrie, 1966; Baum and Eagon, 1967).

---

[8] Kruskal (1983) showed that, despite the name, the Viterbi algorithm actually had a "remarkable history of multiple independent discover[ies]."

10.8   References and Further Readings

Christopher Olah wrote a great visual description of LSTM networks in his blog (Olah, 2015).

More recent variants of LSTMs offer the same advantages and similar performance as the original, while reducing the number of parameters (Cho et al., 2014).

Khandelwal et al. (2018) highlighted some drawbacks of LSTMs, which partially paved the way for the introduction of attention-based architectures such as transformer networks (see Chapter 12).

10.9   Summary

This chapter introduced methods that explicitly capture word-order information. In particular, we discussed several types of RNNs, including stacked (or deep) RNNs, bidirectional RNNs, and LSTM networks. We showed that RNNs can be used in three different ways: (a) as acceptors, where a classification layer is added on top of the last network cell; (b) as transducers, which add a classification layer for each cell; and (c) as encoder-decoders, in which case two RNNs are combined: an encoder that codes an input sequence into a single vector, and a decoder that generates one element at a time from an output sequence. Last, we introduced conditional random fields, which extend transducer RNNs with an extra layer that explicitly models transition probabilities between two cells. This allows us to estimate (and optimize during training) the probability of an entire sequence of labels rather than probabilities of individual labels.

11   Implementing Part-of-Speech Tagging Using Recurrent Neural Networks

The previous chapter was our first exposure to RNNs, which included intuitions for why they are useful for NLP, various architectures, and training algorithms. In this chapter, we will put them to use in order to implement a common sequence modeling task.

11.1   Part-of-Speech Tagging

The task we will use as an example for this chapter is POS tagging, an NLP application that, as we discussed in the previous chapter, benefits from word order. Please see Chapter 16 for a more thorough discussion of POS tagging. The entire code presented in this chapter is available in the `chap11_pos_tagging` Jupyter notebook.

To take a break from NLP applications for English, in this chapter, we use the AnCora corpus (Taulé et al., 2008), which primarily consists of newspaper texts in Spanish and Catalan with different linguistic annotations. In this chapter, we work with the Spanish portion of the corpus and the annotations for Universal POS tags (see Chapter 16 for a description of these tags).

The Spanish portion of the corpus is divided into a training set with 14,305 sentences, a development set with 1,654 sentences, and a test set with 1,721 sentences. The data are distributed in the CoNLL-U format. In this format, all sentences in a dataset are stored in the same file, separated by a blank line. Each individual token in a sentence is represented in a line that contains 10 annotation fields separated by tabs: ID, FORM, LEMMA, UPOS, XPOS, FEATS, HEAD, DEPREL, DEPS, and MISC. A comprehensive explanation of this format and the meaning of the different fields is beyond the goal of this chapter; however, the curious reader can find one at the CoNLL-U website.[1] Here, we are only concerned with the fields FORM (the raw word), and UPOS (the Universal POS tag).

As in previous chapters, we use pandas to preprocess the data. For parsing the CoNLL-U files, we rely on the `conllu` Python module.[2] We implement

---

[1]   https://universaldependencies.org/format.html.
[2]   https://github.com/EmilStenstrom/conllu.

a function called `read_tags` that reads the CoNLL-U file corresponding to a dataset and returns a pandas DataFrame that combines all tokens in a sentence into a single row with two columns, one for the words, and one for the POS tags in the corresponding sentence (the output for the code below is on page 167):

[2]:
```python
from conllu import parse_incr

def read_tags(filename):
    data = {'words': [], 'tags': []}
    with open(filename) as f:
        for sent in parse_incr(f):
            words = [tok['form'] for tok in sent]
            tags = [tok['upos'] for tok in sent]
            data['words'].append(words)
            data['tags'].append(tags)
    return pd.DataFrame(data)

train_df = read_tags('data/UD_Spanish-AnCora/es_ancora-ud-train.
↪conllup')
train_df
```

In order to implement our POS tagging application, we need word embeddings that have been pretrained for Spanish. Here we use the publicly available GloVe embeddings trained on the Spanish Billion Word Corpus[3] by the Departamento de Ciencias de la Computación of Universidad de Chile.[4] In contrast to the GloVe embeddings used in Chapter 9, these do include a header that stores metadata about the embeddings (i.e., size of the vocabulary and the dimension of the embedding vectors), so in this case, we do not use the `no_header=True` argument: so in this case, we do not use the `no_header=True` argument:

[4]:
```python
from gensim.models import KeyedVectors
glove = KeyedVectors.load_word2vec_format('glove-sbwc.i25.vec')
glove.vectors.shape
```

[4]: (855380, 300)

Another difference between these GloVe embeddings and the ones we used in Chapter 9 is that these already include an embedding for unknown words. Therefore, there is no need to introduce our own.

---

[3] https://crscardellino.ar/SBWCE.
[4] https://github.com/dccuchile/spanish-word-embeddings#
glove-embeddings-from-sbwc.

	words	tags
0	[El, presidente, de, el, órgano, regulador, de...	[DET, NOUN, ADP, DET, NOUN, ADJ, ADP, DET, PRO...
1	[Sobre, la, oferta, de, interconexión, con, Te...	[ADP, DET, NOUN, ADP, NOUN, ADP, PROPN, ADP, D...
2	[Afirmó, que, sigue, el, criterio, europeo, y,...	[VERB, SCONJ, VERB, DET, NOUN, ADJ, CCONJ, SCO...
3	[La, inversion, en, investigación, básica, es,...	[DET, NOUN, ADP, NOUN, ADJ, AUX, DET, NOUN, AD...
4	[Durante, la, presentación, de, el, libro, ", ...	[ADP, DET, NOUN, ADP, DET, NOUN, PUNCT, DET, P...
...	...	...
14300	[Y, todas, las, miradas, convergen, en, la, lu...	[CCONJ, DET, DET, NOUN, VERB, ADP, DET, NOUN, ...
14301	[Conviene, que, ahora, ,, en, plena, apoteosis...	[VERB, SCONJ, ADV, PUNCT, ADP, ADJ, NOUN, ADP,...
14302	[Cambiar, las, formas, parece, de, rigor, ,, p...	[VERB, DET, NOUN, VERB, ADP, NOUN, PUNCT, CCON...
14303	[Carlos, y, Fayna, se, enzarzan, en, una, bron...	[PROPN, CCONJ, PROPN, PRON, VERB, ADP, DET, NO...
14304	[Él, llega, a, tirar, la, sobre, la, cama, y, ...	[PRON, VERB, ADP, VERB, PRON, ADP, DET, NOUN, ...

14305 rows × 2 columns

[5]:
```
# these embeddings already include <unk>
unk_tok = '<unk>'
unk_id = glove.key_to_index[unk_tok]
unk_tok, unk_id
```

[5]: ('<unk>', 855379)

However, we do need to include a new embedding for padding, which we will use later to guarantee that all sentences in the same minibatch have the same length. We add a vector of zeros for the padding token in the same way as before:

[6]:
```
# add padding embedding
pad_tok = '<pad>'
pad_emb = np.zeros(300)
glove.add_vector(pad_tok, pad_emb)
pad_tok_id = glove.key_to_index[pad_tok]
pad_tok, pad_tok_id
```

[6]: ('<pad>', 855380)

Next, we need to preprocess our tokens to match the vocabulary of the embeddings. In particular, these embeddings were trained on words that were lowercased and on sequences of digits that were replaced with a single 0. We will apply the same modifications to our tokens:

[7]:
```
def preprocess(words):
    result = []
    for w in words:
        w = w.lower()
        if w.isdecimal():
            w = '0'
        result.append(w)
    return result

train_df['words'] = train_df['words'].progress_map(preprocess)
train_df
```

(From now on, we will omit the pandas tables for readability, but, as usual, the corresponding Jupyter notebook contains all necessary information.)

Next, we add a new column to the dataframe that stores the *word IDs* corresponding to the embedding vocabulary. Note that at this point, we are not padding the sequences of word IDs. We will address padding later.

```
[8]:    def get_ids(tokens, key_to_index, unk_id=None):
            return [key_to_index.get(tok, unk_id) for tok in tokens]

        def get_word_ids(tokens):
            return get_ids(tokens, glove.key_to_index, unk_id)

        # add new column to the dataframe
        train_df['word ids'] = train_df['words'].
         ↪progress_map(get_word_ids)
        train_df
```

We also need to generate the IDs for the POS tags. To this end, we first need to construct a vocabulary of POS tags. Once again, we generate a list of tags using explode(), which linearizes our sequence of tags, and remove repeated tags using unique(). We also add a special tag for the padding token:

```
[9]:    pad_tag = '<pad>'
        index_to_tag = train_df['tags'].explode().unique().tolist()
        index_to_tag += [pad_tag]
        tag_to_index = {t:i for i,t in enumerate(index_to_tag)}
        pad_tag_id = tag_to_index[pad_tag]
        pad_tag, pad_tag_id
```

```
[9]:    ('<pad>', 17)
```

We now use this POS tag vocabulary to construct a new dataframe column that stores the POS tag IDs:

```
[11]:   def get_tag_ids(tags):
            return get_ids(tags, tag_to_index)

        train_df['tag ids'] = train_df['tags'].progress_map(get_tag_ids)
        train_df
```

The implementation of the Dataset class that stores our POS dataset is trivial: we simply return the lists of word and tag IDs, converted to PyTorch tensors.

```
[13]:   from torch.utils.data import Dataset

        class MyDataset(Dataset):
            def __init__(self, x, y):
                self.x = x
                self.y = y

            def __len__(self):
```

```
        return len(self.y)

    def __getitem__(self, index):
        x = torch.tensor(self.x[index])
        y = torch.tensor(self.y[index])
        return x, y
```

Now it's time to handle padding. This time, we will use some features of PyTorch that we have not seen before. The DataLoader object can receive an optional argument, collate_fn, which expects a function that can be used to form a minibatch. We will implement this function using PyTorch's torch.nn.utils.rnn.pad_sequence() function, which, unsurprisingly, pads a group of tensors. We will take advantage of this function to pad the tensors while forming the minibatch itself. The advantage of this strategy is that, rather than needing to pad all the examples to be the same length as the largest sentence *in the corpus*, we will instead pad them to the same length as the largest sentence *in the minibatch*. The latter strategy reduces the amount of padding necessary, which should yield more efficient code.

[14]:
```
from torch.nn.utils.rnn import pad_sequence

def collate_fn(batch):
    # separate xs and ys
    xs, ys = zip(*batch)
    # get lengths
    lengths = [len(x) for x in xs]
    # pad sequences
    x_padded = pad_sequence(
        xs,
        batch_first=True,
        padding_value=pad_tok_id)
    y_padded = pad_sequence(
        ys,
        batch_first=True,
        padding_value=pad_tag_id)
    # return padded
    return x_padded, y_padded, lengths
```

The collate_fn() function takes a single argument, batch, which is a list of tuples. Each tuple has two elements: the list of word IDs and the list of tag IDs corresponding to a single example. We first unzip this list of tuples into two lists; the first list has all the word IDs, and the second has the tag IDs. An explanation of how zip(*batch) works is provided in Appendix A. Next, we compute the lengths of each of the examples in the batch, which we will use later to inform the RNN where padding starts for each example. We then use the pad_sequence() function to add padding. This function will find the longest sequence in the batch and pad all examples accordingly using the provided padding value. This method is designed to work with PyTorch's RNNs, which by default assume the batch index is in the second dimension. However, we will be organizing our tensors such that the batch index is always in the first dimension, which we feel to be more intuitive. For this reason, we also need to provide the batch_first=True argument to pad_sequence. Finally, we return the padded data, as well as the original lengths of the examples.

Next, we implement our POS tagging model class. The model consists of: (a) an embedding layer for our Spanish pretrained embeddings; (b) an LSTM that can be set to be uni- or bidirectional (see Figure 10.3; the RNN is configured to be bidirectional by setting the bidirectional argument to True in the LSTM constructor), with a configurable number of layers (see Figure 10.2; the number of layers is set through the num_layers argument of the constructor) and (c) a linear layer on top of each hidden state, which is used to predict the scores for each of the POS tags for the corresponding token.

The forward() method receives the padded minibatch and the list of lengths for the (unpadded) examples in this minibatch. The first step in the function is to retrieve the embeddings for all words referenced in this minibatch. We then apply dropout over these embedding vectors. Next, before passing the data to the LSTM, we *pack* the padded data. Note that the PyTorch PackedSequence class, which is the output of the pack_padded_sequence() function, stores a batch of sequences that had different lengths before padding.[5] One important advantage of using PackedSequence is that its internal data structure removes the padding tokens (which is why we had to keep track of the example lengths before padding in x_lengths), and, thus, the RNN will not backpropagate over the padded elements.[6]

Once we have a PackedSequence, we pass it to the LSTM. Since the output of the LSTM is also packed, we then unpack it using pad_packed_sequence(). Next we apply dropout to this unpacked LSTM output. Finally, we pass this to the linear layer to predict the tag scores for the tokens.

[15]:
```
from torch import nn
from torch.nn.utils.rnn import pack_padded_sequence,␣
↪pad_packed_sequence
```

---

[5] https://pytorch.org/docs/stable/generated/torch.nn.utils.rnn
.PackedSequence.html.

[6] The astute reader might ask at this point, "Why did we pad the minibatch examples in the first place if we are removing the padding later?" The padding is needed because this allows us to store the minibatch as a single three-dimensional tensor.

```python
class MyModel(nn.Module):
    def __init__(self, vectors, hidden_size, num_layers,↵
↪bidirectional, dropout, output_size):
        super().__init__()
        # ensure vectors is a tensor
        if not torch.is_tensor(vectors):
            vectors = torch.tensor(vectors)
        # init embedding layer
        self.embedding = nn.Embedding.
↪from_pretrained(embeddings=vectors)
        # init lstm
        self.lstm = nn.LSTM(
            input_size=vectors.shape[1],
            hidden_size=hidden_size,
            num_layers=num_layers,
            bidirectional=bidirectional,
            dropout=dropout,
            batch_first=True)
        # init dropout
        self.dropout = nn.Dropout(dropout)
        # init classifier
        self.classifier = nn.Linear(
            in_features=hidden_size * 2 if bidirectional else↵
↪hidden_size,
            out_features=output_size)

    def forward(self, x_padded, x_lengths):
        # get embeddings
        output = self.embedding(x_padded)
        output = self.dropout(output)
        # pack data before lstm
        packed = pack_padded_sequence(output, x_lengths,↵
↪batch_first=True, enforce_sorted=False)
        packed, _ = self.lstm(packed)
        # unpack data before rest of model
        output, _ = pad_packed_sequence(packed,↵
↪batch_first=True)
        output = self.dropout(output)
        output = self.classifier(output)
        return output
```

Despite the small number of lines of code, the code of the forward() method, which switches between embedding vectors, padded tensors, and packed sequences, is not trivial. To clarify it, let us walk through an example. Imagine that the input to the forward() method is a batch, x_padded, with shape (10, 20), corresponding to 10 examples, each with 20 word IDs (some of which are padding). Then we retrieve the embeddings. Assuming our word embeddings – that is, the input vectors $\mathbf{x}_i$ in Chapter 10, are of dimension 300, the new tensor will have a shape of (10, 20, 300), corresponding to 10 examples, each with 20 embeddings, each with dimension 300. After dropout, the shape hasn't changed, but some of the elements have been zeroed out. After unpacking the output of the LSTM, we will have a tensor of shape (10, 20, hidden_size), where hidden_size is the size of the LSTM hidden state – that is, the $\mathbf{h}_t$ vector in Equation 10.6, (hidden_size is a hyper parameter we will set later on). After passing this tensor to the linear layer, we will obtain a tensor of shape (10, 20, tag_vocab_size), where tag_vocab_size is the number of POS tags in our vocabulary. Thus, for each token in each example, we will have a distribution of POS tag scores. For each token, the assigned POS tag will be the one corresponding to the highest score.

We next initialize all the hyper parameters and all the required components:

```
[16]:  from torch import optim
       from torch.utils.data import DataLoader
       from sklearn.metrics import accuracy_score

       # hyperparameters
       lr = 1e-3
       weight_decay = 1e-5
       batch_size = 100
       shuffle = True
       n_epochs = 10
       vectors = glove.vectors
       hidden_size = 100
       num_layers = 2
       bidirectional = True
       dropout = 0.1
       output_size = len(index_to_tag)

       # initialize model, loss function, optimizer, data loader
       model = MyModel(vectors, hidden_size, num_layers,
        ↪bidirectional, dropout, output_size).to(device)
       loss_func = nn.CrossEntropyLoss()
       optimizer = optim.Adam(model.parameters(), lr=lr,
        ↪weight_decay=weight_decay)
       train_ds = MyDataset(train_df['word ids'], train_df['tag ids'])
       train_dl = DataLoader(train_ds, batch_size=batch_size,
        ↪shuffle=shuffle, collate_fn=collate_fn)
       dev_ds = MyDataset(dev_df['word ids'], dev_df['tag ids'])
       dev_dl = DataLoader(dev_ds, batch_size=batch_size,
        ↪shuffle=shuffle, collate_fn=collate_fn)
```

```
train_loss, train_acc = [], []
dev_loss, dev_acc = [], []
```

The training procedure is very similar to the one implemented in Chapter 7. One notable difference is that the output of this model has three dimensions instead of two: number of examples, number of tokens, and number of POS tag scores. Thus, we have to reshape the output in order to pass it to the loss function. Additionally, we need to discard the padding before computing the loss. We reshape the gold tag IDs, using the `torch.flatten()` function, in order to transform the two-dimensional tensor of shape (n_examples, n_tokens) to a one-dimensional tensor with n_examples * n_tokens elements. The predictions are reshaped using the `view(-1, output_size)` method. By passing two arguments, we are stipulating that we want two dimensions. The second dimension will be of size `output_size`. The -1 indicates that the first dimension should be inferred from the size of the tensor. This means that for a tensor of shape (n_examples, n_tokens, output_size), we will get a tensor of shape (n_examples * n_tokens, output_size). Then, we use a Boolean mask to discard the elements corresponding to the padding. This way, the loss function will consider each actual word individually, as if the whole batch was just one big sentence. Note that treating a minibatch as a single virtual sentence does affect the evaluation results.

[17]:
```
# train the model
for epoch in range(n_epochs):
    losses, acc = [], []
    model.train()
    for x_padded, y_padded, lengths in tqdm(train_dl,
    ↪desc=f'epoch {epoch+1} (train)'):
        # clear gradients
        model.zero_grad()
        # send batch to right device
        x_padded = x_padded.to(device)
        y_padded = y_padded.to(device)
        # predict label scores
        y_pred = model(x_padded, lengths)
        # reshape tensors for loss function
        y_true = torch.flatten(y_padded)
        y_pred = y_pred.view(-1, output_size)
        # discard padding
        mask = y_true != pad_tag_id
        y_true = y_true[mask]
        y_pred = y_pred[mask]
        # compute loss
```

```
            loss = loss_func(y_pred, y_true)
            # accumulate for plotting
            gold = y_true.detach().cpu().numpy()
            pred = np.argmax(y_pred.detach().cpu().numpy(), axis=1)
            losses.append(loss.detach().cpu().item())
            acc.append(accuracy_score(gold, pred))
            # backpropagate
            loss.backward()
            # optimize model parameters
            optimizer.step()
        train_loss.append(np.mean(losses))
        train_acc.append(np.mean(acc))

    model.eval()
    with torch.no_grad():
        losses, acc = [], []
        for x_padded, y_padded, lengths in tqdm(dev_dl,
    ↪desc=f'epoch {epoch+1} (dev)'):
            x_padded = x_padded.to(device)
            y_padded = y_padded.to(device)
            y_pred = model(x_padded, lengths)
            y_true = torch.flatten(y_padded)
            y_pred = y_pred.view(-1, output_size)
            mask = y_true != pad_tag_id
            y_true = y_true[mask]
            y_pred = y_pred[mask]
            loss = loss_func(y_pred, y_true)
            gold = y_true.cpu().numpy()
            pred = np.argmax(y_pred.cpu().numpy(), axis=1)
            losses.append(loss.cpu().item())
            acc.append(accuracy_score(gold, pred))
        dev_loss.append(np.mean(losses))
        dev_acc.append(np.mean(acc))
```

Last, we evaluate the performance of our POS tagger on the test set, similarly to how we have done it before:

```
[21]:   from sklearn.metrics import classification_report

        model.eval()

        test_ds = MyDataset(test_df['word ids'], test_df['tag ids'])
        test_dl = DataLoader(test_ds, batch_size=batch_size,
          ↪shuffle=shuffle, collate_fn=collate_fn)
```

```
all_y_true = []
all_y_pred = []

with torch.no_grad():
    for x_padded, y_padded, lengths in tqdm(test_dl):
        x_padded = x_padded.to(device)
        y_pred = model(x_padded, lengths)
        y_true = torch.flatten(y_padded)
        y_pred = y_pred.view(-1, output_size)
        mask = y_true != pad_tag_id
        y_true = y_true[mask]
        y_pred = torch.argmax(y_pred[mask], dim=1)
        all_y_true.append(y_true.cpu().numpy())
        all_y_pred.append(y_pred.cpu().numpy())

y_true = np.concatenate(all_y_true)
y_pred = np.concatenate(all_y_pred)
target_names = index_to_tag[:-2]
print(classification_report(y_true, y_pred,
 ↪target_names=target_names))
```

	precision	recall	f1-score	support
DET	0.99	1.00	0.99	8040
NOUN	0.95	0.96	0.95	9533
ADP	1.00	1.00	1.00	8332
ADJ	0.92	0.93	0.92	3468
PROPN	0.91	0.89	0.90	4101
PRON	0.97	0.95	0.96	2484
VERB	0.98	0.98	0.98	4544
SCONJ	0.93	0.96	0.95	1210
PUNCT	0.98	0.99	0.98	6314
AUX	0.97	0.99	0.98	1396
CCONJ	1.00	1.00	1.00	1439
ADV	0.97	0.96	0.97	1710
NUM	0.97	0.85	0.91	958
PART	0.79	0.61	0.69	18
SYM	0.70	0.38	0.49	37
INTJ	0.80	0.50	0.62	16
accuracy			0.97	53600
macro avg	0.93	0.87	0.89	53600
weighted avg	0.97	0.97	0.97	53600

The results indicate that our POS tagger obtains an overall accuracy of 97%, which is in line with state-of-the-art approaches! This is encouraging considering that our approach does not include the CRF layer we discussed in Chapter 10. We challenge the reader to add this layer and experiment with this architecture for other sequence tasks such as named entity recognition.[7]

11.2   Summary

In this chapter, we have implemented a Spanish POS tagger using a bidirectional LSTM and a set of pretrained, static word embeddings. Through this process, we have also introduced several new PyTorch features such as the pad_sequence, pack_padded_sequence, and pad_packed_sequence functions, which allow us to work more efficiently with variable length sequences for RNNs.

---

[7] See, for example, the LSTM-CRF implementation from the PyTorch tutorial: https://pytorch.org/tutorials/beginner/nlp/advanced_tutorial.html.

12 Contextualized Embeddings and Transformer Networks

The distributional similarity algorithms discussed in Chapter 8 conflate all senses of a word into a single numerical representation (or embedding). For example, the word *bank* receives a single representation, regardless of its financial (e.g., as in *the bank gives out loans*) or geological (e.g., *bank of the river*) sense. This chapter introduces a solution for this limitation, in the form of a new neural architecture called transformer networks (TNs) (Vaswani et al., 2017; Devlin et al., 2018), which learns *contextualized embeddings* of words, which, as the name indicates, change depending on the context. That is, the word *bank* receives a different numerical representation for each of its instances in these two texts because the contexts in which they occur are different.

An important note: in this chapter, we discuss only the first half of the TN architecture, the *encoder*, which generates contextualized embeddings for an input text. Reusing the terminology from Chapter 10, this enables two ways to use a TN: as a transducer, where we build applications that rely on the contextualized embeddings of each word in an input sequence (e.g., POS tagging), or as an acceptor, in which case we build applications over a single contextualized representation of the entire input text (e.g., text classification). We will discuss the second half of the TN architecture, the *decoder*, which generates one element at a time from an output sequence, in Chapter 14.

While TNs have a relatively complex structure, the intuition behind them is simple: each word receives a contextualized embedding that is a *weighted average* of some context-independent input embeddings (i.e., embeddings that are conceptually similar to the word2vec embeddings from the previous chapter).[1,2] Figure 12.1 visualizes this intuition. As the figure shows, the contextualized embedding of the word *bank* combines all four of the input embeddings. Because this includes the input embedding for *river*, the resulting contextualized embedding of *bank* will lean toward a semantic representation closer to geology than finance. As the figure indicates, the same process applies

---

[1] In reality, TNs operate over tokens that may be subword units. We will clarify this distinction in Section 12.2, but, until then, we continue to use "word" for readability.

[2] Or, rather, each word receives a weighted average of *projections* of input embeddings into a new feature space. We will deal with these details later in this chapter.

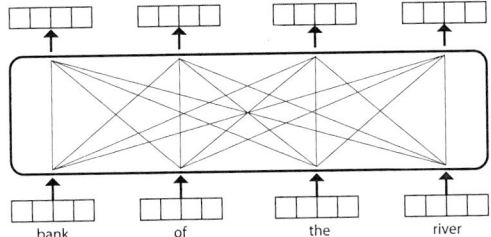

Figure 12.1 Intuition behind transformer networks: Each output embedding is a weighted average of all input embeddings in the context

to all other words in the text. Importantly, the weights used for the weighted averages that generate the output embeddings are specific to each word. This allows TNs to generate distinct output embeddings for the different words in the text to be processed.

In reality, TNs consist of multiple layers, where each layer implements a weighted average, as discussed earlier. This stacked architecture is summarized in Figure 12.2. As the figure shows, the output embeddings for layer i become the input embeddings for layer $i + 1$. The number of layers typically ranges between 2 and 24. Relying on multiple layers allows TNs to learn more complex functions for assembling the eventual output embeddings, which has been empirically shown to yield more meaningful representations.

In the next section, we will discuss the architecture of the individual TN layer, which is the key TN building block. Then, we will discuss the tokenization strategy used by TNs, which is different from what we discussed so far in the book. We will conclude the technical description of TNs in this chapter with a discussion of their training procedure.

12.1   Architecture of a Transformer Layer

Figure 12.3 shows the internal architecture of an individual TN layer. As shown, each layer implements a sequence of five operations. The first operation adds positional information to the input embedding of each word in the input text. That is, at this stage, the embedding of each word changes depending on its position in the input text. For example, each occurrence of the word *shipping* in the tongue twister *A ship shipping ship shipping shipping ships* will receive a different embedding because they occur at different positions in the text.

The second operation implements the key functionality of the TN layer summarized in Figure 12.1 – that is, generating embeddings that are a weighted average of the embeddings produced by the previous operation. In deep

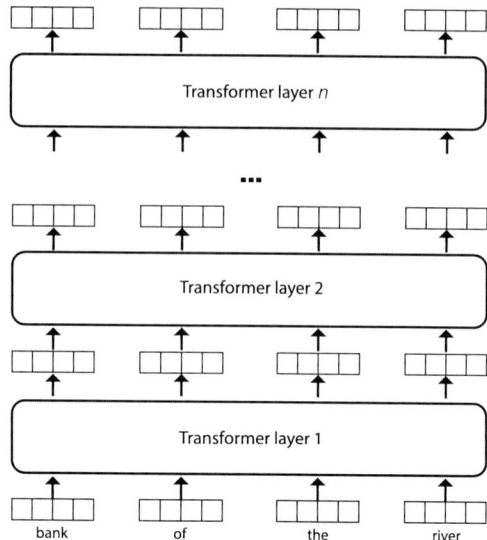

Figure 12.2 A transformer network consists of multiple layers, where each layer performs a weighted average of its input embeddings

learning parlance, this weighted average is called "self attention."[3] The term "attention" is used to indicate that this component identifies the important parts of the data and "pays attention" to them more (through higher weights in the weighted average). "Self" indicates that this component operates over its own input text (we will see other types of attention that operate over other texts in the following chapters).

The next three operations are not that important conceptually, but they have been shown to have a significant contribution empirically. For example, the "Add and normalize" components sum up input and output embeddings from the previous component in the pipeline (e.g., the input embeddings to the self-attention layer and the corresponding output embeddings computed through the weighted average), and normalize the results to avoid values that might be too large, which may negatively impact gradient descent. The feed-forward components encode each received numerical representation into a new vector, which allows the TN layer to learn more complex functions.

We detail all these operations next.

---

[3] This terminology is inspired from cognitive science and the actual cognitive attention. To the knowledge of the authors, it has not been proven that this connection between the weighted average performed by TNs (or other deep learning architectures) and the cognitive attention process is justified. However, because "attention" is widely used in the deep learning literature, we will continue to use this terminology throughout the book.

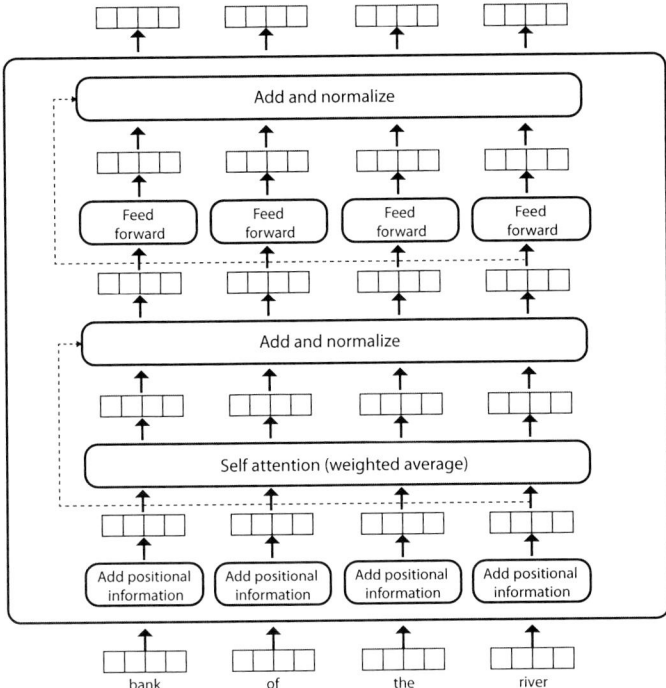

Figure 12.3 Architecture of an individual transformer layer

12.1.1 Positional Embeddings

Our astute reader has likely observed that the weighted average summarized in Figure 12.1 suggests a "bag-of-words" model. That is, TNs seem to produce contextualized embeddings that are independent of the order of the words in the input text. This is clearly less than ideal. For example, in the text *Bank of America financed a repair of the river bank* the machine should be able to figure out that the first instance of *bank* refers to a financial institution because it is closer to the word *financed* and farther away from the word *river*. Modeling such proximity information requires keeping track of positional information – that is, where words occur in the input texts.

Before TNs, neural approaches typically modeled positional information as a separate numerical representation. That is, each value i in a list of possible word positions in a text (say, 1 to 100), is associated with a numerical representation $\mathbf{p}_i$, which is initialized with random values. Then, the input embedding of word w_i becomes a concatenation of two vectors: $\mathbf{x}_i = [\mathbf{w}_i; \mathbf{p}_i]$, where $\mathbf{w}_i$ is a regular, static word embedding, similar to the ones computed by word2vec. The training process of a neural network on top of such embeddings backpropagates gradients all the way back to these vectors. Thus, during training, the network also learns numerical representations for word positions through the **p** vectors. The advantage of this strategy to model positional information is flexibility: the network learns numerical representations for position values that are customized for the task at hand. The disadvantages of this approach are additional cost caused by the extra back-propagation operations, and the inability to handle position values not seen in training. For example, assume that all training sentences for some NLP task contain fewer than 100 words. The network trained on these data will not know what numerical representations to assign to words that occur at positions larger than 100 during evaluation.

Transformer networks address these drawbacks by using hard-coded functions to generate numerical representations of word positions. That is, for a word at position i in the input text, TNs generate a vector $\mathbf{p}_i$, where the value at position j in this vector is computed using a function that depends on both i and j: $f(i,j)$. The actual function f used is not that important; suffice to say that the resulting vector $\mathbf{p}_i$ encodes positional information because it is unique for each word position i.[4] For example, Alammar (2018) calculated the following positional vectors for a toy example containing a text with three words and $|\mathbf{p}| = 4$:

$$\mathbf{p}_0 = [0, 0, 1, 1]$$
$$\mathbf{p}_1 = [0.84, 0.0001, 0.54, 1]$$
$$\mathbf{p}_2 = [0.91, 0.0002, -0.42, 1]$$

This example highlights that each positional embedding is indeed unique for a given word position i in the text.

Once the positional vectors are calculated, the actual input embedding $\mathbf{x}_i$ for word w_i becomes: $\mathbf{x}_i = \mathbf{w}_i + \mathbf{p}_i$. This approach mitigates the two disadvantages mentioned before: the function f is hard-coded, and, thus, there is no need to learn it. f generates different values for any word position i, and thus works for previously unseen word positions. The TNs' creators mention that the proposed method performs similarly in practice as the more expensive and less flexible strategy discussed at the beginning of this subsection.

12.1.2   Self Attention

As indicated earlier, this self-attention layer is the key building block in TNs. For each word w_i, this layer produces an output embedding, $\mathbf{z}_i$, that is a combination of all input embeddings $\mathbf{x}$ (i.e., the embeddings produced by the previous component that infuses positional information), for all the words in the input text.

---

[4]   We refer the reader to the TNs paper for details on this function (Vaswani et al., 2017).

To generate the output embeddings (i.e., the vector z_i for every word w_i), the self-attention layer uses three vectors for each word w_i: a query vector q_i, a key vector k_i, and a value vector v_i. These vectors are nothing magical: they are just arrays of real values – for example, a q_i vector might look like $(0.58, -045, 0.34, \ldots)$. We will discuss later in this chapter how the values in these vectors are learned; for now, let us assume that these three vectors exist for every word in the input text, and they are populated with some meaningful values.

At a high level, the query and key vectors are used to generate unique attention weights for each pair of words w_i and w_j. The intuition behind this attention mechanism can be explained with a real-world analogy: suppose you are at bakery and are interested in buying a bagel. Here, the query is "bagel," and the keys are the names of all products in the bakery – for example, "bread," or "croissant." The value of each product is its price. Our goal in this shopping experience is to pay more attention to the prices (or values) of the products (keys) that are closest to our interest (query).[5] In our context, the dot product $q_i \cdot k_j$ indicates how important word w_j is for the output embedding of word w_i. The value vector v_i is a projection (or transformation) of the input vector x_i into a new feature space. Each output embedding z_i will be a weighted average of these value vectors.[6]

To formalize a bit more, given the query, key, and value vectors for all words in the input text (i.e., q_i, k_i, v_i for word w_i), the self-attention algorithm operates as follows:

1. For each pair of words, w_i and w_j, compute the attention weight a_{ij} using the q_i and k_j vectors. In particular:
   (a) Initialize the attention weights a_{ij} with the product of the corresponding query and key vectors: $a_{ij} = q_i \cdot k_j$.
   (b) Divide these values by the square root of the length of the key vector: $a_{ij} = a_{ij}/\sqrt{|k_1|}$ (all k_i vectors have the same size, so we arbitrarily use k_1 here).
2. For each word w_i, apply softmax on all its attention weights, a_{ij}.
3. For each word w_i, multiply the aforementioned attention weights with the corresponding value vectors, for all words w_j. Then sum up all these weighted vectors to produce the output vector for w_i: $z_i = \sum_j a_{ij} v_j$.

---

[5] We thank Sandeep Suntwal for this analogy, which was influenced by the discussion in (Kiat, 2021).
[6] Thus, Figure 12.1, oversimplifies the self-attention component when it suggests that the weighted average operates over the input vectors. We hope the reader pardons this temporary approximation, which was introduced for the sake of pedagogy.

Table 12.1 A self-attention walkthrough example for computing the contextual embedding z_1 for the word *bank* in the text *bank of the river*

1 (a)	Compute all the $\mathbf{q}_1 \cdot \mathbf{k}_i$ dot products for the four words in the text: $a_{11} = \mathbf{q}_1 \cdot \mathbf{k}_1 = 40$ $a_{12} = \mathbf{q}_1 \cdot \mathbf{k}_2 = 16$ $a_{13} = \mathbf{q}_1 \cdot \mathbf{k}_3 = 8$ $a_{14} = \mathbf{q}_1 \cdot \mathbf{k}_4 = 32$
1 (b)	Divide all the above values by $\sqrt{\|\mathbf{k}_1\|} = 8$: $a_{11} = 5$ $a_{12} = 2$ $a_{13} = 1$ $a_{14} = 4$
2	Apply softmax on the above 4 values: $a_{11} = 0.70$ $a_{12} = 0.03$ $a_{13} = 0.01$ $a_{14} = 0.26$
3	Compute the contextualized embedding z_1 for *bank*, as a weighted average of the value vectors: $z_1 = 0.70\mathbf{v}_1 + 0.03\mathbf{v}_2 + 0.01\mathbf{v}_3 + 0.26\mathbf{v}_4$

Table 12.1 shows an artificial walkthrough example for this algorithm.[7] Step 1(a) shows the result of the $\mathbf{q}_1 \cdot \mathbf{k}_j$ multiplications – that is, for all combinations between the word *bank* and the words in the text (*bank* included). These attention weight values indicate that the word *bank* pays attention to itself (i.e., a_{11} is large) and the word *river* (a_{14} is also large), which disambiguates *bank* in the current context, and less attention to the preposition *of* and the determiner *the*. The next step, 1(b), divides these values by the square root of the length of the key vector. This heuristic is necessary to mitigate the *exploding gradient* phenomenon. This phenomenon is the opposite of the vanishing gradient phenomenon we discussed in Chapter 5. That is, large parameter values in the network such as these attention weights yield gradient values that are consequently also large during backpropagation, which cause unstable learning due to too much "jumping around" in the parameter space, or even overflow in parameter values.

Step 2 applies a softmax layer on the resulting weights, which converts them into a probability distribution. This is necessary to: (a) produce a meaningful weighted average in the next step, and (b) to further control for large weight

---

[7] That is, the values in step 1(a) are made-up simple numbers for readability. In reality, all the values shown here are computed using the query, key, and value vectors, which, as we will discuss later in this chapter, are learned during training.

values. Later on in this section, we will see another component that aims to control for unreasonably large parameter values in TN. Last, step 3 computes the output embedding, z_1, as a weighted average of the v vectors for all the words in the context multiplied by their corresponding attention weights generated in the previous step.

This walkthrough example shows that, given query, key, and value vectors for words in the input text, it is relatively trivial to produce contextualized output embeddings. But where do the query, key, and value vectors come from? All these vectors are generated by projecting the input embeddings x into a new feature space. More formally, each self-attention block contains three matrices, W^Q, W^K, and W^V, where W^Q and W^K have dimension $|x| \times |k|$, and W^V has dimension $|x| \times |v|$.[8] Each of the q_i, k_i, and v_i vectors is then computed as:

(i) $q_i = x_i \times W^Q$,
(ii) $k_i = x_i \times W^K$, and
(iii) $v_i = x_i \times W^V$.

Thus, the parameters in a self-attention layer are the three matrices, W^Q, W^K, and W^V. A typical configuration for the self-attention layer has $|x| = 512$, and $|k| = |v| = 64$.

12.1.3 Multiple Heads

Transformer networks further expand the self-attention layer by repeating it multiple times. A typical configuration includes eight different instances of this algorithm. Each of these instances is called a "head." To make sure that the heads capture different information, each head receives different copies of the W^Q, W^K, and W^V matrices, which are all initialized with different values. This allows each layer to produce output embeddings z_i that operate in different feature spaces, and, hopefully, capture complementary information. Then, the actual embedding z_i for the word at position i is computed as the product between the concatenated embeddings produced by each head and a new "output" matrix W^O:

$$z_i = [z_i^1; z_i^2; \ldots z_i^n] \times W^O, \tag{12.1}$$

where the superscript j in z_i^j indicates which head produced it, and n indicates the total number of heads. The dimension of the output matrix W^O is $n|v| \times |x|$, which guarantees that the output embeddings z have the same dimension as the input embeddings x.

Thus, the complete list of parameters for a multiheaded self-attention layer includes n copies of the W^Q, W^K, and W^V matrices, and one copy of the

---

8   Note that the q and k vectors must have the same dimension because of the dot product in the computation of the attention weights.

output matrix $\mathbf{W^O}$. Under a typical configuration of $|\mathbf{x}| = 512$, $|\mathbf{k}| = |\mathbf{v}| = 64$, and $n = 8$, this means that a multiheaded self-attention layer contains 1,048,576 parameter weights.

12.1.4   Add and Normalize and Feed Forward Layers

As already mentioned, the next three components – that is, two instances of the add and normalize layer, and one feed-forward layer, are not that important conceptually, but they do matter empirically.

Each of the add and normalize layers starts by summing up the input and output embeddings produced by the previous component. For example, the add and normalize layer that follows the self-attention layer sums up the $\mathbf{x}_i$ and $\mathbf{z}_i$ for each word i. The motivation for this summation is to make sure that the signal from the input embeddings $\mathbf{x}$ (which we know from Chapter 8 carry important information) does not get lost in the machinery implemented by the self-attention layer. Then, the resulting embedding, $\mathbf{x}_i + \mathbf{z}_i$, is normalized using layer normalization, as discussed in Section 6.8. This latter normalization step is yet one more component that aims to mitigate the exploding gradient problem.

The feed-forward layer between the two add and normalize layers projects the output embeddings into a new feature space, similar to what the value matrix, $\mathbf{W^V}$, is doing to the input embeddings. This introduces more parameters in the transformer layer, which allow TNs to learn more complex functions.

12.2   Subword Tokenization

So far, we have used the term "word" to describe the inputs to the first transformer layer, but that is a misnomer introduced to simplify our presentation. In reality, TNs operate over subword units – that is, their tokens may be word fragments rather than complete words. These subword units are generated automatically using algorithms such as the *byte pair encoding* (BPE) algorithm for word segmentation (Sennrich et al., 2015). In a nutshell, this algorithm creates a symbol vocabulary, which keeps track of the allowed subword units. This dictionary is initialized with individual characters, including a special symbol </w> that indicates the end of a word. Then, it iteratively counts the frequency of symbol pairs in a large text corpus and replaces the most frequent pair with the concatenation of the two symbols, which is also added to the symbol dictionary. Merging is not allowed across word boundaries, which means that the new symbols created during the merge operations are always subwords (up to entire words), and never include parts of different words. The size of the output symbol dictionary is equal to the size of the initial dictionary plus the number of merge operations (because each merge creates a new symbol).

For example, assume that *bank of the river* is part of the training corpus for the BPE algorithm. Then, in the first iteration, this phrase will be segmented into individual characters and end-of-word markers:

b a n k </w> o f </w> t h e </w> r i v e r </w>

Let's say that the most frequent sequence of symbols so far is *t h*. Then, after merging these two symbols and adding *th* the symbol vocabulary, our text becomes:

b a n k </w> o f </w> th e </w> r i v e r </w>

Then, if the most frequent pair now is *th e*, the text becomes:

b a n k </w> o f </w> the </w> r i v e r </w>

and so on, until we reach the desired size for the output symbol dictionary. A typical size for this dictionary is 50,000. Since 50,000 is clearly smaller than the size of any language's vocabulary, it is unavoidable that BPE tokenization will fragment infrequent words into one or more subword units. For example, the word *transformers* might be fragmented into *transform* and *ers*.

Because TNs rely on this subword tokenization, their input embeddings – that is, the embeddings that are fed into the first layer in the architecture (Figure 12.3), no longer align with "traditional" word embedding algorithms such as word2vec, which operate over complete words. For this reason, TNs initialize the input embeddings assigned to subword units randomly, and update them together with the rest of the parameters during training (which we will discuss in the next section).

But why go through this additional trouble of subword tokenization? There are two advantages. First, operating over subwords makes the TN more robust to unknown words. For example, assume that a TN sees the word *transformers* for the first time after training. A traditional word-embedding algorithm may not know how to handle this word (or, at least, how to handle it well), but a TN may still be able to, if it tokenizes it into subwords that were seen in training such as *transform* and *ers*, for which it has trained input embeddings. The second advantage is saving space. That is, a word-embedding algorithm that relies on complete words may quickly reach a vocabulary size in the millions. This is because the vocabulary size keeps growing indefinitely as the underlying text corpus grows (Chapter 5 in (Schütze et al., 2008)). This is a problem: if we have a vocabulary of one million words, and we use vectors of size 512 for the input embeddings, we would need $4 \times 512 \times 1,000,000 = 2,048$ megabytes just to store the input embeddings![9] In contrast, a TN with 50,000 subword units requires only $4 \times 512 \times 50,000 = 102.4$ megabytes for its input embeddings.

---

[9] Assuming we use four bytes to store each real number in the 512-dimensional embedding vector.

12.3   Training a Transformer Network

So far, we have introduced TNs, which allow us to construct contextualized embeddings for a sequence of (subword) tokens. But how do we learn their parameters – for example, the input embeddings for the subword tokens, or the various **W** matrices in each layer? At a high level, the training process for TNs consists of two steps: one unsupervised procedure called *pretraining*, and a supervised one called *fine-tuning* (Devlin et al., 2018). Devlin et al. (2018) called the TN that resulted from this training process BERT, from **B**idirectional **E**ncoder **R**epresentations from **T**ransformers.[10] We discuss the two training procedures next.

12.3.1   Pretraining

The pretraining procedure uses a *masked language model* (MLM) objective (Devlin et al., 2018). That is, during pretraining, we randomly mask tokens in some input text by replacing them with a special token – for example, [MASK] – and ask the TN to guess the token behind the mask. Typically, 15% of the input tokens are masked in the training texts. While this task may sound trivial, several important details need discussing:

(i) First, this task is implemented as a multiclass classifier that uses as input the contextualized embeddings of the masked words and generates one of the subword tokens from the vocabulary created by the BPE algorithm. That is, imagine the transformer as the sequence of layers as summarized in Figure 12.2. Then, if we mask the word *river*, the classifier that guesses the masked word will operate on top of the contextualized embedding produced by layer *n* for the word *river*. The classifier itself is implemented as a softmax layer that produces a probability distribution over all the tokens in the BPE vocabulary, and is trained with the usual cross-entropy loss function, which maximizes the probability of the "correct" token – that is, the token that was masked.

(ii) Pretraining is often referred to as an unsupervised algorithm because it does not require any annotated training data from domain experts (e.g., as we would when training our review classifier). This is not entirely correct: after all, the texts used during pretraining are written by people and thus provide some human supervision. However, what is important is that, for most languages, there is a plethora of texts available on the web that can be easily transformed into training data for TNs using the masked language model trick. For this reason, most MLM pretraining settings

---

[10] This started a somewhat unfortunate trend that overused Muppet names for variants of TN architectures.

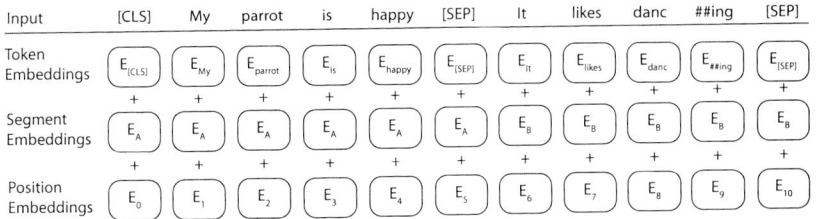

Figure 12.4 Input example for the next sentence prediction pretraining task. [SEP] is a special separator token used to indicate
end of sentence. The [CLS] token stands for *class*, and is used
to train the binary classifier, which indicates if sentence B follows
sentence A in text or not. The ## marker indicates that the corresponding token is a subword token that should be appended to the
token to its left

include billions of words. This allows TNs to capture many language patterns before they are trained on any specific NLP application (see the
fine-tuning subsection)!

(iii) Because the masked tokens can appear anywhere in a given text and
thus there is meaningful context that can be used to guess the masked
token both to the left and to the right of the mask, this pretraining procedure is called a *bidirectional* language model. This is to contrast it with
traditional language modeling (LM) tasks, which normally proceed left
to right. That is, a traditional LM guesses what word follows after the
user types a few words, like the texting application in your phone. The
pretraining procedure does not have this directionality constraint, as it
can "peek" on both sides of the mask.

A second pretraining procedure that was proposed by Devlin et al. (2018)
is *next sentence prediction* (NSP). This task trains TNs to predict which sentence follows a given sentence. Similar to the previous MLM task, this task is
"unsupervised," in the sense that is relies solely on text without any additional
expert annotations. However, unlike MLM, which operates over contextualized
embeddings of individual tokens, this task requires a different setting, which
is exemplified in Figure 12.4. The figure shows that, unlike the original TN,
the input embeddings for NSP sum up three embeddings: token and position
embeddings (similar to the original architecture) and a new embedding that
encodes which segment the current token belongs to (A or B). More importantly, this architecture introduces the virtual [CLS] token, which is inserted
at the beginning of the text, and whose contextualized embedding is used to
train the binary NSP classifier. That is, the actual classifier is a sigmoid on
top of the contextualized embedding for [CLS]. Note that the [CLS] token is
treated like any other token in the text – that is, its attention weights cover all
the tokens in the text. Thus, the classifier that operates on top of the [CLS]

contextualized embedding has indirect access to the entire text through its attention mechanism.

To train the classifier, NSP generates positive examples from actual sentences that follow a given sentence (as in the figure 12.4), and negative examples from sentences randomly sampled from the corpus. The proportion of negative to positive examples is 1:1.

Devlin et al. (2018) report that pretraining BERT with NSP benefits downstream tasks such as question answering. However, other works have "questioned the necessity of the NSP loss" for NLP applications (Liu et al., 2019). That is, Liu et al. (2019) observe that training solely with the MLM objective performs just as well or better than training with both MLM and NSP on downstream tasks.

12.3.2   Fine-tuning

These pretraining procedures allow TNs to capture a variety of language patterns that are application independent. In contrast, fine-tuning trains TNs for specific NLP applications such as text classification, question answering, natural language inference, and so forth. We will discuss these applications in detail in Chapter 16. For now, we will just mention that many of these applications can be modeled with architectures similar to the one shown in Figure 12.4, where one or more texts (separated by [SEP]) are preceded by a [CLS] token, which drives the actual classification. Table 12.2 shows a few example inputs for two NLP applications. As the table shows, many of these tasks can be reduced to classification tasks (again, on top of the [CLS] embedding) that receive as input one or more sentences. For example, for review classification, the input sentences are the actual sentences in the review, and the classification task needs to produce multiple labels such as Positive, Negative, or Neutral. For natural language inference, a hypothesis sentence is followed by a premise sentence. The classification task in this case is to indicate if the premise entails the hypothesis, it contradicts it, or the premise is neutral for the current hypothesis.

During fine-tuning, the training process receives a TN that was pretrained using one of the algorithms discussed in the previous subsection. The training continues with a loss function that is specific to the task at hand. For example, for review classification, the loss will be the standard cross-entropy, which will maximize the probability of the correct review label – for example, Positive for the review in Table 12.2.

12.4   Drawbacks of Transformer Networks

At the time this book was written, TNs dominated the field of NLP. However, despite their success, TNs are not perfect. The first important limitation

Table 12.2 Two examples of natural language processing application inputs formatted for transformer networks. In the first example, the classifier on top of the [CLS] embedding should predict the Positive label; in the second case, the prediction is Entailment

Application	Example input
Review classification	[CLS] although this was obvious ##ly a low budget production the perform ##ances and the songs in this movie are worth seeing . [SEP] one of walken 's few musical roles to date . [SEP]
Natural language inference	[CLS] Some students are reading . [SEP] Two students are reading the deep learning book . [SEP]

is that their self-attention mechanism (Section 12.1.2) has a quadratic runtime. That is, the contextualized representation for each token is computed by iterating over all tokens in the input. This cost is mitigated through the parallelism offered by GPUs. However, when GPUs are not available, TNs are considerably slower than RNNs, whose runtime overhead is linear in the sequence size. More recent transformer variants replaced the original quadratic self-attention mechanism with linear methods (see Section 12.6).

Second, the positional embeddings in the original transformer architecture (Section 12.1.1) encode *absolute* token positions in the text. This makes little sense for most NLP applications. For example, in the case of syntactic parsing, what information does position 17 provide for a token's syntactic role? Recent works have made similar observations and proposed novel architectures that encode the *relative* positions between any two tokens in the self-attention mechanism. This allows transformers to encode information that is more meaningful for language processing. For example, continuing the syntactic parsing use case, the position of a noun relative to a verb provides hints of its subject or object role (see Section 12.6 for details).

The last important drawback of TNs is that they are large. For example, at the time this book was written, most "large" transformer configurations had grown to contain tens or hundreds of billions of parameters, and there were no signs that this trend would stop anytime soon. While this is an impressive technical achievement, it also makes this technology inaccessible to many who do not have access to computers with GPUs and enough memory. See Section 12.6 for potential solutions to this drawback.

12.5 Historical Background

Transformer networks were introduced by Vaswani et al. (2017) for machine translation. Devlin et al. (2018) focused on the encoder component of the transformer architecture. They introduced several pretraining strategies and showed that, when properly pretrained, transformer encoders obtain new state-of-the-art results on multiple NLP tasks such as question answering and language inference. These two papers redefined the NLP landscape in the late 2010s and 2020s. A battery of transformer variants have been proposed since then that continued to improve the performance of the original architecture on many applications and languages. We discuss a few of these methods in the next section.

12.6   References and Further Readings

Liu et al. (2019) revisited the pretraining strategies proposed in the original BERT paper (Devlin et al., 2018) and observed that BERT was under-trained. They showed that performance on several downstream NLP tasks improves: (a) when the transformer encoder is pretrained longer on more data that include longer sequences, and (b) interestingly, when the next sentence prediction objective is removed from pretraining.

A series of recent papers have proposed more efficient alternatives to the original quadratic self-attention mechanism (Beltagy et al., 2020; Kitaev et al., 2020; Wang et al., 2020; Zaheer et al., 2020). For example, Longformer introduces a self-attention mechanism that scales linearly with the size of the input text (Beltagy et al., 2020). To this end, Longformer uses a sliding local window for attention around each token. That is, each token attends only to a constant number of tokens in its neighborhood (rather than the whole text, as in the original algorithm). This local-window strategy is coupled with the original "global" self-attention mechanism that is applied only to the [CLS] token, which is meant to capture a representation of the whole text. Big Bird (Zaheer et al., 2020) follows a similar strategy that combines local and global self-attention, but, to increase global coverage, they replace a random subset of connections from the sliding windows with random connections from the entire input text.

Several efforts have shown that replacing the absolute positional embeddings in the original transformer with relative positional information yields better contextualized representations that improve downstream tasks (He et al., 2020; Raffel et al., 2020; Ontanón et al., 2021). For example, T5 (Raffel et al., 2020) embeds relative position information in the self-attention mechanism itself. In particular: (a) each relative position (e.g., *two tokens to the right*) is mapped to a scalar parameter that is learned with the rest of the network parameters, and (b) this parameter is "added to the corresponding score used for computing the attention weights" between the key and the query (Raffel et al., 2020). The important thing about these relative position representations is that they are *position invariant* – that is, they are the same regardless where the tokens appear in the actual input text.

To mitigate the large size of transformer models, a few papers have "distilled" larger transformer models into smaller ones that preserve most of the original capabilities (Jiao et al., 2019; Sanh et al., 2019). For example, Sanh et al. (2019) have shown that it is possible to "reduce the size of a BERT model by 40%, while retaining 97% of its language understanding capabilities and being 60% faster." They achieved this by training the smaller model to mimic the predictions of the larger BERT during pretraining.

12.7   Summary

This chapter introduced the encoder component of TNs, which produces contextualized representations of words – that is, embeddings that capture the context in which the words appear. The transformer encoder enables one to use this architecture as a transducer (i.e., classifying every word in the input text) or acceptor (i.e., classifying the entire input text). The decoder component of this architecture (which will enable encoder-decoder configurations for TNs) will be discussed in Chapter 14. We also discussed several architectural choices that enabled the tremendous success of TNs – that is, self attention, multiple heads, stacking of multiple layers, and subword tokenization – as well as how transformers can be pretrained on large amounts of data through masked language modeling and next-sentence prediction.

13   Using Transformers with the Hugging Face Library

One of the key advantages of TNs is the ability to take a model that was pretrained over vast quantities of text and fine-tune it for the task at hand. Intuitively, this strategy allows TNs to achieve higher performance on smaller datasets by relying on statistics acquired at scale in an unsupervised way (e.g., through the masked language model training objective). To this end, in this chapter, we will use the Hugging Face library, which has a rich repository of datasets and pretrained models, as well as helper methods and classes that make it easy to target downstream tasks.[1] Using pretrained transformer encoders, we will implement the two tasks that served as use cases in the previous chapters: text classification and POS tagging.

The code presented in this chapter is available in the following notebooks: `chap13_classification_bert` for text classification using the BERT encoder (Devlin et al., 2018), `chap13_classification_distilbert` for text classification using DistilBERT, a more compact encoder (Sanh et al., 2019) (Section 13.2), and `chap13_pos_tagging` for Spanish POS tagging using a multilingual transformer encoder (Section 13.3).

Note that transformer models are much larger than the models we have used so far in this book, and therefore take longer to train. Models such as DistilBERT ameliorate this issue, but even then, access to a GPU is crucial. Depending on the model and amount of data, a GPU can bring down training time from hours to minutes. If you do not have access to a GPU, we recommend using an online service such as Google Colab, which provides free access to GPUs and is easy to use (Johnson, 2021).[2]

13.1   Tokenization

As discussed in Section 12.2, transformers rely on subword tokens. This strategy provides an elegant way to handle unknown and low-frequency words by splitting them into more frequent subword parts. At the same time, these

---

[1] https://huggingface.co/docs/transformers/main/en/index.
[2] https://colab.google.

tokenization algorithms maintain frequently occurring words as standalone tokens, so the signal for these common words is preserved.

To make this more concrete, we show how tokenizers are employed in the Hugging Face library. First, we load the tokenizer that corresponds to the transformer we intend to use. This is important for two reasons: (a) different transformers rely on different tokenization algorithms, and (b) even for the ones that use the same algorithm, their tokenizer vocabularies are likely to be different if they were pretrained on different corpora. Next, we tokenize some example text and display some of the resulting attributes with pandas:

```
[1]:  from transformers import AutoTokenizer
      import pandas as pd

      # load tokenizer
      transformer_name = 'bert-base-cased'
      tokenizer = AutoTokenizer.from_pretrained(transformer_name)

      # tokenize text
      text = 'I am the walrus.'
      output = tokenizer(text)

      # display results
      pd.DataFrame(
          [output.tokens(), output.word_ids(), output.input_ids],
          index=['tokens', 'word_ids', 'input_ids'],
      )
```

	0	1	2	3	4	5	6	7	8
tokens	[CLS]	I	am	the	wa	##l	##rus	.	[SEP]
word_ids	None	0	1	2	3	3	3	4	None
input_ids	101	146	1821	1103	20049	1233	6208	119	102

As shown here, the tokenizer splits the text into tokens and adds two special tokens: the [CLS] token at the beginning of the token sequence, and the [SEP] token at the end. Also, note that the ## characters at the beginning of some tokens indicate that they are not standalone words, but rather subwords that continue a word previously started. The output in our example shows that the word *walrus* was split into three subwords. Note, however, that this is specific to this particular tokenization algorithm, and other tokenizers may indicate word continuation in different ways. A better way to detect word continuations is using the word_ids() method of the tokenizer output, which assigns the same ID to all token part of the same word. For example, all fragments of the word *walrus* share the word ID 3. Last, the `input_ids` attribute provides the token IDs used internally by the transformer to map tokens to embeddings.

To briefly demonstrate how different tokenizers produce different outputs, here is the same text tokenized with the tokenizer corresponding to `xlm-roberta-base`:

	0	**1**	**2**	**3**	**4**	**5**	**6**	**7**
tokens	\<s\>	_I	_am	_the	_wal	rus	.	\</s\>
word_ids	None	0	1	2	3	3	3	None
input_ids		0	87	444	70	32973	6563	5	2

Note how the [CLS] and [SEP] special tokens have been replaced with \<s\> and \</s\> respectively. Also, spaces have been replaced with the Unicode character _ (U+2581, LOWER ONE EIGHTH BLOCK). Tokens that start with that character are considered word beginnings and the rest are word continuations, as can be confirmed by looking at the word IDs. This illustrates the importance of using the tokenizer that corresponds to the transformer you intend to use.

13.2   Text Classification

For our text classification example, we will continue using the AG News dataset from previous chapters. We will load, preprocess, and split the dataset into pandas DataFrames in the same way as before. Now however, rather than continuing with pandas, we will create a Hugging Face dataset from the dataframes. Hugging Face datasets are convenient because of their built-in support of batching, efficient data transformations, and caching. In particular, we convert each dataframe into a Hugging Face dataset. The various datasets are managed with a `DatasetDict`. Note that this is the same data structure seen when downloading a Hugging Face dataset from their hub.[3] The keys in this dictionary are usually *train*, *validation*, and *test*:[4]

```
[5]:  from datasets import Dataset, DatasetDict

      ds = DatasetDict()
```

---

[3] https://huggingface.co/datasets.
[4] These correspond to the more common terms *train*, *development*, and *test* we have used throughout the book so far. In this chapter, we use the Hugging Face naming conventions for consistency.

```
ds['train'] = Dataset.from_pandas(train_df)
ds['validation'] = Dataset.from_pandas(eval_df)
ds['test'] = Dataset.from_pandas(test_df)
ds
```

```
[5]: DatasetDict({
         train: Dataset({
             features: ['label', 'title', 'description', 'text'],
             num_rows: 108000
         })
         validation: Dataset({
             features: ['label', 'title', 'description', 'text'],
             num_rows: 12000
         })
         test: Dataset({
             features: ['label', 'title', 'description', 'text'],
             num_rows: 7600
         })
     })
```

Once our dataset is loaded, we load a tokenizer. Different pretrained models
are tokenized differently, and it is important to select the tokenizer that corresponds to the model we will use so that the inputs are consistent with model
expectations. In our example, we will use the bert-base-cased pretrained
model and tokenizer:

```
[6]: from transformers import AutoTokenizer

     transformer_name = 'bert-base-cased'
     tokenizer = AutoTokenizer.from_pretrained(transformer_name)
```

Datasets have a map() method that transforms the dataset by applying a
function to each example. The method returns a new dataset with the transformation applied. We use the map() method to tokenize our dataset. To this
end, we define a function that tokenizes an example using the tokenizer we
loaded previously. Note that tokenizers support many options that you may
need depending on your situation. However, since this is a simple scenario, all
we need to do is provide the text to tokenize and specify how to handle texts
that exceed the maximum number of tokens permitted by the pretrained model.
Here we have our tokenizer truncate any inputs that are too long by specifying
the truncation=True parameter. The output of this function will be added
to the new dataset as extra columns. Further, we also want to *remove* some of
the columns that are no longer needed, simplifying subsequent steps. For this,

we use the `remove_columns` argument, listing the columns that we want to discard.

Additionally, the dataset's map() method can batch the dataset; we enable this option with the `batched=True` argument (the output for the code below is on page 199):

```
[7]:  def tokenize(batch):
          return tokenizer(batch['text'], truncation=True)

      train_ds = ds['train'].map(
          tokenize,
          batched=True,
          remove_columns=['title', 'description', 'text'],
      )
      eval_ds = ds['validation'].map(
          tokenize,
          batched=True,
          remove_columns=['title', 'description', 'text'],
      )
      train_ds.to_pandas()
```

Next, we implement a classifier for our task. Hugging Face provides a variety of models corresponding to several types of downstream tasks. However, for pedagogical purposes, we implement one from scratch. In particular, our model class inherits from `BertPreTrainedModel`, which provides several useful methods such as `init_weights()` and `from_pretrained()` methods, which we will use later.[5] The model constructor takes a configuration object as its only parameter. Configuration objects contain all the hyper parameters used by the corresponding pretrained models. We will show later how the configuration model is retrieved and customized.

Models that implement specific downstream tasks are usually composed of a pretrained model (sometimes referred as the body), and one or more task-specific layers (usually referred as the head). Here, we initialize a `BertModel` using the provided configuration, as well as a dropout layer and a task-specific linear layer used for classifying the BERT output. Each of these layers is initialized by calling the `init_weights()` method inherited from `BertPreTrainedModel`.

The `forward()` method, which implements the task-specific forward pass, implements the complete encoder pipeline summarized in Figures 12.1 and 12.2 in the previous chapter. This method takes as arguments the outputs of the tokenizer, and, optionally, the gold labels corresponding to the input data

---

[5]   Note that the text-classification notebooks provided in this chapter have two versions: one that uses a BERT encoder, and one that uses a more compact encoder (DistilBERT). Despite its smaller size, the latter encoder obtains the same performance on our text classification dataset. Since the two notebooks are similar, here we discuss just the code that uses the BERT encoder.

	label	input_ids	token_type_ids	attention_mask
0	3	[101, 3270, 11906, 1522, 1146, 7106, 1111, 251...	[0, 0, 0, 0, 0, 0, 0, 0, 0, 0, 0, 0, 0, 0, ...	[1, 1, 1, 1, 1, 1, 1, 1, 1, 1, 1, 1, 1, 1, ...
1	0	[101, 4222, 11404, 1174, 117, 1476, 1130, 2696...	[0, 0, 0, 0, 0, 0, 0, 0, 0, 0, 0, 0, 0, 0, ...	[1, 1, 1, 1, 1, 1, 1, 1, 1, 1, 1, 1, 1, 1, ...
2	0	[101, 158, 119, 156, 119, 12068, 5084, 1116, 9...	[0, 0, 0, 0, 0, 0, 0, 0, 0, 0, 0, 0, 0, 0, ...	[1, 1, 1, 1, 1, 1, 1, 1, 1, 1, 1, 1, 1, 1, ...
3	2	[101, 22087, 8223, 1611, 1106, 4417, 5572, 324...	[0, 0, 0, 0, 0, 0, 0, 0, 0, 0, 0, 0, 0, 0, ...	[1, 1, 1, 1, 1, 1, 1, 1, 1, 1, 1, 1, 1, 1, ...
4	0	[101, 7270, 118, 2733, 1383, 1111, 12448, 7430...	[0, 0, 0, 0, 0, 0, 0, 0, 0, 0, 0, 0, 0, 0, ...	[1, 1, 1, 1, 1, 1, 1, 1, 1, 1, 1, 1, 1, 1, ...
...	...	...	...	...
107995	0	[101, 6096, 117, 10378, 3969, 5977, 1111, 8988...	[0, 0, 0, 0, 0, 0, 0, 0, 0, 0, 0, 0, 0, 0, ...	[1, 1, 1, 1, 1, 1, 1, 1, 1, 1, 1, 1, 1, 1, ...
107996	0	[101, 16409, 118, 16587, 159, 4064, 1106, 1564...	[0, 0, 0, 0, 0, 0, 0, 0, 0, 0, 0, 0, 0, 0, ...	[1, 1, 1, 1, 1, 1, 1, 1, 1, 1, 1, 1, 1, 1, ...
107997	0	[101, 9569, 5480, 10582, 2087, 1867, 158, 119...	[0, 0, 0, 0, 0, 0, 0, 0, 0, 0, 0, 0, 0, 0, ...	[1, 1, 1, 1, 1, 1, 1, 1, 1, 1, 1, 1, 1, 1, ...
107998	0	[101, 11560, 3881, 108, 3614, 132, 3498, 2944,...	[0, 0, 0, 0, 0, 0, 0, 0, 0, 0, 0, 0, 0, 0, ...	[1, 1, 1, 1, 1, 1, 1, 1, 1, 1, 1, 1, 1, 1, ...
107999	3	[101, 1130, 139, 24683, 131, 21107, 2050, 1739...	[0, 0, 0, 0, 0, 0, 0, 0, 0, 0, 0, 0, 0, 0, ...	[1, 1, 1, 1, 1, 1, 1, 1, 1, 1, 1, 1, 1, 1, ...

108000 rows × 4 columns

points. Our implementation of the forward pass sends the input tokens to the BERT model to produce the contextualized representations for all tokens. This output has several components, including the last_hidden_state that contains the final hidden-state embedding for each token. For our task, we will represent the whole sequence using the embedding for the [CLS] token that occurs at the start of each example. We retrieve it by selecting the first element of each output sequence in the batch (i.e., last_hidden_state[:, 0, :]).

As in the previous chapters, we apply dropout to our sequence representation, and then pass it through our linear classification layer. If gold labels are provided (i.e., we are training), we now compute the loss using the cross-entropy loss. The output of the forward pass is wrapped in a Hugging Face SequenceClassifierOutput object[6] and returned:

```
[8]:  from torch import nn
      from transformers.modeling_outputs import␣
      ↪SequenceClassifierOutput
      from transformers.models.bert.modeling_bert import BertModel,␣
      ↪BertPreTrainedModel

      class BertForSequenceClassification(BertPreTrainedModel):
          def __init__(self, config):
              super().__init__(config)
              self.num_labels = config.num_labels
              self.bert = BertModel(config)
              self.dropout = nn.Dropout(config.hidden_dropout_prob)
              self.classifier = nn.Linear(config.hidden_size, config.
      ↪num_labels)
              self.init_weights()

          def forward(self, input_ids=None, attention_mask=None,␣
      ↪token_type_ids=None, labels=None, **kwargs):
              outputs = self.bert(
                  input_ids,
                  attention_mask=attention_mask,
                  token_type_ids=token_type_ids,
                  **kwargs,
              )
              cls_outputs = outputs.last_hidden_state[:, 0, :]
              cls_outputs = self.dropout(cls_outputs)
              logits = self.classifier(cls_outputs)
              loss = None
```

---

[6] Hugging Face utilizes a set of output objects to standardize model output for a given task. These objects typically include additional information – for example, attention weights, which can be used for visualizing or debugging model behavior.

```
if labels is not None:
    loss_fn = nn.CrossEntropyLoss()
    loss = loss_fn(logits, labels)
return SequenceClassifierOutput(
    loss=loss,
    logits=logits,
    hidden_states=outputs.hidden_states,
    attentions=outputs.attentions,
)
```

Next, we load the configuration of the pretrained model and instantiate our model. The `AutoConfig` class can load the configuration for any pretrained model, retrieving it from Hugging Face if needed. Then we use the configuration to instantiate our model using the `from_pretrained()` method. With this call, the pretrained model will be loaded, which includes downloading if necessary:

```
[9]:  from transformers import AutoConfig

      config = AutoConfig.from_pretrained(
          transformer_name,
          num_labels=len(labels),
      )

      model = (
          BertForSequenceClassification
          .from_pretrained(transformer_name, config=config)
      )
```

Hugging Face provides a `Trainer` class that greatly simplifies the training process. This class not only implements the training loop we have been using in the previous chapters, but also handles other useful steps such as saving checkpoints (i.e., intermediate models after a number of minibatches have been processed during training), and tracking custom measures about model performance. In order to create a `Trainer`, we first need to specify its configuration in a `TrainingArguments` object. In ours, we specify certain hyper parameters such as batch size, weight decay, and number of epochs, as well as where to store model checkpoints:

```
[11]:  from transformers import TrainingArguments

       num_epochs = 2
       batch_size = 24
       weight_decay = 0.01
```

```
model_name = f'{transformer_name}-sequence-classification'

training_args = TrainingArguments(
    output_dir=model_name,
    log_level='error',
    num_train_epochs=num_epochs,
    per_device_train_batch_size=batch_size,
    per_device_eval_batch_size=batch_size,
    evaluation_strategy='epoch',
    weight_decay=weight_decay,
)
```

The TrainingArguments class provides a wide variety of arguments that we have not shown.[7] These arguments usually have appropriate default values, so it is often fine to omit them. For example, we did not use the label_names argument, which specifies the key that corresponds to the training labels. When omitted, it defaults to keys such as label, labels, and label_ids.[8] In this chapter, we used label.

Note that we also specify how often we would like to see the performance of the current model (at the end of each epoch) with evaluation_strategy='epoch'. This means that after each epoch we print the current loss on the training partition and on the evaluation dataset, if one is available. Additionally, we can report custom metrics at this time. For this purpose, we use the compute_metrics parameter of the Trainer, which expects a function that receives a transformers.EvalPredictions object containing the label IDs and the predicted logits. The expected return type is a dictionary whose keys correspond to different metrics, each of which will be displayed as a separate result column.

```
[12]:  from sklearn.metrics import accuracy_score

       def compute_metrics(eval_pred):
           y_true = eval_pred.label_ids
           y_pred = np.argmax(eval_pred.predictions, axis=-1)
           return {'accuracy': accuracy_score(y_true, y_pred)}
```

Using the TrainingArguments and compute_metrics function, we create our Trainer. Note that when you provide a tokenizer, the trainer will

---

[7] https://huggingface.co/docs/transformers/main/en/main_classes/trainer#transformers.TrainingArguments.

[8] In the case of extractive question answering (see Chapter 16), the start_positions and end_positions store the start/end positions of the correct answers.

automatically pad the sequences in each batch. Also, the trainer will automatically use any GPU that is available, unless specifically disabled in the `TrainingArguments`.

```
[13]: from transformers import Trainer

      trainer = Trainer(
          model=model,
          args=training_args,
          compute_metrics=compute_metrics,
          train_dataset=train_ds,
          eval_dataset=eval_ds,
          tokenizer=tokenizer,
      )
```

Training our model takes a single call to the `train()` method of the Trainer object. As specified in the our instance of `TrainingArguments`, the training and validation losses, as well as the accuracy, are reported every epoch.

```
[14]: trainer.train()
```

Epoch	Training Loss	Validation Loss	Accuracy
1	0.187800	0.172629	0.941667
2	0.104000	0.183001	0.946250

As in the other chapters, we can write custom code to obtain the model's predictions on the test data. However, the Trainer class provides a predict() method that drastically simplifies this:

```
[17]: from sklearn.metrics import classification_report

      output = trainer.predict(test_ds)
      y_true = output.label_ids
      y_pred = np.argmax(output.predictions, axis=-1)
      print(classification_report(y_true, y_pred,
          ↪target_names=labels))
```

	precision	recall	f1-score	support
World	0.96	0.95	0.96	1900
Sports	0.99	0.99	0.99	1900
Business	0.93	0.91	0.92	1900
Sci/Tech	0.91	0.94	0.92	1900
accuracy			0.95	7600
macro avg	0.95	0.95	0.95	7600
weighted avg	0.95	0.95	0.95	7600

As shown in the table, this model achieves an accuracy of 95%, which is the highest performance we have achieved so far on this dataset.

13.3   Part-of-Speech Tagging

To showcase part-of-speech tagging using transformers, we continue with the Spanish section of the AnCora corpus introduced in Chapter 11. Recall that the dataset is stored in the CoNLL-U format. We load this format in the same way as before, but then we convert the loaded dataset into a Hugging Face DatasetDict:

```
[5]: from datasets import Dataset, DatasetDict

     ds = DatasetDict()
     ds['train'] = Dataset.from_pandas(train_df)
     ds['validation'] = Dataset.from_pandas(valid_df)
     ds['test'] = Dataset.from_pandas(test_df)
     ds
```

```
[5]: DatasetDict({
         train: Dataset({
             features: ['words', 'tags'],
             num_rows: 14305
         })
         validation: Dataset({
             features: ['words', 'tags'],
             num_rows: 1654
         })
         test: Dataset({
             features: ['words', 'tags'],
             num_rows: 1721
         })
     })
```

Importantly, because the CoNLL-U dataset is already tokenized, we use the is_split_into_words=True tokenizer argument to ensure that the tokenizer respects the existing word boundaries during its subword tokenization. Further, while we want to predict one POS tag per word, any given word may be split into smaller pieces by our tokenizer. Thus, we need to align the tokenizer output to the CoNLL-U words. The original BERT paper (Devlin et al., 2018) addresses this by only using the embedding corresponding to the first subtoken for each word. We follow the same approach for consistency. For the subwords that do not correspond to the beginning of a word, we use a special value that indicates that we are not interested in their predictions. The CrossEntropyLoss has a parameter called ignore_index for this purpose. The default value for this parameter is −100, which we use as the label for the subwords we wish to ignore during training:

```
[9]:   # default value for CrossEntropyLoss ignore_index parameter
       ignore_index = -100

       def tokenize_and_align_labels(batch):
           labels = []
           # tokenize batch
           tokenized_inputs = tokenizer(
               batch['words'],
               truncation=True,
               is_split_into_words=True,
           )
           # iterate over batch elements
           for i, tags in enumerate(batch['tags']):
               label_ids = []
               previous_word_id = None
               # get word ids for current batch element
               word_ids = tokenized_inputs.word_ids(batch_index=i)
               # iterate over tokens in batch element
               for word_id in word_ids:
                   if word_id is None or word_id == previous_word_id:
                       # ignore seen word ids
                       label_ids.append(ignore_index)
                   else:
                       # get tag id for corresponding word
                       tag_id = tag_to_index[tags[word_id]]
                       label_ids.append(tag_id)
                   # remember this word id
                   previous_word_id = word_id
               # save label ids for current batch element
               labels.append(label_ids)
```

```
# store labels together with the tokenizer output
tokenized_inputs['labels'] = labels
return tokenized_inputs
```

Next, we use this function to preprocess the train and validation folds in our
DatasetDict (the output for the code below is on page 207):

```
[10]: train_ds = ds['train'].map(
          tokenize_and_align_labels,
          batched=True,
      )
      eval_ds = ds['validation'].map(
          tokenize_and_align_labels,
          batched=True,
      )
      train_ds.to_pandas()
```

Next, we implement our model class that uses a transformer encoder as a
transducer. Because our downstream task consists of POS tagging for Spanish, we need a transformer model that was pretrained on Spanish texts. Here,
we chose XLM-RoBERTa (Conneau et al., 2019) as our base model. XLM-
Roberta is a RoBERTa model (Liu et al., 2019) that has been pretrained on
100 different languages, including Spanish. Of note, XLM-RoBERTa does not
require us to specify what language we are working on. Similar to BERT, it
only requires the input_ids.

We discussed in the text classification section that Hugging Face provides implementations for text classification models. This is also true for
token classification problems that require transducers. In particular, the
XLMRobertaForTokenClassification model provided by Hugging Face
does everything needed for this task. However, as before, here we implement
it ourselves for pedagogical purposes.

The model architecture is similar to our text classification example. It consists of a transformer, a dropout layer, and a linear layer used for classification.
The number of labels that determines the output dimension of the linear layer is
equal to the number of POS tags. The primary difference between the text classification example and this token classification model is that with the former
we produced one label for each text document, while here we produce one
label *for each token* in the input text. Specifically, in our text classification
model the output shape was two-dimensional: (batch_size, num_labels).
Here, our output is three-dimensional: (batch_size, sequence_size,
num_labels). So, while much of the forward method is familiar to us, when
we are required to compute the loss, we need to reshape the logits and the
labels before passing them to the CrossEntropyLoss, since it expects two-
dimensional input and one-dimensional labels. For this purpose, we use the

	words	tags	input_ids	attention_mask	labels
0	[El, presidente, de, el, órgano, regulador, de...	[DET, NOUN, ADP, DET, NOUN, ADJ, ADP, DET, PRO...	[0, 540, 9692, 8, 88, 103633, 15913, 1846, 8, ...	[1, 1, 1, 1, 1, 1, 1, 1, 1, 1, 1, 1, 1, 1, 1, 1, ...	[-100, 0, 1, 2, 0, 1, 3, -100, 2, 0, 4, -100,...
1	[Sobre, la, oferta, de, interconexión, con, Te...	[ADP, DET, NOUN, ADP, NOUN, ADP, PROPN, ADP, D...	[0, 44125, 21, 19806, 8, 1940, 2271, 3355, 194...	[1, 1, 1, 1, 1, 1, 1, 1, 1, 1, 1, 1, 1, 1, 1, 1, ...	[-100, 2, 0, 1, 2, 1, -100, -100, -100, 2, 4, ...
2	[Afirmó, que, sigue, el, criterio, europeo, y,...	[VERB, SCONJ, VERB, DET, NOUN, ADJ, CCONJ, SCO...	[0, 62, 38949, 849, 41, 58453, 88, 166220, 620...	[1, 1, 1, 1, 1, 1, 1, 1, 1, 1, 1, 1, 1, 1, 1, 1, ...	[-100, 6, -100, -100, 7, 6, 0, 1, 3, 10, 7, 6...
3	[Le, inversión, en, investigación, básica, es,...	[DET, NOUN, ADP, NOUN, ADJ, AUX, DET, NOUN, AD...	[0, 239, 98649, 22, 31674, 124528, 198, 88, 46...	[1, 1, 1, 1, 1, 1, 1, 1, 1, 1, 1, 1, 1, 1, 1, 1, ...	[-100, 0, 1, 2, 1, 3, 9, 0, 1, 2, 0, 1, 10, 0,...
4	[Durante, la, presentación, de, el, libro, ", ...	[ADP, DET, NOUN, ADP, DET, NOUN, PUNCT, DET, P...	[0, 24292, 21, 43945, 8, 88, 7750, 44, 239, 78...	[1, 1, 1, 1, 1, 1, 1, 1, 1, 1, 1, 1, 1, 1, 1, 1, ...	[-100, 2, 0, 1, 8, 0, 4, -100, 2, 4, ...
⋮					
14300	[Y, todas, las, miradas, convergen, en, la, lu...	[CCONJ, DET, DET, NOUN, VERB, ADP, DET, NOUN, ...	[0, 990, 5136, 576, 100688, 7, 158, 814, 1409...	[1, 1, 1, 1, 1, 1, 1, 1, 1, 1, 1, 1, 1, 1, 1, 1, ...	[-100, 10, 0, 0, 1, -100, 6, -100, -100, 2, 0,...
14301	[Conviene, que, ahora, , en, plena, apoteosis....	[VERB, SCONJ, ADV, PUNCT, ADP, ADJ, NOUN, ADP...	[0, 1657, 7772, 13, 41, 18451, 6, 4, 22, 31161...	[1, 1, 1, 1, 1, 1, 1, 1, 1, 1, 1, 1, 1, 1, 1, 1, ...	[-100, 6, -100, -100, 7, 11, 8, -100, 2, 3, 1...
14302	[Cambiar, las, formas, parece, de, rigor, , p...	[VERB, DET, NOUN, VERB, ADP, NOUN, PUNCT, CCON...	[0, 313, 61055, 42, 576, 26497, 12295, 8, 7599...	[1, 1, 1, 1, 1, 1, 1, 1, 1, 1, 1, 1, 1, 1, 1, 1, ...	[-100, 6, -100, -100, 0, 1, 6, 2, 1, 8, -100, ...
14303	Carlos, y, Fayna, se, enzarzan, en, una, bron...	[PROPN, CCONJ, PROPN, PRON, VERB, ADP, DET, NO...	[0, 24856, 113, 114162, 76, 40, 22, 6383, 5935...	[1, 1, 1, 1, 1, 1, 1, 1, 1, 1, 1, 1, 1, 1, 1, 1, ...	[-100, 4, 10, 4, -100, 5, 6, -100, -100, 2, 0,...
14304	[É, llega, a, tirar, la, sobre, la, cama, y, ...	[PRON, VERB, ADP, VERB, PRON, ADP, DET, NOUN, ...	[0, 124043, 47612, 10, 61846, 21, 1028, 21, 39...	[1, 1, 1, 1, 1, 1, 1, 1, 1, 1, 1, 1, 1, 1, 1, 1, ...	[-100, 5, 6, 2, 6, 5, 2, 0, 1, 10, 5, 6, 0, 1...

14305 rows × 5 columns

view() method to reshape the tensors. This method is efficient because it does not copy the tensor data. Instead it provides a new view of the same data that behave like a tensor with a different shape.[9] As mentioned before, the number of arguments passed to this method determines the number of dimensions in the output tensor. Here, for our logits, we pass two arguments and so our new view will have two dimensions. The second will be the size of self.num_labels, while the first (because we pass −1) will be inferred based on the original tensor shape. For our labels, on the other hand, we only provide one argument and so the new view will have one dimension, inferred by the original shape:

[11]:
```python
from torch import nn
from transformers.modeling_outputs import TokenClassifierOutput
from transformers.models.roberta.modeling_roberta import
 ↪RobertaModel, RobertaPreTrainedModel

class XLMRobertaForTokenClassification(RobertaPreTrainedModel):
    def __init__(self, config):
        super().__init__(config)
        self.num_labels = config.num_labels
        self.roberta = RobertaModel(config,
 ↪add_pooling_layer=False)
        self.dropout = nn.Dropout(config.hidden_dropout_prob)
        self.classifier = nn.Linear(config.hidden_size, config.
 ↪num_labels)
        self.init_weights()

    def forward(self, input_ids=None, attention_mask=None,
 ↪token_type_ids=None, labels=None, **kwargs):
        outputs = self.roberta(
            input_ids,
            attention_mask=attention_mask,
            token_type_ids=token_type_ids,
            **kwargs,
        )
        sequence_output = self.dropout(outputs[0])
        logits = self.classifier(sequence_output)
        loss = None
        if labels is not None:
            loss_fn = nn.CrossEntropyLoss()
```

[9]   Similar to NumPy, PyTorch tensors are represented internally by a block of memory storing the data and some metadata that describes how the data should be read – for example, type, shape, and stride. The view() method returns a new tensor with new metadata but pointing to the same memory block.

```
        inputs = logits.view(-1, self.num_labels)
        targets = labels.view(-1)
        loss = loss_fn(inputs, targets)
    return TokenClassifierOutput(
        loss=loss,
        logits=logits,
        hidden_states=outputs.hidden_states,
        attentions=outputs.attentions,
    )
```

Next, we instantiate our model using the XLM-RoBERTa configuration:

```
[12]: from transformers import AutoConfig

      transformer_name = 'xlm-roberta-base'

      config = AutoConfig.from_pretrained(
          transformer_name,
          num_labels=len(index_to_tag),
      )

      model = (
          XLMRobertaForTokenClassification
          .from_pretrained(transformer_name, config=config)
      )
```

As before, we create a TrainingArguments object and define a compute_metrics function in order to customize a Trainer:

```
[13]: from transformers import TrainingArguments

      num_epochs = 2
      batch_size = 24
      weight_decay = 0.01
      model_name = f'{transformer_name}-finetuned-pos-es'

      training_args = TrainingArguments(
          output_dir=model_name,
          log_level='error',
          num_train_epochs=num_epochs,
          per_device_train_batch_size=batch_size,
          per_device_eval_batch_size=batch_size,
          evaluation_strategy='epoch',
          weight_decay=weight_decay,
      )
```

While the `TrainingArguments` code has no substantial changes, we need to adjust the `compute_metrics` function to account for the fact that our model uses subword tokens rather than complete words. Recall that only the first subword token per word was assigned a POS tag. This function discards the labels corresponding to the ignored subword tokens and evaluates the rest, returning the accuracy score:

```
[14]:  from sklearn.metrics import accuracy_score

       def compute_metrics(eval_pred):
           # gold labels
           label_ids = eval_pred.label_ids
           # predictions
           pred_ids = np.argmax(eval_pred.predictions, axis=-1)
           # collect gold/predicted labels, ignoring ignore_index label
           y_true, y_pred = [], []
           batch_size, seq_len = pred_ids.shape
           for i in range(batch_size):
               for j in range(seq_len):
                   if label_ids[i, j] != ignore_index:
                       y_true.append(index_to_tag[label_ids[i][j]])
                       y_pred.append(index_to_tag[pred_ids[i][j]])
           # return computed metrics
           return {'accuracy': accuracy_score(y_true, y_pred)}
```

The last component required for the `Trainer` is a collator. Since this time we are batching sequences of tokens, we need a collator that can pad them dynamically when constructing the batches. The `transformers` library includes a `DataCollatorForTokenClassification` specifically for this purpose. Once we have our collator and our trainer object, we can train our model:

```
[15]:  from transformers import Trainer
       from transformers import DataCollatorForTokenClassification

       data_collator = DataCollatorForTokenClassification(tokenizer)

       trainer = Trainer(
           model=model,
           args=training_args,
           data_collator=data_collator,
           compute_metrics=compute_metrics,
           train_dataset=train_ds,
           eval_dataset=eval_ds,
           tokenizer=tokenizer,
       )
```

```
trainer.train()
```

Next, we evaluate our newly trained model on the test dataset. For this purpose, we preprocess the data in the same way we did for the train and validation
partitions. Then, for convenience, we use the trainer's predict() method to
generate the predicted logits using our model:

[16]:
```
test_ds = ds['test'].map(
    tokenize_and_align_labels,
    batched=True,
)
output = trainer.predict(test_ds)
```

As before, we use scikit-learn's classification_report() function
to display the results of the evaluation. This function expects two one-
dimensional lists of labels, so we need to follow a similar approach to the
one we employed for text classification. Note that output.label_ids and
output.predictions are NumPy arrays rather than PyTorch tensors. This
time we use NumPy's reshape() method to reshape the arrays. This method
is similar to PyTorch's view() method that we used previously, except that
view() may copy the array's data in some situations. We discard the labels
corresponding to ignored subword tokens, and then we print the classification
report:

[17]:
```
from sklearn.metrics import classification_report

num_labels = model.num_labels
label_ids = output.label_ids.reshape(-1)
predictions = output.predictions.reshape(-1, num_labels)
predictions = np.argmax(predictions, axis=-1)
mask = label_ids != ignore_index

y_true = label_ids[mask]
y_pred = predictions[mask]
target_names = tags[:-1]

report = classification_report(
    y_true, y_pred,
    target_names=target_names
)
print(report)
```

	precision	recall	f1-score	support
DET	1.00	1.00	1.00	8040
NOUN	0.99	0.99	0.99	9533
ADP	1.00	1.00	1.00	8332
ADJ	0.98	0.97	0.97	3468
PROPN	0.99	0.99	0.99	4101
PRON	0.99	0.99	0.99	2484
VERB	0.99	0.99	0.99	4544
SCONJ	0.97	0.98	0.98	1210
PUNCT	1.00	1.00	1.00	6314
AUX	0.99	0.99	0.99	1396
CCONJ	1.00	1.00	1.00	1439
ADV	0.99	0.99	0.99	1710
NUM	0.97	0.98	0.97	958
PART	0.93	0.78	0.85	18
SYM	0.97	0.95	0.96	37
INTJ	0.86	0.75	0.80	16
accuracy			0.99	53600
macro avg	0.98	0.96	0.97	53600
weighted avg	0.99	0.99	0.99	53600

Our model based on XLM-RoBERTa achieves 99% accuracy. This is considerably better than the LSTM-based model developed in Chapter 11. In order to understand the differences between the two methods, we produce in what follows a confusion matrix for the results of each model. Rows in the confusion matrix represent the true labels and columns represent the predicted labels. In these confusion matrices, each cell x_{ij} corresponds to the *proportion* of values with label i that were assigned the label j.[10] For a perfect model, all cells in the diagonal would have value 1 and all other cells would have value 0. The code used to generate the confusion matrix is also shown. The confusion matrices for the LSTM and transformer are shown in Figure 13.1 and Figure 13.2, respectively.

```
[18]:  import matplotlib.pyplot as plt
       from sklearn.metrics import ConfusionMatrixDisplay,␣
        ↪confusion_matrix

       cm = confusion_matrix(y_true, y_pred, normalize='true')
       disp = ConfusionMatrixDisplay(
```

[10] This is the case because we used the normalize='true' parameter of the confusion_matrix() function.

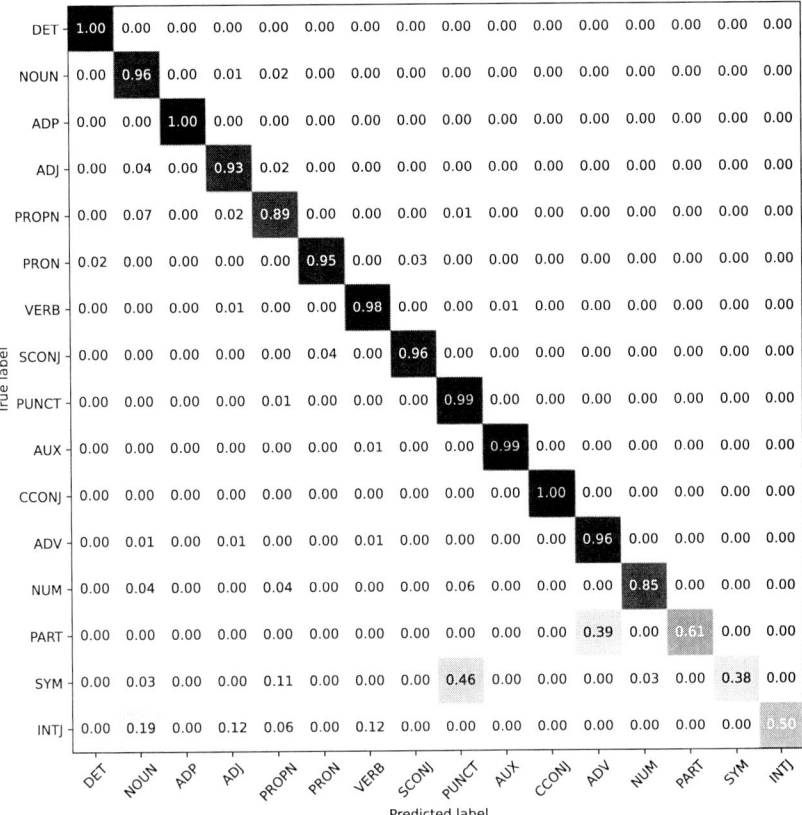

Figure 13.1 Confusion matrix corresponding to the long short-term memory–based part-of-speech tagger developed in Chapter 11

```
    confusion_matrix=cm,
    display_labels=target_names,
)

fig, ax = plt.subplots(figsize=(10,10))
disp.plot(
    cmap='Blues',
    values_format='.2f',
    colorbar=False,
    ax=ax,
    xticks_rotation=45,
)
```

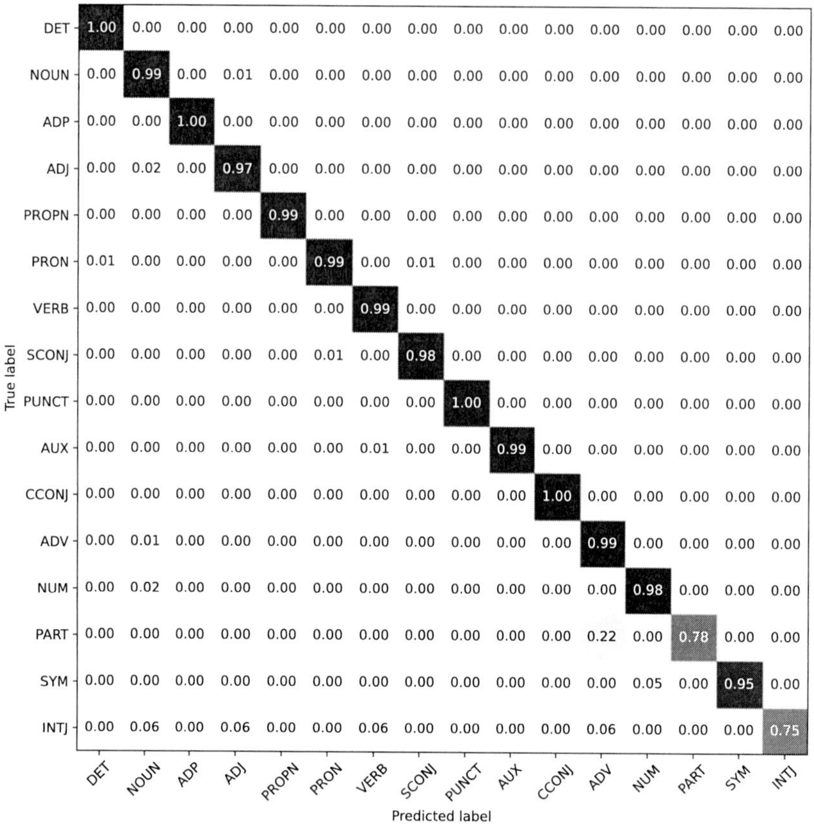

Figure 13.2 Confusion matrix corresponding to the transformer-based part-of-speech tagger

The two confusion matrices highlight a couple of important observations. First, the transformer model is considerably better at predicting POS tags with infrequent support in the dataset. For example, the accuracy for predicting the SYM POS tag increased from 38% in the LSTM model to 95% in the transformer model! Equally impressive, the transformer improved the performance of tags that are extremely common and thus provide plenty of opportunity to both approaches to learn a good model. For example, the accuracy of tagging NOUN, the second most common POS tag in the dataset, increased from 96% in the LSTM model to 99% in the transformer model.

Manning (2011) provides an interesting point of comparison, concluding that the "limit of human consistency on part-of-speech tagging is 97%." At first glance, this suggests that TNs perform the POS task at least as well as people do. However, the human consistency number in (Manning, 2011) was reported on a different dataset, resulting in a potential apples/oranges comparison to be avoided. Further, it relies on a measure that is somewhat stricter than the simple accuracy we discuss in this chapter. Last, the POS dataset we use in this section contains training and evaluation partitions that come from the same domain and genre. It is likely that machine performance in the wild – that is, on documents very different from the training data, will be lower.

13.4   Summary

In this chapter, we presented two applications driven by the encoder component of a TN. First, we used the transformer encoder as an acceptor and implemented a text classification application for English news. Second, we used the encoder as a transducer to develop a Spanish POS tagger. Both tasks were implemented using pretrained transformer models from the Hugging Face library. For both applications, the transformer-based methods outperform considerably all approaches introduced in the previous chapters, highlighting the value of the transformer architecture.

14   Encoder-Decoder Methods

In Chapters 10 and 12, we focused on two common usages of RNNs and TNs: acceptors and transducers. In the former case, we use one single vector that captures the entire text to train classification tasks that require access to the complete text such as review classification. This single vector is typically the last hidden state in an RNN or the [CLS] embedding in a TN. In contrast, transducers train tasks that operate at word level (e.g., POS tagging) by operating over the representations produced for each word (by either RNNs or TNs). Both acceptors and transducers are very useful for many NLP applications (see Chapter 16 for details). However, these two directions ignore one of the most important applications in NLP: machine translation. To address this task, we discuss in this chapter a third architecture for both RNNs and TNs: encoder-decoder methods.

At their core, encoder-decoder architectures encode an input sequence (similar to the acceptor architecture), and use the representation of the input text to create (or decode) the output sequence. Figure 14.1 shows an example of this approach for machine translation, where both encoder and decoder are implemented using RNNs.

Encoder-decoder approaches are a relatively new direction in the field of machine translation that started around 2014. Before then, machine translation operated using explicit features extracted from the languages to be translated at various levels of complexity. The Vauquois triangle (Figure 14.2) depicts a hierarchy of these "traditional" approaches to machine translation based on their sophistication (Jurafsky and Martin, 2009). As the figure indicates, these approaches range from simplistic methods that rely on lexical conversion between languages to more complex methods that incorporate syntax and semantics, to the ultimate idealized goal of translating to/from an universal interlingua. If we have to position encoder-decoder methods in this hierarchy, they probably fit somewhere between the semantic transfer methods (because the representations generated by RNNs and TNs capture (some) syntax and semantics) and approaches that rely on the universal interlingua. One may argue, as food for thought, that the representation that encodes the input sequence (i.e., the hidden state **c** of the </s> token of the English sentence in Figure 14.1) *is* an interlingua, but an interlingua for machines not people.

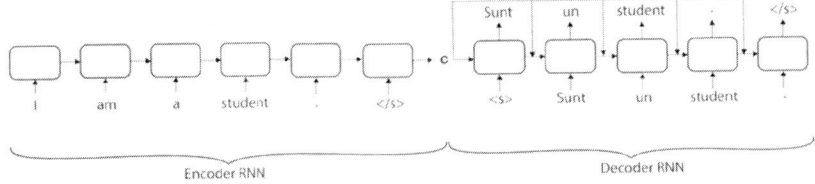

Figure 14.1 An encoder-decoder example of machine translation from English to Romanian, where both encoder and decoder are implemented using recurrent neural networks. Two virtual tokens, </s> and <s>, indicate end of sentence and beginning of sentence, respectively. The decoder uses the representation generated for the entire input sequence – that is, the hidden state vector **c** of the </s> token in the English sentence, to generate the equivalent Romanian words

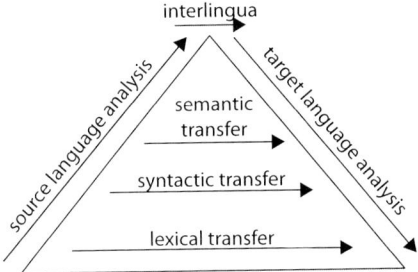

Figure 14.2 The Vauquois triangle that describes the hierarchy of machine translation approaches

Regardless of where encoder-decoder methods fit in the Vauquois triangle, they did simplify the machine translation task tremendously (to the point where we can now discuss them in this introductory book!), while, at the same, they improved overall machine translation results. In the remainder of this chapter, we describe three encoder decoder architectures that apply to RNNs and TNs.

14.1 BLEU: An Evaluation Measure for Machine Translation

Before we discuss specific neural architectures for machine translation, we briefly overview the most widely used evaluation measure for machine translation performance. This measure is called BLEU, from *bi*lingual *e*valuation *u*nderstudy (Papineni et al., 2002).

Intuitively, BLEU measures the word overlap between a candidate translation produced by the machine and a reference translation produced by a human.

Table 14.1 A simple example of the BLEU evaluation measure. The underlined words indicate matches between the candidate translation and the reference. The BLEU score for this candidate translation is 3/6

| Candidate (machine) | The feline sits <u>on the mat</u>. |
| Reference (human) | There is a cat <u>on the mat</u>. |

Table 14.2 The BLEU measure allows multiple reference translations. In such cases, the highest overlap is used. In this example, the BLEU score is 4/6 due to the higher overlap with the second reference translation

Candidate (machine)	<u>The</u> feline sits <u>on the mat</u>.
Reference 1 (human)	There is a cat <u>on the mat</u>.
Reference 2 (human)	<u>The</u> cat is <u>on the mat</u>.

More formally, BLEU "counts up the number of candidate translation words (unigrams) which occur in any reference translation and then divides by the total number of words in the candidate translation" (Papineni et al., 2002). Table 14.1 shows a simple example of this computation.

To account for multiple surface forms describing the same content, BLEU allows multiple reference translations for a given sentence. If more than one reference translation is available, BLEU picks the highest overlap score. Table 14.2 shows an example of this situation.

Note that simple overlap can be easily abused by generating the same word from the reference translation over and over again. BLEU controls for this situation by allowing each word from the reference translation to be used just once during the computation of the overlap. Table 14.3 shows an example of this situation.

Last, to account for word order, BLEU can be computed for n-grams, for any value of n. For example, for the candidate and reference translation in Table 14.1, the BLEU score over bigrams is 2/5 because two of the five bigrams in the candidate translation match the reference bigrams (*on the* and *the mat*).

Table 14.3 Simple overlap can be abused by repeatedly generating the same word from the reference translation. BLEU prevents this by allowing each word from a reference translation to be used just once. Naive overlap would score this candidate translation 6/6; BLEU scores it 2/6

Candidate (machine)	The the the the the the.
Reference (human)	The cat is on the mat.

Sidebar 14.1 Other evaluation measures for machine translation

The development of automated evaluation measures for machine translation remains an active research area. Other measures such as METEOR (Banerjee and Lavie, 2005) and BLEURT (Sellam et al., 2020), which allow candidate words to match reference words based not only on their surface forms but also their stemmed forms or meanings and also offer better control for word order, have been proposed in the literature. However, BLEU has passed the test of time successfully as it remains the most widely used machine translation evaluation measure to date.[1]

14.2 A First Sequence-to-Sequence Architecture

The first sequence-to-sequence architecture was proposed by Sutskever et al. (2014) in the context of machine translation. The intuition behind this approach (exemplified in Figure 14.1) is simple:

(i) We first encode the input sequence (i.e., the text in the source language), using an RNN such as a left-to-right LSTM. The hidden state of the last word in the input text – that is, the **c** vector produced for the </s> token in Figure 14.1 – becomes the representation of the entire input.

(ii) Starting from this representation, we then generate (or decode) one word at a time from the target language using a second RNN. Each decoder cell receives three inputs: (a) the input embedding of the previously decoded word (e.g., the embedding of the word *Sunt* for the cell that decodes *un*), (b) the hidden state of the previous RNN cell, which is combined with (c) the encoding of the input sentence **c** (e.g., through concatenation). Thus, each decoding cell "knows" where it stands in the currently decoded text, as well as what the representation of the source text is. Decoding continues

---

[1] https://naacl2018.wordpress.com/2018/03/22/test-of-time-award-papers.

until the stop symbol (</s>) is generated for the target text. Importantly, the decoded text does not need to have the same length as the input text.

This entire architecture is trained to maximize the accuracy of the decoded text. That is, each cell in the decoding RNN is encouraged to produce the correct target word at that position from the vocabulary of the target language. More formally, the overall loss function is the sum of multiple cross-entropy losses (see Section 6.4), one for each word in the gold target sequence.

Sutskever et al. (2014) showed that this approach performs better than previous machine translation methods that relied on explicit feature representations. However, this performance improvement was observed only after several important "tricks" were implemented:

- According to the authors, the most important modification was reversing the word order in the input sequence. For example, instead of using the input sequence *I am a student .* from Figure 14.1, this method uses the reversed text: *. student a am I*. Thus, the decoder starts decoding words in the target language using the hidden state of the first word in the source sentence, *I*. The intuition behind this somewhat strange idea is that reversing the input sequence yields a better initialization for the decoder. Since it is probable that the sentence in the target language will start with the subject, it helps the decoder if its starting point is the hidden state of the equivalent subject word in the source language.[2] As we discussed in Section 6.8, how we initialize the training process for neural networks matters. This trick aims to find a better initial state for the decoder using the aforementioned linguistic observation.
- The authors also observed that it is important to use two different LSTMs that do not share parameters: one for the encoder, and one for the decoder. This makes intuitive sense: since the two RNNs handle texts in different languages, it is sensible that their parameters be different.
- Further, the authors used stacked (or deep) LSTMs (see Figure 10.2) with four layers, each generating hidden state vectors of dimension 1,000, for both encoder and decoder. This suggests that to capture the subtleties of language translation, one needs networks that encode complex nonlinear functions and have many parameters. To speed up the training process, each LSTM layer is executed in parallel on a different GPU. This speeds up the training runtime from 1,700 words per second (with a single GPU) to 6,300 words per second (with four GPUs).
- Last, the authors used two strategies to hedge the risk of the decoder committing to an incorrect translation early. First, for each decoding position, the decoder maintains the best B hypotheses up to that point. For example,

---

[2] According to Crystal (1997), the word order in 75% of the world's languages is either subject-verb-object or subject-object-verb.

say that for $B = 2$ the two best hypotheses after decoding one word are: w_1 and w_2. At position two, the decoder estimates the probability of all words in the target language for each of the two hypotheses and keeps the two word sequences with the highest overall probability.[3] For example, the top two hypotheses at position two could be: w_1, w_3 and w_1, w_4. This process continues until end-of-sentence symbols are generated. This hedging algorithm is called *beam search*, from the analogy of a flashlight that has a light beam that only partially illuminates an unknown, dark space. Second, the authors use an ensemble of five different instantiations of their system, where each of these instances is trained using a different random number generator, which changes the initialization of the network parameters as well as the order in which they process the data during training. The same beam-search decoding strategy is applied here, with the only change being that hypotheses are generated by multiple translation systems.

Sutskever et al. (2014) showed that their overall machine translation approach outperforms previous approaches based on "traditional" statistical ML approaches. For example, using an ensemble of five LSTMs and a beam search with $B = 12$ for decoding, the authors report a BLEU score of 34.8% for the translation of English to French, while a state-of-the-art traditional machine translation obtained only 33.3% BLEU. The fact that this relatively simple approach outperforms the more complex traditional translation system is impressive. However, this performance was only observed after the improvements just described were implemented. Without them, performance is considerably more modest. For example, a configuration that uses a single LSTM and processes the source English text in its natural order obtains only 26.2% BLEU.

14.3 Sequence-to-Sequence with Attention

One critical limitation of the previous approach is that "a neural network needs to be able to compress all the necessary information of a source sentence into a fixed-length vector" (Bahdanau et al., 2015) – that is, into the hidden state of the last token in the source sentence, as shown in Figure 14.1. Sutskever et al.'s (2014) workaround for this problem was to reverse the source sentence, which yields a better initialization for the decoder. However, this signal becomes less and less meaningful as we keep decoding the target sentence, which makes it hard for the decoder to "cope with long sentences, especially those that are longer than the sentences in the training corpus" (Bahdanau et al., 2015).

A first solution for this problem was proposed by Bahdanau et al. (2015). The intuition behind the proposed solution is simple: instead of providing the

---

[3] Sutskever et al. (2014) used a vocabulary of 80,000 words for the target language.

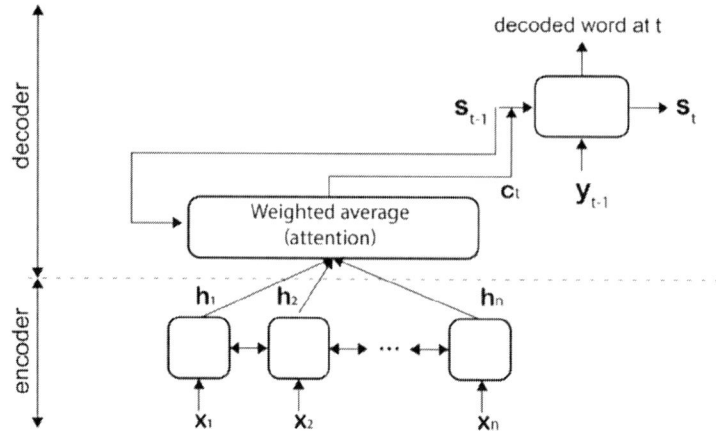

Figure 14.3 The architecture of a single decoder cell in a sequence-to-sequence architecture with attention. The encoder for this architecture is a bidirectional recurrent neural network that uses the input word embeddings, x_1 to x_n, to produce a sequence of hidden states, h_1 to h_n. The decoder is a left-to-right recurrent neural network. To avoid confusion between the source and target languages, we use y_t to indicate the input representation of the target word decoded at position t, and s_t to indicate the hidden state produced by the decoder cell at position t. c_t indicates the custom encoding vector of the source text for position t in the decoder

decoder a single vector **c** that encodes the entire source sentence, the decoder receives a *different* encoding of the source text for *each* decoded word. For instance, following the example in Figure 14.1, when decoding the word *student* in Romanian, the decoder benefits from focusing on (or paying more attention to) the word *student* in the source English text, rather than other parts of the text.

Figure 14.3 shows a description of this attention-based architecture. The first observation about this architecture is that its encoder changes from a unidirectional RNN to a bidirectional one. Thus, the hidden state h_i generated for each word at position i in the source text encodes context from both the left and the right of the current word. Second, each cell at position t in the decoder RNN receives the hidden state s_{t-1} generated by the previous decoder cell (similar to the architecture of Sutskever et al. [2014]), as well as a vector c_t that provides an encoding of the source text. Importantly, each c_t vector is *customized* for this position in the output sequence. For example, when decoding the first word in the output sequence for the example shown in Figure 14.1 (i.e., the word *Sunt*, which is the first-person singular form of the verb *be* in Romanian) the corresponding c_1 vector should be closer to the hidden state representation of the word *am* in the input sequence. Similarly, the $\mathbf{c}_3$ vector for the output word *student* should be closer to the hidden state vector of *student* in the input English text. This is the key contribution of this architecture.

Formally, the $\mathbf{c}_t$ vector is computed as a weighted average of all hidden states produced by the encoder on the source text, where the weights α used in the average change for each position t during decoding. That is:

$$c_t = \sum_i \alpha_{ti} \mathbf{h}_i \tag{14.1}$$

$$\alpha_{ti} = \text{softmax}(\mathbf{s}_{t-1}^T \mathbf{W}_a \mathbf{h}_i). \tag{14.2}$$

Equation 14.1 computes the average of all hidden states in the input text, where each hidden state is weighted by a parameter α. These weights are computed in Equation 14.2 for each input token i and each decoded token t. The attention matrix $\mathbf{W}_a$ allows us to combine the last hidden state of the decoder ($\mathbf{s}_{t-1}$) with the hidden state of the corresponding input token ($\mathbf{h}_i$), to produce a single scalar value that indicates how much attention the decoded token t should pay to the input token i. The ^{T} symbol in the equation indicates the transpose operation, which rotates the $\mathbf{s}_{t-1}$ vector $90°$ such that its rows become columns. This means that, in order for the multiplications in Equation 14.2 to be valid, the matrix $\mathbf{W}_a$ must have the same number of rows as the encoder's hidden state and the same number of columns as the number of rows of the decoder's hidden state. The attention matrix $\mathbf{W}_a$ is shared between the decoder cells, and is trained jointly with the parameters of the encoder and decoder. Last, the softmax function simply converts all α weights to a probability distribution to be used in the actual weighted average described in Equation 14.1.

Sidebar 14.2 *Other strategies for modeling attention*

Note that Equation 14.2 is not the only way to compute attention weights. Luong et al. (2015) describe two other strategies that produce good results empirically. The first is to simply skip the attention matrix $\mathbf{W}_a$ completely and compute attention weights as dot products of the two corresponding hidden states – that is, $\alpha_{ti} = \text{softmax}(\mathbf{s}_{t-1}^T \mathbf{h}_i)$. For this to work, the hidden states of the encoder and decoder must have the same dimensions. The second strategy concatenates the two hidden states into a single vector ($[\mathbf{s}_{t-1}; \mathbf{h}_i]$). This vector then becomes the input to an arbitrary FFNN with a single output neuron that computes the attention weight. The advantage of the first strategy is that it has fewer parameters to learn (because there is no $\mathbf{W}_a$ matrix) and thus might be trained more quickly from smaller datasets. The disadvantage is that it is less flexible than the method from Equation 14.2, which can learn more complex combination formulas due to the additional attention matrix. The second strategy is more flexible than Equation 14.2 due to the arbitrary neural network, but may introduce more parameters than it, which may complicate training.

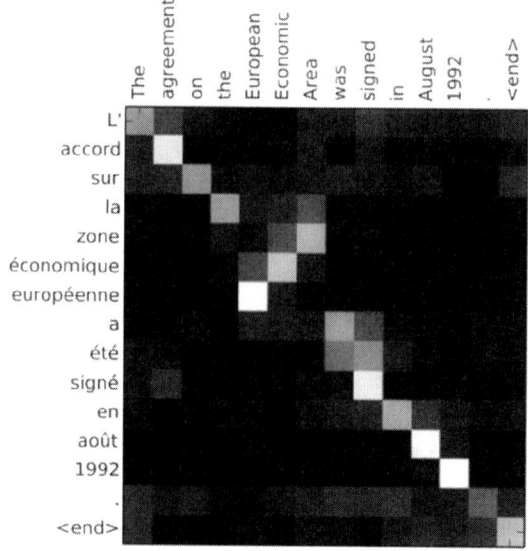

Figure 14.4 Example of attention weights from Bahdanau et al., 2015. The x-axis corresponds to words in the source language (English); the y-axis contains the decoder words from the target language (French). Each cell visualizes an attention weight α between the corresponding words, where black indicates 0 and white indicates 1

Bahdanau et al. (2015) have shown that their attention-based sequence-to-sequence method produces better results than the previous method of Sutskever et al. (2014). For example, they report a relative improvement of 50% in BLEU score in an English-to-French machine translation experiment over a vanilla sequence-to-sequence method without attention. A qualitative analysis of the attention weights learned (of which Figure 14.4 shows an example) indicates that the attention weights indeed learn to align decoded words with the corresponding source words, even when the word order in the target language changes. For example, Figure 14.4 shows that the French word *zone* is aligned mostly with the English word *Area* even though the word order is different – that is, in French, modifiers of nouns follow the noun rather than precede it, as in English.

14.4   Transformer-Based Encoder-Decoder Architectures

The previous approach combines three relatively complex components: two RNNs (one for the encoder and one for the decoder) and the attention mechanism that connects the two. Vaswani et al. (2017) showed that this architecture can be simplified based on the claim that "attention is all you need." That is, they replaced the encoder/decoder RNNs with distinct transformer layers that have their own self-attention, which are then connected through another attention mechanism. For the encoding side, they used exactly the same transformer layers we discussed in Chapter 12. The decoding blocks follow a similar architecture with a few important additions (see Figure 14.5). Intuitively, this approach replaces the RNN/attention combination of Bahdanau et al. (2015) with three attention mechanisms:

Encoder self-attention: The input text is encoded with the same stack of transformer layers introduced in Chapter 12, which means that the output embeddings $\mathbf{z}$ for the source text are computed using the self-attention mechanism we discussed before.

Decoder self-attention: Similarly, the decoded text is fed to a distinct stack of transformer layers that contain the same self-attention mechanism as the encoder. However, since the decoding operation works from left to right, the decoder only has information about the words decoded so far. For instance, when decoding the third word in the example shown in Figure 14.5, the self-attention mechanism can only use the words previously decoded (*bank* and *of*). The other entries reserved for the remaining text to be decoded are masked – that is, they are not used in any of the operations in the decoder layer.

Encoder-decoder attention: Last, the transformer encoder-decoder architecture includes an attention mechanism between the encoder and decoder layers, which, intuitively, serves the same purpose as the attention approach introduced in the previous section. This attention method is implemented very similarly to the self-attention mechanism described in Section 12.1.2. The primary difference is that the query, key, and value vectors ($\mathbf{q}_i$, $\mathbf{k}_j$, and $\mathbf{v}_j$, respectively) are used to compute the attention weights $\mathbf{a}_{ij}$ that align the decoded word at position i with the input word at position j. Due to this, the query vector $\mathbf{q}_i$ is computed using the representation produced by the decoder for the word at position i (i.e., the vector generated by the previous add-and-normalize operation – see Figure 14.5), while the key and value vectors $\mathbf{k}_j$ and $\mathbf{v}_j$ are computed using the output embedding $\mathbf{z}_j$ produced by the corresponding encoding layer for the input token at position j.

Importantly, the three sets of parameter matrices that contain the query, key, and value vectors behind all these attention types are learned during training (just like any other parameters in a neural network!). and then applied at testing time to incorporate attention information in the encoding of new input texts and the decoding of the corresponding output texts.

Vaswani et al. (2017) showed that this encoder-decoder transformer network, which relies heavily on attention while skipping sequence modeling, performs

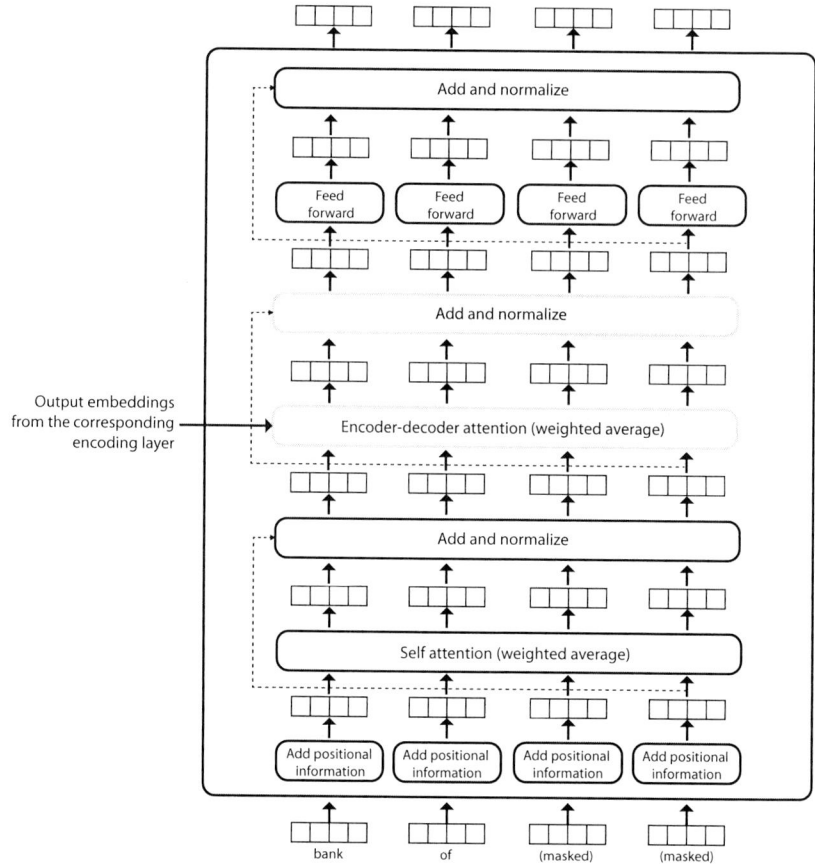

Figure 14.5 Architecture of an individual transformer decoder layer. The decoder layer follows closely the architecture of the encoder layer (see Figure 12.3), but it includes two new components (shown in grey in the figure): a component that implements an attention mechanism between the encoded and the decoded texts, and an additional add-and-normalize layer that normalizes the outputs of the encoder-decoder attention component

better for machine translation than approaches that rely on RNNs for encoding and decoding, including methods that couple RNNs with attention, similar to the strategy we discussed in the previous section. However, for this improvement to be considerable, the authors had to deploy a "big" transformer – that is, one that had six stacked layers for both encoder and decoder, produced a contextualized embedding of dimension 1,024 per token, and was trained for 300K steps (or batches).

14.5   Drawbacks of Encoder-Decoder Methods

At the time of writing, in 2023, TNs had emerged as the predominant approach for encoder-decoder methods. This means that the same drawbacks we discussed in Section 12.4 apply here. That is, TNs are large – for example, some models have hundreds of billions of parameters – and they are slow due to their attention mechanism, which is usually quadratic in the length of the texts. The situation is exacerbated for encoder-decoder methods, which have *two* transformer networks, one for the encoder and one for the decoder, and *three* attention mechanisms, as we discussed in Section 14.4. These drawbacks give rise to an equity concern: ML groups that do not have access to the necessary hardware resources to train these large networks are precluded from participating in the research discussed in this chapter.

One further drawback that applies to all encoder-decoder methods discussed in this chapter is that they require parallel texts for training – for example, to train an English-Romanian machine translation system, one needs a corpus of aligned sentences for the two languages. This limitation induces a different equity issue: cultures that rely on low-resource languages are largely isolated from the rest of the world. This is mitigated to a certain extent by recent directions in machine translation that rely solely on monolingual corpora for training (Artetxe et al., 2018). However, as expected, these methods do not perform as well as machine translation approaches that train using parallel corpora.

14.6   Historical Background

Encoder-decoder approaches are a relatively recent addition to the ML landscape. Kalchbrenner and Blunsom (2013) were the first to introduce a neural machine translation method that mapped the entire input sentence into a single vector. However, their method relied on vanilla RNNs, and suffered from the same vanilla RNN limitations we discussed in Chapter 10. Sutskever et al. (2014) replaced the building blocks in the aforementioned approach with LSTMs, and showed that their sequence-to-sequence machine translation approach achieves state-of-the-art results. Bahdanau et al. (2015) introduced the attention mechanism we discussed in this chapter, which removed the limitation that all decoded words must be generated using the same representation of the entire input text. This motivated the design of TNs (Vaswani et al., 2017), which have since dominated the NLP field.

14.7   References and Further Readings

While this chapter described key encoder-decoder approaches, one aspect we did not discuss is how to adjust the pretraining objectives to address the distinct capabilities required by encoder-decoder tasks such as machine translation. For example, the masked language model pretraining objective we discussed in Chapter 12 is not multilingual. Further, it does not capture how to align tokens from the source language to tokens in the target language. We discuss pretraining objectives tailored for machine translation in Chapter 16 where we discuss NLP applications (Raffel et al., 2020; Xue et al., 2020).

14.8   Summary

This chapter introduced three encoder-decoder architectures, which enable important NLP applications such as machine translation. In particular, we discussed the sequence-to-sequence method of Sutskever et al. (2014), which couples an encoder LSTM with a decoder LSTM. We followed this method with the approach of Bahdanau et al. (2015), which extend the previous decoder with an attention component, which produces a different encoding of the source text for each decoded word. Last, we introduced the complete encoder-decoder TN, which relies on three attention mechanisms: one within the encoder (which we discussed in Chapter 12), a similar one that operates over decoded words, and, importantly, an attention component that connects the input words with the decoded ones (which serves the same purpose as the attention mechanism of Bahdanau et al. [2015]).

15   Implementing Encoder-Decoder Methods

In this chapter, we implement a machine translation application as an example of an encoder-decoder task. In particular, we build on pretrained encoder-decoder transformer models, which exist in the Hugging Face library for a wide variety of language pairs. We first show how to use one of these models outside of the box in order to perform translation for one of the language pairs it has been exposed to during pretraining: English to Romanian. Afterward, we fine-tune the model to a new language combination it has not seen before: Romanian to English. In both use cases, we use the T5 encoder-decoder model, which has been pretrained for several tasks, including machine translation (Raffel et al., 2020). Please see Chapter 16 for a description of T5's pretraining process. The data for this task come from the WMT 2016 dataset (Bojar et al., 2016), which consists of English sentences aligned pairwise to German, Czech, Russian, Finnish, Romanian, and Turkish. In this chapter, we use only the English-Romanian texts (in both directions).

15.1   Translating English to Romanian

As a first example, we use T5 to translate from English to Romanian, which is one of the language pairs it has been exposed to during pretraining. The code discussed in this section is available in the notebook chap15_trans lation_en_to_ro.

Even though in this exercise we are not fine-tuning the model, we still need to define a few hyper parameters to frame the task and help the model understand how to work with the data:

```
[2]: transformer_name = 't5-small'
     source_lang = 'en'
     target_lang = 'ro'
     max_source_length = 1024
     max_target_length = 128
     task_prefix = 'translate English to Romanian: '
     batch_size = 100
```

These settings indicate that we use the t5-small model, a smaller T5 variant, to minimize the amount of memory required. The source_lang and target_lang variables define the direction of translation – that is, from English to Romanian. To keep our computing requirements small, we limit the length of our input and output. That is, English text longer than max_source_length tokens will be truncated. Further, we limit our generated Romanian text to max_target_length. We chose a maximum target length of 128 tokens to limit the computational cost incurred during text generation (recall that the text is generated one token at a time).

The T5 models are trained to support multiple tasks such as translation and summarization (please see Chapter 16 for details). Thus, during training and inference, the user must specify which task the model should perform using a text prefix. Here, we use the prefix "translate English to Romanian:" to indicate that the input text is in English and should be translated to Romanian.

Next, we load the model and the corresponding tokenizer and move them to the GPU if one is available:

[3]:
```
from transformers import AutoTokenizer, AutoModelForSeq2SeqLM

tokenizer = AutoTokenizer.from_pretrained(transformer_name)
model = AutoModelForSeq2SeqLM.from_pretrained(transformer_name)
model = model.to(device)
```

We use the datasets library to load our translation dataset. Note that the first time one calls load_dataset() the dataset will be downloaded automatically from the Hugging Face repository.[1] The load_dataset() function takes a dataset name and configuration, which in our case are wmt16 and ro-en, respectively. Since in this example we are only evaluating the model, we only load the *test* partition (or split) of the dataset:

[4]:
```
from datasets import load_dataset

test_ds = load_dataset('wmt16', 'ro-en', split='test')
test_ds
```

[4]:
```
Dataset({
    features: ['translation'],
    num_rows: 1999
})
```

---

[1] https://huggingface.co/datasets/wmt16.

The dataset consists of a single column called translation. Each element in this column is a dictionary that contains the aligned pair. The dictionary keys are the abbreviated language names and the values are the corresponding sentences. An example of one of these dictionaries is shown next:

```
[5]:  test_ds['translation'][0]
```

```
[5]:  {'en': 'UN Chief Says There Is No Military Solution in Syria',
       'ro': 'Şeful ONU declară că nu există soluţii militare în Siria'}
```

We encapsulate the logic for translating the English text into Romanian in a function called translate(). Inside this function, for a batch of aligned pairs, we select the English sentence as our input and prepend the task prefix. Then we tokenize these inputs, including the prefix, specifying that sentences longer than max_source_length should be truncated, the batch should be padded, and the tokenizer should return PyTorch tensors.

Once the tokenizer output has been moved to the GPU, we pass it to the model's generate() method. This is the first time we have seen this method, because only decoder and encoder-decoder models support it. This method generates an output sequence by predicting one token at a time, stopping when either the *end-of-sequence* token is produced or when the sequence reaches a maximum length. Several generation techniques are supported, such as beam search, in which several alternate translations are maintained by the model so that it is able to select an overall best translation from several options. For efficiency, we use a greedy approach, which chooses the best token at each step of the generation. This is equivalent to using a beam search with a beam of size one.

Since the model generates its predictions as a sequence of token IDs, we need to convert them back into the corresponding tokens to be able to read the translated text. We do this using the tokenizer's batch_decode() method. Finally, we return the gold and predicted Romanian sentences in a dictionary:

```
[6]:  def translate(batch):
          # get source language examples and prepend task prefix
          inputs = [x[source_lang] for x in batch["translation"]]
          inputs = [task_prefix + x for x in inputs]

          # tokenize inputs
          encoded = tokenizer(
              inputs,
              max_length=max_source_length,
              truncation=True,
              padding=True,
              return_tensors='pt',
```

```
        )

        # move data to gpu if possible
        input_ids = encoded.input_ids.to(device)
        attention_mask = encoded.attention_mask.to(device)

        # generate translated sentences
        output = model.generate(
            input_ids=input_ids,
            attention_mask=attention_mask,
            num_beams=1,
            max_length=max_target_length,
        )

        # generate predicted sentences from predicted token ids
        decoded = tokenizer.batch_decode(
            output,
            skip_special_tokens=True,
        )

        # get gold sentences in target language
        targets = [x[target_lang] for x in batch["translation"]]

        # return gold and predicted sentences
        return {
            'reference': targets,
            'prediction': decoded,
        }
```

Next, we apply our `translate()` function to our `Dataset` to translate all the sentences (the output for the code below is on page 233):

```
[7]:  results = test_ds.map(
          translate,
          batched=True,
          batch_size=batch_size,
          remove_columns=test_ds.column_names,
      )

      results.to_pandas()
```

We evaluate the quality of these translations using the BLEU metric, which we introduced in Chapter 14. To this end, we load an existing implementation

	reference	prediction
0	Șeful ONU declară că nu există soluții militar...	eful ONU declară că nu există o soluție milita...
1	Secretarul General Ban Ki-moon afirmă că răspu...	Secretarul General Ban Ki-moon declară că răsp...
2	Șeful ONU a solicitat din nou tuturor părților...	eful U.N. a cerut din nou tuturor partidelor, ...
3	Ban a declarat miercuri în cadrul unei conferi...	Ban a declarat la o conferință de presă susțin...
4	Ban și-a exprimat regretul că divizările în co...	El și-a exprimat regretul că diviziunile din c...
...	...	...
1994	Nu sunt bani puțini.	Banii sunt suficienți.
1995	Uneori mi-e rușine să ridic banii de la casierie.	Uneori mi-e rușine să iau banii de la biroul c...
1996	La sfârșitul mandatului voi face un raport cu ...	La sfârșitul biroului voi raporta tot ceea ce ...
1997	S-a întâmplat să ridic într-o lună și 30.000 d...	Într-o lună am adunat 30 000 de lei cu ramburs...
1998	"Să spună un parlamentar că nu-i ajung banii e...	"A spune că un parlamentar nu are suficienți b...

1999 rows × 2 columns

of BLEU from the datasets library as a Metric object.[2] Metric objects have a method called add(), which is used to accumulate the predictions and gold labels, one example at a time. After accumulating all examples, the compute() method returns the results of the evaluation. Note that for each predicted sentence, BLEU expects a list of reference sentences (as there are often many correct ways of translating a given text). Since we only have one reference, we wrap it in a list before passing it to the metric:

```
[8]: from datasets import load_metric

metric = load_metric('sacrebleu')

for r in results:
    prediction = r['prediction']
    reference = [r['reference']]
    metric.add(prediction=prediction, reference=reference)

metric.compute()
```

```
[8]: {'score': 25.18405390123436,
     'counts': [27521, 14902, 8681, 5141],
     'totals': [49236, 47237, 45240, 43245],
     'precisions': [55.89609229019417,
      31.547304020153693,
      19.188770999115828,
      11.888079546768413],
     'bp': 1.0,
     'sys_len': 49236,
     'ref_len': 48945}
```

The score corresponds to the BLEU score. The rest of the items correspond to the components required to compute the score. That is, the counts, totals, and precisions correspond to the counts, totals, and precisions for 1-, 2-, 3-, and 4-grams. The bp is the brevity penalty. The sys_len and ref_len correspond to the predictions and reference lengths.

The BLEU score of 25.2% is slightly lower than the state of the art, but we are being penalized by the peculiarities of diacritic usage in Romanian characters. For example, the letters ș and ț (corresponding to the sounds *sh* and *ts* in English) are usually spelled with a comma below the characters *s* and *t*, which is the standard imposed by the Romanian Academy. However, in "the

---

[2] https://huggingface.co/docs/datasets/v2.4.0/en/package_reference/main_
classes#datasets.Metric.

wild," these characters are often written using a cedilla instead of a comma – for example, *ţ* instead of *ț* (or, using the names of these Unicode characters, LATIN SMALL LETTER T WITH CEDILLA instead of LATIN SMALL LETTER T WITH COMMA BELOW). Further, some of these characters with diacritics are often omitted altogether in the T5 output. The T5 output contains an example for each of these two situations (e.g., *soluţi(e)* instead of *soluţi(i)*, and *eful* instead of *Şeful*):

```
[9]: results[0]
```

```
[9]: {'reference': 'Şeful ONU declară că nu există soluţii militare␣
     ↪în Siria',
      'predicted': 'eful ONU declară că nu există o soluţie militară␣
     ↪în Siria'}
```

To avoid being penalized at scoring time for these arbitrary discrepancies, post-processing scripts are sometimes used to normalize diacritic usage.[3] Usage of such post-processing scripts can improve the BLEU score substantially. However, this is beyond the scope of this chapter.

15.2   Implementation of Greedy Generation

To gain a better intuition of how the encoder-decoder model generates its output sequence, we show next an implementation of the greedy version of the generate() method just used. Note that our function implements text tokenization as well. Thus, this function takes as an argument a single English text (i.e., no batching) and returns the corresponding Romanian text:

```
[10]: def greedy_translation(text):
          # prepend task prefix
          text = task_prefix + text

          # tokenize input
          encoded = tokenizer(
              text,
              max_length=max_source_length,
              truncation=True,
              return_tensors='pt',
          )

          # encoder input ids
          encoder_input_ids = encoded.input_ids.to(device)
```

---

[3] https://github.com/huggingface/transformers/blob/main/examples/legacy/
seq2seq/romanian_postprocessing.md.

```
# decoder input ids, initialized with start token id
start = model.config.decoder_start_token_id
decoder_input_ids = torch.LongTensor([[start]]).to(device)

# this loop replaces the generic generate() method
# we generate tokens here, one at a time
for _ in range(max_target_length):
    # get model predictions
    output = model(
        encoder_input_ids,
        decoder_input_ids=decoder_input_ids,
    )
    # get logits for last token
    next_token_logits = output.logits[0, -1, :]
    # select most probable token
    next_token_id = torch.argmax(next_token_logits)
    # append new token to decoder_input_ids
    output_id = torch.LongTensor([[next_token_id]]).to(device)
    decoder_input_ids = torch.cat(
        [decoder_input_ids, output_id],
        dim=-1,
    )
    # break loop if predicted token is the end of sequence
    if next_token_id == tokenizer.eos_token_id:
        break

# return text corresponding to predicted token ids
return tokenizer.decode(
    decoder_input_ids[0],
    skip_special_tokens=True,
)
```

This function interacts directly with the encoder and decoder components of the T5 model, so we must construct the input for both. The encoder's input is constructed by prepending the task prefix to the English text and tokenizing it. On the other hand, the decoder's input is constructed incrementally by accumulating the tokens predicted so far in order to predict the next token in the sequence. At the beginning, before any tokens are predicted, the decoder's input is initialized with a single token that corresponds to the beginning of the sequence. We retrieve this token, called decoder_start_token_id, from the model's configuration object.

The tokens are predicted one at a time, until the model produces eos_token_id, which indicates that the sequence is finished. However, in case the model does not produce this end-of-sequence token within a reasonable number of steps, we also enforce a maximum number of predicted tokens, determined by the max_target_length parameter we defined previously. The T5 model's forward() method, called indirectly through its __call__() method, takes the inputs for both the encoder and the decoder. The output returned by this method corresponds to all the tokens in the decoder's input plus an extra one: the newly predicted token. To select the best prediction, we retrieve the logits from the output and select the logits corresponding to the last token in the sequence (recall that the output shape is [batch size, sequence length, vocabulary size]). From these selected logits, we use the `argmax()` to select the token id corresponding to the highest-scoring vocabulary item. We append this new token ID to the decoder's input and repeat the process until we encounter the end-of-sequence token or the decoded text reaches the maximum length.

Once we are finished generating token IDs, we retrieve the corresponding text by calling the tokenizer's `decode()` method. This method is identical to the `batch_decode()` method we used previously, except that it only decodes a single example.

Next is an usage example for the `greedy_translation()` function:

```
[11]:  greedy_translation("this is a test")
```

```
[11]:  'Acesta este un test'
```

15.3 Fine-Tuning Romanian to English Translation

In this section, we fine-tune a T5 model on the translation of Romanian to English, a language pair that was not included in the T5 pretraining. To confirm that these data were not included in pretraining, we evaluated the performance of the vanilla `t5-small` model on the translation from Romanian to English using code equivalent to the code discussed in the previous section (see the `chap15_translation_ro_to_en` notebook). The resulting BLEU score was only 3.2%, which is substantially lower than the score we obtained when translating English to Romanian (25.2%).

Note that the transformers library includes scripts to fine-tune a translation model directly from the command line.[4] For didactic purposes, we will not use these scripts in this section, but instead write the fine-tuning code explicitly.

For this exercise, we continue using the WMT16 dataset, but this time we load the *train* and *validation* splits. We employ the same `t5-small` model that we used previously. The code from the last section to load the model, tokenizer, and dataset does not need to change for this use-case, so we do not repeat it here. However, as before, the complete code is available in a Jupyter notebook (`chap15_translation_ro_to_en_finetune`).

We begin by tokenizing the source (Romanian) and target language (English) texts. As in the last section, we need to prepend the task prefix to the source texts prior to tokenizing. This time, since we are translating in the

---

[4] https://github.com/huggingface/transformers/tree/main/examples/pytorch/translation.

opposite direction, we use the prefix `"translate Romanian to English:"`, and we prepend it to the Romanian text.

Each call to the tokenizer with a batch of texts produces `input_ids` and an `attention_mask`. This output is what we need for the Romanian text, which will serve as the input to the model. To generate the `labels` – that is, the correct translated tokens – we use the `input_ids` corresponding to the English text. Recall that `"labels"` is the default key name expected by trainers in Hugging Face.

```python
[5]: def tokenize(batch):
        # get source sentences and prepend task prefix
        sources = [x[source_lang] for x in batch["translation"]]
        sources = [task_prefix + x for x in sources]
        # tokenize source sentences
        output = tokenizer(
            sources,
            max_length=max_source_length,
            truncation=True,
        )

        # get target sentences
        targets = [x[target_lang] for x in batch["translation"]]
        # tokenize target sentences
        labels = tokenizer(
            targets,
            max_length=max_target_length,
            truncation=True,
        )
        # add targets to output
        output["labels"] = labels["input_ids"]

        return output
```

We apply our `tokenize()` function to both the train and validation splits (the output for the code below is on page 240):

```python
[6]: train_dataset = wmt16['train']
     eval_dataset = wmt16['validation']
     column_names = train_dataset.column_names

     train_dataset = train_dataset.map(
         tokenize,
         batched=True,
         remove_columns=column_names,
     )
```

```
eval_dataset = eval_dataset.map(
    tokenize,
    batched=True,
    remove_columns=column_names,
)


train_dataset.to_pandas()
```

Recall that in order to construct a trainer, we need a data collator for batching, a function to compute the metrics of interest, and a Training Arguments object. In this section, we use a data collator called Data CollatorForSeq2Seq, which is included in the transformers library specifically for sequence-to-sequence models. The collator pads the batches using the label_pad_token_id, which we have set to -100, as we did in Chapter 13 (this is the default ignore_index value used by Cross EntropyLoss):

```
[8]:  from transformers import DataCollatorForSeq2Seq

      label_pad_token_id = -100

      data_collator = DataCollatorForSeq2Seq(
          tokenizer,
          model=model,
          label_pad_token_id=label_pad_token_id,
      )
```

The compute_metrics() function computes the BLEU score. It uses the tokenizer to decode the token IDs into text, for both the predicted and gold labels, ignoring padding:

```
[9]:  from datasets import load_metric

      metric = load_metric('sacrebleu')

      def compute_metrics(eval_preds):
          preds, labels = eval_preds
          # get text for predictions
          predictions = tokenizer.batch_decode(
              preds,
              skip_special_tokens=True,
          )
          # replace -100 in labels with pad token
          labels = np.where(
```

	input_ids	attention_mask	labels
0	[13959, 3871, 29, 12, 1566, 10, 4961, 106, 204,...	[1, 1, 1, 1, 1, 1, 1, 1, 1, 1, 1, 1, 1, 1, 1,...	[19428, 13, 12876, 10, 217, 13687, 7, 1]
1	[13959, 3871, 29, 12, 1566, 10, 5085, 5840, 49,...	[1, 1, 1, 1, 1, 1, 1, 1, 1, 1, 1, 1, 1, 1, 1,...	[2276, 8843, 138, 13, 13687, 7, 13, 1767, 3823...
2	[13959, 3871, 29, 12, 1566, 10, 4961, 106, 204,...	[1, 1, 1, 1, 1, 1, 1, 1, 1, 1, 1, 1, 1, 1, 1,...	[19428, 13, 12876, 10, 217, 13687, 7, 1]
3	[13959, 3871, 29, 12, 1566, 10, 781, 8750, 9, ...	[1, 1, 1, 1, 1, 1, 1, 1, 1, 1, 1, 1, 1, 1, 1,...	[781, 2420, 13, 17500, 10, 217, 13687, 7, 1]
4	[13959, 3871, 29, 12, 1566, 10, 374, 6225, 49,...	[1, 1, 1, 1, 1, 1, 1, 1, 1, 1, 1, 1, 1, 1, 1,...	[11167, 7, 1204, 10, 217, 13687, 7, 1]
...	...	...	...
610315	[13959, 3871, 29, 12, 1566, 10, 4540, 4031, 9,...	[1, 1, 1, 1, 1, 1, 1, 1, 1, 1, 1, 1, 1, 1, 1,...	[4540, 4031, 9, 7, 1672, 7, 2262, 900, 17, 38,...
610316	[13959, 3871, 29, 12, 1566, 10, 2364, 4540, 40...	[1, 1, 1, 1, 1, 1, 1, 1, 1, 1, 1, 1, 1, 1, 1,...	[242, 4540, 4031, 9, 7, 6, 8, 516, 65, 66, 8,...
610317	[13959, 3871, 29, 12, 1566, 10, 2262, 900, 17,...	[1, 1, 1, 1, 1, 1, 1, 1, 1, 1, 1, 1, 1, 1, 1,...	[2262, 900, 17, 641, 65, 46, 3761, 6, 1069, 31...
610318	[13959, 3871, 29, 12, 1566, 10, 3, 25882, 759,...	[1, 1, 1, 1, 1, 1, 1, 1, 1, 1, 1, 1, 1, 1, 1,...	[9810, 157, 31, 7, 516, 92, 3088, 21, 46, 3839...
610319	[13959, 3871, 29, 12, 1566, 10, 18420, 83, 362...	[1, 1, 1, 1, 1, 1, 1, 1, 1, 1, 1, 1, 1, 1, 1,...	[3625, 32, 5788, 35, 15, 3844, 31, 7, 3, 16143...

610320 rows × 3 columns

```
        labels != label_pad_token_id,
        labels,
        tokenizer.pad_token_id,
    )
    # get text for gold labels
    references = tokenizer.batch_decode(
        labels,
        skip_special_tokens=True,
    )
    # metric expects list of references for each prediction
    references = [[ref] for ref in references]

    # compute bleu score
    results = metric.compute(
        predictions=predictions,
        references=references,
    )

    return {'bleu': results['score']}
```

We use the Seq2SeqTrainingArguments class, which adds the predict_with_generate parameter to the regular TrainingArguments class. This is needed to indicate that the trainer should use the generate() method for inference in order to compute the metrics (BLEU in this case):

[10]:
```
from transformers import Seq2SeqTrainingArguments

training_args = Seq2SeqTrainingArguments(
    output_dir=output_dir,
    per_device_train_batch_size=batch_size,
    per_device_eval_batch_size=batch_size,
    save_steps=save_steps,
    predict_with_generate=True,
    evaluation_strategy='steps',
    eval_steps=save_steps,
    learning_rate=learning_rate,
    num_train_epochs=num_train_epochs,
)
```

Finally, we construct the trainer using the Seq2SeqTrainer class, which is a subclass of Trainer that adds the ability to compute scores such as BLEU during training by calling generate() during evaluation:

```
[11]:  from transformers import Seq2SeqTrainer

       trainer = Seq2SeqTrainer(
           model=model,
           args=training_args,
           train_dataset=train_dataset,
           eval_dataset=eval_dataset,
           tokenizer=tokenizer,
           data_collator=data_collator,
           compute_metrics=compute_metrics,
       )
```

Fine-tuning a translation model takes considerably longer than training or fine-tuning the models we have developed so far in this book. To account for this, here we add support for resuming training from a *checkpoint* – that is, a model that was saved after training on a number of examples. Similar to how one can resume a video game, this allows one to pick up from the last "save point," in case training was interrupted and needs to be resumed:

```
[12]:  import os
       from transformers.trainer_utils import get_last_checkpoint

       last_checkpoint = None
       if os.path.isdir(output_dir):
           last_checkpoint = get_last_checkpoint(output_dir)

       if last_checkpoint is not None:
           print(f'Checkpoint detected at {last_checkpoint}.')
```

When calling the trainer's `train()` method, we either provide a model checkpoint or `None`. In the former case, the trainer will continue training from the provided checkpoint. In the latter case, the trainer will begin training from scratch. Once the training has completed, we save the trained model and tokenizer using the trainer's `save_model()` method into the output directory. Note that the first instruction may take a considerable amount of time. For example, on a machine with an NVIDIA TITAN RTX GPU, training took approximately five hours. On a computer without a GPU, the same training process may take more than 100 hours.

```
[13]:  train_result = trainer.
       ↪train(resume_from_checkpoint=last_checkpoint)
       trainer.save_model()
```

We then compute and save the metrics corresponding to the training partition. This is not required, but it is helpful to keep a record of the model's performance on the training data. Note that the metrics do not automatically include the number of examples in the training partition, so we add them explicitly:

```
[14]: metrics = train_result.metrics
      metrics['train_samples'] = len(train_dataset)
      trainer.save_metrics('train', metrics)
      trainer.save_state()
```

Next, we evaluate our final model on the validation data and save the corresponding metrics. These metrics indicate that our BLEU score on the validation data is 35.2%, which is evidence that fine-tuning has helped dramatically:

```
[15]: metrics = trainer.evaluate(
          max_length=max_target_length,
          num_beams=num_beams,
          metric_key_prefix='eval',
      )

      metrics['eval_samples'] = len(eval_dataset)

      trainer.log_metrics('eval', metrics)
      trainer.save_metrics('eval', metrics)
```

```
***** eval metrics *****
  epoch                    =        3.0
  eval_bleu                =    35.1923
  eval_loss                =     1.4452
  eval_runtime             = 0:01:50.71
  eval_samples             =       1999
  eval_samples_per_second  =     18.055
  eval_steps_per_second    =      4.516
```

Last, we save a *model card* into our output directory. A model card is akin to an automatically generated README file that includes information about the model used, the data, settings used, and performance throughout the training process. This file is helpful for reproducibility as it contains all of this key information in one place. These cards are often uploaded to the Hugging Face Hub together with the model itself.[5]

---

[5] We do not discuss the model uploading process here. Please see the documentation on model sharing at: https://huggingface.co/docs/transformers/v4.14.1/model_sharing.

```
[16]:  kwargs = {
           'finetuned_from': transformer_name,
           'tasks': 'translation',
           'dataset_tags': dataset_name,
           'dataset_args': dataset_config_name,
           'dataset': f'{dataset_name} {dataset_config_name}',
           'language': [source_lang, target_lang],
       }
       trainer.create_model_card(**kwargs)
```

15.4   Using a Previously Saved Model

Models that have been saved locally can be loaded using the same
from_pretrained() methods we have used before. In particular, instead of
providing a model name, we provide the path to the local directory where the
model is stored, using the local_files_only parameter to indicate that we
want to load the model from the local file system instead of downloading it
from the Hugging Face Hub (Make sure you use an output directory that is
valid on your machine!):

```
[3]:  from transformers import AutoTokenizer, AutoModelForSeq2SeqLM

      output_dir = '/media/data2/t5-translation-example'
      tokenizer = AutoTokenizer.from_pretrained(
          output_dir,
          local_files_only=True,
      )
      model = AutoModelForSeq2SeqLM.from_pretrained(
          output_dir,
          local_files_only=True,
      )
      model = model.to(device)
```

Once our fine-tuned model is loaded, we use it the same way as
before. That is, we use the translate() function to generate translations for our test partition. Then we use the BLEU metric to score this
output. From this metric, we obtain the final BLEU score of 33.4%,
which is markedly better than our initial score (i.e., without fine-tuning) of
3.2%! The code corresponding to this section is available in the notebook
chap15_translation_ro_to_en_finetuned.

15.5   Summary

In this chapter, we used a complete encoder-decoder TN to implement a
machine translation application. Importantly, transformers with a decoder

component have a `generate()` method that simplifies the generation process and provides multiple options for decoding. We encourage you to explore these options! For example, try comparing the quality of the output with the resources required to produce it (e.g., runtime overhead) when the size of the search beam increases. Additionally, we saw how to fine-tune an encoder-decoder model on a new language pair that it has not seen during its pretraining. This exercise included using checkpoints to support resuming training in case of unexpected interruptions, saving our fine-tuned model, and loading it for later use.

16 Neural Architectures for Natural Language Processing Applications

So far, we have discussed neural architectures such as FFNNs, RNNs, and TNs, and have exemplified how they can be used to implement a few NLP applications such as text classification and POS tagging. In this chapter, we describe several common applications (including the ones we touched on before) and multiple possible neural approaches for each. It is important to note that research on how to improve these methods is ongoing and exciting new methods continue to be proposed. However, given the scope of this book, we focus on simple neural approaches that work well and should be familiar to anybody beginning research in NLP or interested in deploying robust strategies in industry.

16.1   Text Classification

Text classification is one of the most widely used NLP applications due to its many real-world uses such as sentiment analysis (Pang et al., 2008; Socher et al., 2013), product reviews (Maas et al., 2011), news classification (Zhang et al., 2015), or classifying user intent in search queries (Li and Roth, 2002).

Because of its ubiquity in the real world and also its relative simplicity, we used text classification as a walkthrough example in the first chapters of the book (until Chapter 9, and then again in Chapter 13). However, to facilitate nonlinear reading, we summarize next three common neural architectures for text classification.

The first is deep averaging networks (DAN) (Iyyer et al., 2015), illustrated in Figure 16.1. This architecture is the neural equivalent of a bag-of-words approach for text classification. That is, this approach takes as input a set of static embeddings (generated offline using an algorithm similar to those covered in Chapter 8), which are then simply averaged into a single vector. This average can theoretically be improved. For example, one could remove words whose embeddings are likely to be too noisy (because they appear in too many ambiguous contexts) such as prepositions and determiners. One could apply the attention mechanism discussed in Section 14.3 to turn the simple average into a weighted one. What is important is that this averaging mechanism produces a single vector regardless of the number of input words. This is useful,

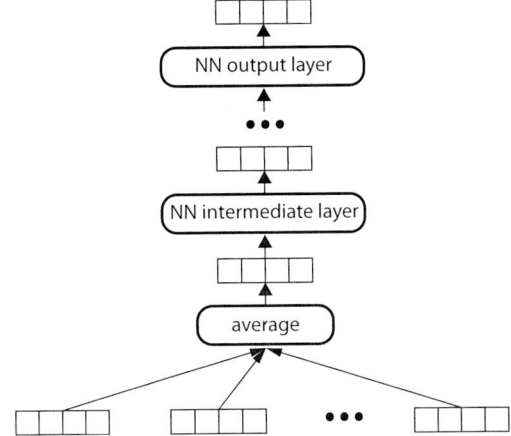

Figure 16.1 Deep averaging network (DAN) for text classification

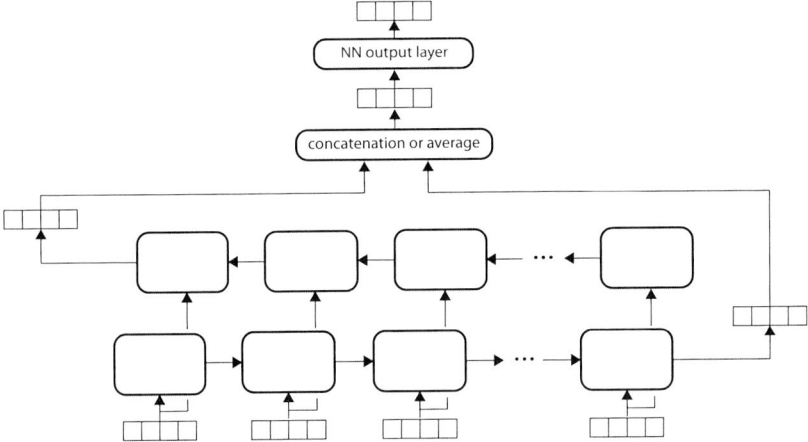

Figure 16.2 An acceptor bidirectional recurrent neural network for
text classification

as it makes this architecture applicable to texts of any length. The vector that
stores the average of the input embeddings is followed by one or more intermediate neural layers, which apply Equation 5.1 (hence the "deep" in the name),
and one output layer, which produces the output vector with one neuron per
class. Please see Section 9.2 for a complete PyTorch implementation of this
architecture.

The second architecture for text classification, illustrated in Figure 16.2,
uses an acceptor bidirectional RNN (typically a bidirectional LSTM). In this

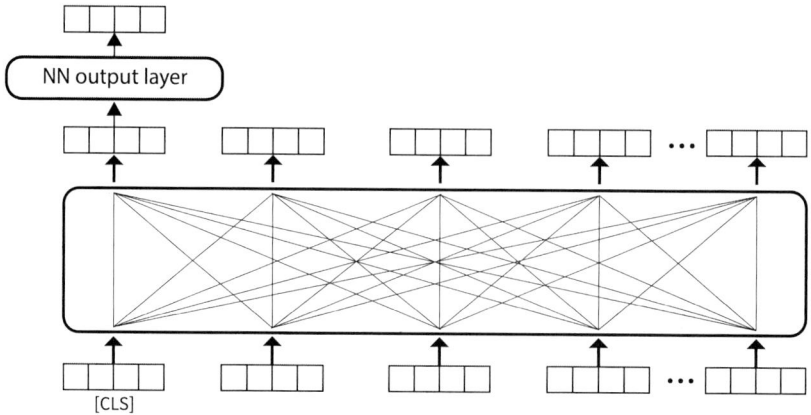

Figure 16.3 Transformer network for text classification

architecture, we use only the right-most hidden state of the left-to-right RNN and the left-most hidden state of the right-to-left RNN. These two vectors are then combined, either through concatenation or averaging, into a new vector, which is then fed to an output layer that produces an output vector similar to the one in the previous architecture. We challenge the curious reader to adapt the transducer RNN implemented in Chapter 11 to this acceptor configuration! However, be advised that in the experience of the authors, RNNs seldom outperform simpler bag-of-words architectures for text classification.

Last, Figure 16.3 shows how the (encoder of a) TN can be used for text classification. As the figure illustrates, the TN can be used immediately for text classification by adding an output layer on top of the contextualized embedding of the virtual [CLS] token. Please see Chapter 13 for an implementation of this architecture. As our results in this chapter indicate, this simple extension of the transformer encoder achieves state-of-the-art results in our multiclass text classification use case.

16.2   Part-of-Speech Tagging

As we mentioned in Chapter 10, parts of speech are categories of words that share similar grammatical properties – for example, nouns or verbs. Part-of-speech tagging is the task that assigns these tags to words in a text. Because POS tags are familiar to most readers, POS tagging is commonly used as an example of an NLP task that requires the modeling of sequences. For example, for both English and Spanish, once we see a determiner, it is much more likely that the next word will be a noun. In Spanish and Romanian, adjectives are likely to appear after a noun. And so on. Importantly, the usefulness of

POS tagging goes considerably beyond this didactic use case. Part-of-speech information has been shown to improve several downstream tasks such as syntactic parsing (Collins, 1996; Kiperwasser and Goldberg, 2016; Vacareanu et al., 2020a) and information extraction (Mintz et al., 2009; Valenzuela-Escárcega et al., 2018).

While linguistic studies of parts of speech go back at least 2,500 years (Matilal, 1990), large datasets annotated with standardized POS tags are relatively new. In 1964, the Brown corpus was made public with approximately 1 million English words annotated with 87 POS tags (Francis, 1964). Three decades later, (Marcinkiewicz, 1994) released the Penn Treebank corpus, which contains more than 4.5 million words of American English annotated with a somewhat simpler set of 36 POS tags and 12 other tags, for punctuation and currency symbols. These datasets elicited several decades of work on ML approaches for POS tagging, to the point where Manning (2011) suggested that "there is limited further mileage to be had . . . from better machine learning."

However, it was observed that the Penn Treebank POS tag set is too fine-grained to be applicable across languages (Petrov et al., 2011). For example, the Penn Treebank uses six POS tags for verbs, which distinguish between different verb forms such as gerund and past participle, some of which do not exist in many other languages. This research motivated the creation of a "universal" set of coarser POS tags that are more easily applicable across languages. Ultimately, this effort grew into the Universal Dependencies project, which is a "framework for consistent annotation of grammar (parts of speech, morphological features, and syntactic dependencies) across different human languages" (Nivre et al., 2016, 2020).[1] Table 16.1 lists the universal POS tags used in the latest iteration of Universal Dependencies. As the table indicates, these POS tags are split into three categories: (a) open-class tags, which contain an unbounded number of words as new words belonging to these categories are continuously added to the language (e.g., new protein names are continuously created as our understanding of molecular biology improves); (b) closed-class tags, which contain a fixed number of words (e.g., the number of prepositions in most languages is finite and reasonably small); and (c) other tags, which contain punctuation signs, symbols such as currency, and other words that do not fit anywhere else. Importantly, these tags are coarser than the older Brown of Penn Treebank tags. For example, there is a single POS tag for verbs rather than six as in the Penn Treebank.

Part-of-speech tagging is commonly implemented with transducer RNNs (Figure 16.4) or TN transducers (Figure 16.5). Both these methods follow typical transducer settings, as described in Chapters 10 and 12. That is, they include an output layer on top of the hidden states corresponding to each word, which projects the corresponding representations into a vector with one neuron

---

[1] https://universaldependencies.org.

Table 16.1 Universal part-of-speech tags

Name	Definition	Description	Example English words
		Open-class tags	
ADJ	Adjective	Words that modify nouns and specify their properties	*blue, big*
ADV	Adverb	Words that typically modify verbs, but can also modify adjectives, and other adverbs	*rapidly, very*
INTJ	Interjection	Words typically used as exclamation or part of an exclamation	*ouch, hello*
NOUN	Noun	Words denoting a person, place, thing, animal or idea	*dog, honesty*
PROPN	Proper noun	(Part of the) name of an individual, place, or object	*Jane, IBM*
VERB	Verb	Words that typically indicate events and actions, and can serve as predicate in a sentence	*eat, run*
		Closed-class tags	
ADP	Adposition	Words that occur before (preposition) or after (postposition) a complement composed of a noun phrase, noun, pronoun, or clause that functions as a noun phrase	*in, at*
AUX	Auxiliary	Words that accompany lexical verbs to add grammatical distinctions	*has, should*
CCONJ	Coordinating conjunction	Words that link other words or larger constituents	*and, but*
DET	Determiner	Words that modify nouns or noun phrases and express the reference of the noun phrase	*the, all*
NUM	Numeral	Words functioning typically as a determiner, adjective or pronoun, that express a number	*1, 3.14*
PART	Particle	Words that must be associated with another word or phrase to impart meaning, and that do not satisfy definitions of other POS tags	*'s, not*
PRON	Pronoun	Words that substitute for nouns or noun phrases	*she, theirs*
SCONJ	Subordinating conjunction	Conjunctions that link constructions by making one of them a constituent of the other	*that, if*
		Other	
PUNCT	Punctuation	Non-alphabetical characters that delimit linguistic units	*. , ,*
SYM	Symbol	Entities that differ from ordinary words by form and/or function	*$, :)*
X	Other	Words that cannot be assigned a real POS category	*blah, xyz*

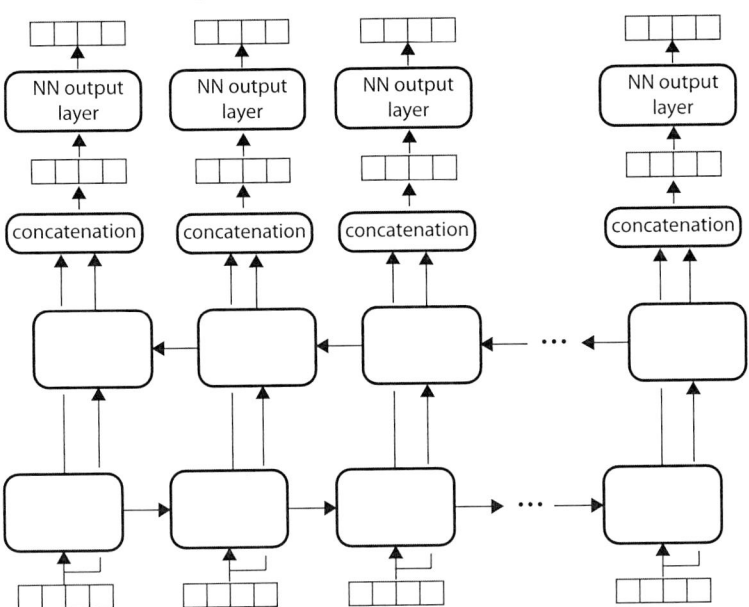

Figure 16.4 A bidirectional transducer recurrent neural network for sequence modeling

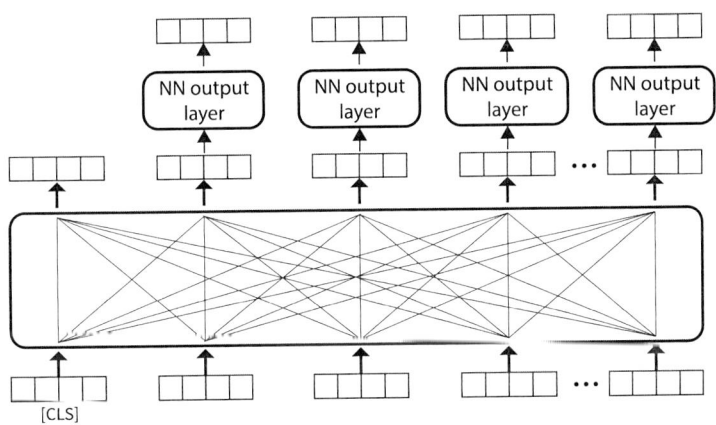

Figure 16.5 Transformer network transducer for sequence modeling

per POS tag. We provide implementations of both these architectures in Chapters 11 and 13, respectively. One subtle but important difference between the two architectures is that TNs operate on top of subword units that are generated by the BPE tokenizer or a similar algorithm (see Section 12.2 for details on the

BPE algorithm) rather than the actual words annotated in the dataset. Thus, one has to map the POS tags from words to subwords before training, and in the opposite direction (from subwords to words) when applying the resulting model to new texts. The strategy Devlin et al. (2018) recommended is to transfer the word labels to the corresponding first subwords, and ignore the other subwords. For example, assume the word *playing* is annotated as VERB in some training sentence and the underlying transformer breaks it down into two subword units: *play* and *ing*. In this case, the first subword, *play*, receives the label VERB. The second subword, *ing*, will not be included in the loss computed for the corresponding sequence (i.e., the sum of the cross-entropy losses for the annotated subwords).[2] At testing time, the opposite process is required – for example, the label assigned to the first subword (*play*) becomes the label of the whole word (*playing*). All other subword labels are ignored.

Last, the two architectures shown in Figures 16.4 and 16.5 do not use the conditional random fields layer discussed in Section 10.5. Recall that, unlike the simple architectures discussed earlier, which maximize the probability of the correct label assignment *for each individual token*, the CRF maximizes the probability of *the entire correct sequence of labels*. The latter probability combines label emission probabilities for individual tokens with transition probabilities between pairs of consecutive labels. In the authors' experience, this CRF layer does not bring a considerable performance boost on large datasets – for example, the Penn Treebank, where there are enough data to implicitly capture transition information in the underlying RNN or TN. But the CRF layer does help on smaller datasets common for underrepresented languages or under-resourced tasks such as the ones discussed in the next section. We challenge the intrepid reader to extend the two architectures implemented in Chapters 11 and 13 with the CRF layer from Section 10.5!

16.3   Named Entity Recognition

Named entities are phrases that contain names of people, organizations, locations, and so forth (Tjong Kim Sang, 2002; Tjong Kim Sang and De Meulder, 2003). For example, in the sentence *Ion Jinga moves to New York to start his UN position*, *Ion Jinga* is a person name, *New York* is a location name, and, last, *UN* is an organization name. The task of named entity recognition (NER) is to identify the boundaries of such names and label them accordingly. Note that NER is not always as simple as this example suggests. For example, in the artificial sentence *George Washington is revered not only by Washington officials, but also by most of the population in Washington, DC*, the same word (*Washington*) appears as part of a person's name (*George Washington*),

---

[2]   As we showed in Chapter 13, this can be achieved in PyTorch by setting the label of the token to be ignored to -100.

organization name (the second instance of *Washington* represents the US federal government), and location name (*Washington, DC*).

Further, there is no wide agreement on what types of names a NER system should recognize. For example, the shared tasks organized by the Computational Natural Language Learning (CoNLL) conference (Tjong Kim Sang, 2002; Tjong Kim Sang and De Meulder, 2003) recognize four types of names: persons, organizations, locations, and miscellaneous, which is a catchall for other names such as gentilics (e.g., *Spanish*), names of works of art (e.g., *Purple Haze*), names of diseases (e.g., *Bovine Spongiform Encephalopathy*), and names of gods (e.g., *Zeus*). Stanford's CoreNLP NER system (Finkel et al., 2005) diverges from this by including also numeric entities such as dates, times of day, and monetary values. Ling and Weld's (2012) *fine grained entity recognition* (FIGER), expands the set of names into a taxonomy containing 112 types. For example, locations are refined into cities, counties, countries, and so forth. Other efforts adapted the NER tasks to domain-specific problems such as the recognition of gene names (Smith et al., 2008). Nevertheless, in open-domain settings, it seems that the CoNLL representation is the one most widely used. For this reason, we will use it as the walkthrough example in this section.

Similar to POS tagging, the NER task requires the modeling of context. For example, one can immediately infer that the blank in *mayor of____* is probably a city name based on the two preceding words. Thus, NER commonly uses very similar architectures to the ones we discussed for POS tagging (Figures 16.4 and 16.5). However, the NER methods built on top of these architectures tend to differ from POS tagging in two significant ways.

First, NER methods are more likely to use a CRF layer on top of the underlying RNN or TN than POS tagging approaches. This is possibly because the common NER datasets such as the ones from the CoNLL 2002 and 2003 shared tasks are considerably smaller than POS tagging datasets. For example, the English NER from 2003 CoNLL shared task is approximately four times smaller than the Penn Treebank POS dataset. Because of their smaller sizes, it is possible that the RNNs and TNs struggle more to capture information on the valid transitions allowed. On the other hand, the CRFs, which model transition information explicitly, are more likely to learn it faster and better. For example, the authors' implementation of the popular LSTM-CRF architecture (Lample et al., 2016) obtains a 4% F_1 performance boost from the CRF component on the 2003 CoNLL English NER task, but only a couple of decimal points for POS tagging on the larger Penn Treebank dataset.

Second, while POS tags apply to individual words, named entities are commonly multi-word phrases, which must be recognized as a single unit. To capture multi-word names, Ramshaw and Marcus (1999) proposed the IOB (or BIO) notation, which introduces three types of labels: (a) O indicates words that are outside named entity mentions, (b) B-CLASS indicates that this word begins

Table 16.2 Example annotations for the BIO, IO, and BILOU annotation schemas, and the CoNLL named entity types. Because in the IO representation the only label prefix is I-, it sometimes is omitted completely – for example, I-PER becomes PER. We show the prefix here for clarity

Sentence	BIO Annotation	IO Annotation	BILOU Annotation
Jane	B-PER	I-PER	B-PER
Smith	I-PER	I-PER	L-PER
and	O	O	O
her	O	O	O
spouse	O	O	O
John	B-PER	I-PER	U-PER
like	O	O	O
to	O	O	O
visit	O	O	O
Trinidad	B-LOC	I-LOC	B-LOC
and	I-LOC	I-LOC	I-LOC
Tobago	I-LOC	I-LOC	L-LOC
.	O	O	O

a new named entity mention of type CLASS (e.g., B-PER indicates the beginning of a new person name in the CoNLL datasets), and I-CLASS indicates that this word is inside (or continues) a named entity mention of type CLASS (e.g., I-PER indicates a word that continues a person's name in CoNLL). The second column in Table 16.2 shows the BIO labels assigned to the sentence listed in the first column. Finkel et al. (2005) observed that the distinction between B- and I- prefixes is linguistically relevant only when two named entity mentions are immediately next to each other in text, and it is important to understand where the second mention starts. However, because such situations are extremely unlikely (at least in English and most European languages), they merged the two prefixes to simplify the NER task. For example, for CoNLL named entity types, merging B- and I- prefixes reduces the label space from $1 + 4 \times 2 = 9$ to $1 + 4 \times 1 = 5$. The third column in Table 16.2 shows the IO labels assigned to the same sentence. Interestingly, Ratinov and Roth (2009) argued for increasing the number of labels as this would produce "a more expressive model." They proposed the BILOU representation, which adds two new label types over BIO: (a) L-CLASS, which indicates that last token in a multi-word named entity mention of type CLASS, and (b) U-CLASS, which indicates that the corresponding word is the unique token in this named entity mention. An example of this annotation is shown in the fourth column in Table 16.2. Using an NER approach without the CRF layer, Ratinov and Roth (2009) reported that the BILOU schema yields better results than BIO on two

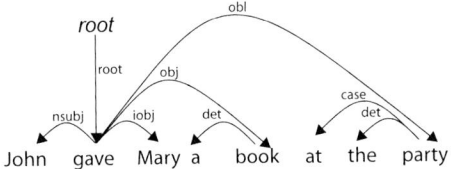

Figure 16.6 A sample sentence parsed with universal dependencies

NER datasets. However, in general, the jury is still out on the best NER annotation schema. It is likely that the best choice is impacted by the underlying
NER architecture, the size of the training dataset, and the types of annotated
named entities.[3]

16.4 Dependency Parsing

Teaching computers to understand the syntactic structure of sentences is crucial for many NLP applications including sentiment analysis (Greene and
Resnik, 2009), information extraction (Bunescu and Mooney, 2005; Moschitti, 2006), question answering (Cui et al., 2005; Surdeanu et al., 2011), and
machine translation (Charniak et al., 2003; Galley and Manning, 2009). While
sentential syntax can be represented using different formalisms, in this book,
we focus on *dependency grammars*, which describe these syntactic structures
using directed, binary dependency relations between words. For example, in
the sentence *John loves Mary*, there is a subject dependency between the verb
loves and the noun *John*, and an object dependency between *loves* and *Mary*.

As did the POS tag representations from Section 16.2, language-specific
dependency representations (De Marneffe and Manning, 2008; Surdeanu et al.,
2008; Hajic et al., 2009) evolved into language-independent *universal dependencies* (Nivre et al., 2016, 2020). For example, the second version of the
Universal Dependencies project uses a set of thirty-seven dependency types to
represent syntactic structures for more than 150 languages (Nivre et al., 2020).

Table 16.3 shows a few of the universal dependency types. Figure 16.6 shows
a sentence parsed with these dependencies. In general, universal dependencies (as well as most other dependency representations) have several important
properties:

- They are *binary* relations between a dominant word (commonly called *head*)
  and another word that immediately depends on it (known as *modifier* or
  dependent). To capture the fact that one word dominates the other, these

---

[3] This debate yielded some lively coffee breaks at NLP conferences.

Table 16.3 Some universal dependency types from https://universaldepend
encies.org/u/dep/all.html. See this URL for the complete list of dependency
types

Name	Definition	Description
		Core arguments
nsubj	Nominal subject	Relation between the verb of a clause and the nominal that is the syntactic subject and the proto-agent of the clause.
obj	Object	Relation between the verb and its object – that is, the entity being acted upon or that undergoes a change of state or motion.
iobj	Indirect object	Relation between the verb and its indirect object, which is a core argument but not a subject or object.
		Non-core dependents
obl	Oblique	Relation between a verb/adjective/adverb and a nominal (noun, pronoun, noun phrase) that functions as a non-core (oblique) argument or adjunct.
		Nominal dependents
det	Determiner	Relation that holds between a nominal and its determiner.
case	Case	Relation used for any case-marking element that is treated as a separate syntactic word (e.g., prepositions, postpositions). Case-marking elements are treated as dependents of the noun they attach to or introduce.

    relations are *directed* from the head to the modifier. For example, Figure 16.6 depicts a directed relation from the verb *gave* and its object, *book*.

- These relations are *labeled* with the syntactic function they capture. For example, the relation between *gave* and *book* is labeled obj to indicate that the noun serves as the object of the verb.
- In general, these relations may create various sentential structures such as directed graphs or trees. However, in practice, most datasets annotate the syntactic structures of sentences as dependency trees. This means that each word has *exactly one head* – that is, it serves as modifier exactly once in a sentence. For example, the word *book* modifies *gave* and no other word in this sentence. To enforce this rule, the dominant word in the sentence – that is, the word that does not depend grammatically on another word (typically the main verb in the sentence such as *gave* in Figure 16.6) – is marked as modifier for a virtual token usually called *root*, which, as its name indicates, becomes the root of the dependency tree.

Most dependency-parsing algorithms fall into two classes: *transition-based* or *graph-based*. The former class carries this name because these algorithms rely on a stack of input words from which the syntactic structure of the sentence

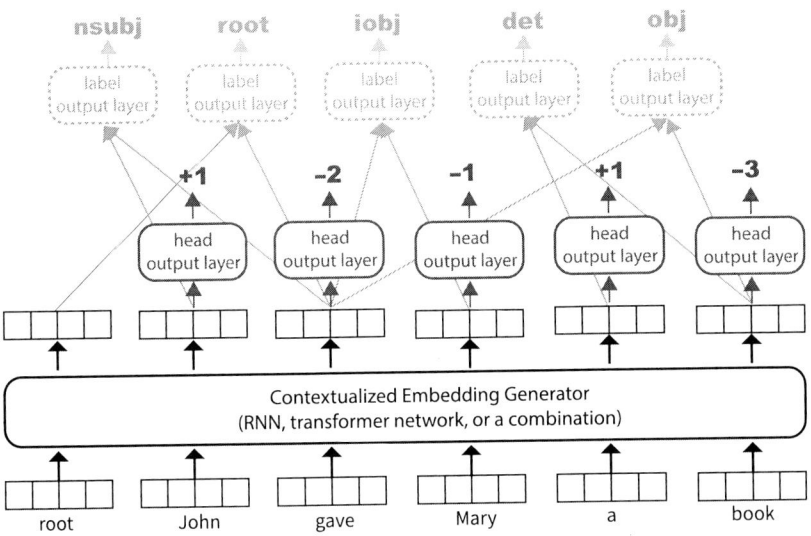

Figure 16.7 Dependency parsing as sequence modeling. The classifier that predicts the relative position of the head token is shown in the boxes with continuous lines; the classifier that predicts dependency labels is shown in the boxes with dashed lines

is incrementally built through a series of transition operations (Nivre, 2003; Ballesteros et al., 2017). For example, one such operation constructs a dependency relation from the word on top of the stack to the second word on the stack, and removes the second word from the stack. The latter class uses graph algorithms to extract the syntactic structure of a sentence from larger graphs that encode likely candidates. For example, McDonald et al.'s (2005) parsing algorithm searches within the graph that contains all possible dependency relations for a given sentence for the tree that maximizes some goodness score and covers all the sentence words.[4]

Recently, a third class of algorithms emerged that treat dependency parsing as a sequence-labeling task, where the key labels to be predicted are the relative positions of the head words (Fernández-González and Gómez-Rodríguez, 2019; Vacareanu et al., 2020a). Because these approaches are extremely simple and yet perform very well, we summarize the algorithm of Vacareanu et al. (2020a) in Figure 16.7, and describe it in detail next. As the figure shows, this approach uses two classifiers on top of contextualized embeddings: one that predicts the relative position of the head for each token in the sentence (shown in the boxes drawn with continuous lines in the figure) and one that predicts

---

[4] We refer the reader interested in more details on transition- and graph-based dependency-parsing algorithms to the excellent book of Jurafsky and Martin (2022).

the label of dependencies, given the predicted heads (shown in the boxes drawn with dashed lines in the figure). This seemingly simple idea requires a few subtle implementation details:

- Because the dominant word in the sentence (*gave* in the figure 16.7) is headed by it, the special token *root* is added at the first position in the sentence.
- The contextualized embeddings are generated by a component that is the encoder of a TN, an RNN, or a combination. Combining both has several advantages. First, TNs do a better job capturing long-distance information in the sentence because their attention mechanism treats all tokens equally regardless of distance while RNNs tend to be "fuzzy far away" (Khandelwal et al., 2018). This is necessary to model syntactic dependencies that connect words that are far apart in the sentence. On the other hand, RNNs are "sharp nearby" (Khandelwal et al., 2018) – that is, they capture sequence information in local neighborhoods – which is necessary for the vast majority of syntactic dependencies. To take advantage of both approaches, Vacareanu et al. (2020a) feed the output embeddings produced by a transformer encoder into a bidirectional LSTM whose hidden states serve as the input to the actual parsing classifiers. Second, connecting a TN with an RNN provides the opportunity to use other information beyond word/token representations. For example, Vacareanu et al. (2020a) construct the input embeddings for their bidirectional LSTM by concatenating the transformer's output embeddings with (learned) embeddings of the corresponding POS tags. This is useful because POS tags provide critical hints for dependency parsing. For example, for the English language, a determiner POS tag almost always indicates that the head of the corresponding token appears to its right in the sentence and that the dependency label is det.
- Other than these two differences, the classifier that predicts the relative position of the head words operates just like any of the sequence models described in the previous two sections. That is, it has an output layer on top of each contextualized representation, which predicts the position of its head. For example, the position of the head of the word *John* in the sentence from Figure 16.7 is $+1$ (because *gave* immediately follows *John* in the sentence); the position of *gave*'s head is -2 because the special token *root* appears two positions to the left of *gave*. Vacareanu et al. (2020a) chose a range of (-50, 50) for the possible relative positions to predict because this range accounts for 99.9% of the English dependencies in the Universal Dependencies training dataset.
- The dependency label classifier uses an output layer that operates on top of the concatenated contextualized embeddings of the head and modifier words. At training time, Vacareanu et al. (2020a) used the correct heads; at testing time, the head classifier is applied first to identify the most likely heads for all sentence words. Importantly, both classifiers share the same

component for the generation of contextualized embeddings, which means that this expensive operation is applied just once.

Despite the simplicity of the approach, Vacareanu et al. (2020a) report state-of-the-art performance on 15 distinct languages from the Universal Dependencies project. However, this simplicity comes with a few limitations:

- Because each word has exactly one head, this algorithm can only construct dependency trees (rather than more complex structures such as directed graphs). However, the basic dependency syntax introduced here uses trees for the vast majority of languages included in the Universal Dependencies effort, so this is not a major limitation in practice.[5]
- Because the classifier that predicts the positions of the heads is not aware of the actual sentence length, it may predict heads that lie outside of the sentence boundaries. For example, this classifier may predict $+2$ for the word a in the sentence in Figure 16.7, which would yield an absolute head position that is outside of the actual sentence. Nevertheless, the fix for this problem is simple: for each token, pick the head position with the highest score that is also within the sentence boundaries.
- The most important drawback of this algorithm is that it can create cycles. As a simple example, consider that the head classifier incorrectly predicts *John* as the head of *gave* (instead of *root*) in Figure 16.7. This mistake would create a cycle between *John* and *gave*. This is an important limitation, as many NLP applications rely on the traversal of dependency trees (Chambers et al., 2007; Valenzuela-Escárcega et al., 2015). The workaround for this is twofold:
  - First, construct a dependency *graph* by keeping all head predictions (or the top k in a pragmatic implementation) for each word in the sentence. For example, the top two heads predicted for *gave* may be *John* and *root* (in descending order of their score). Both these heads generate edges in this graph: *John* $\rightarrow$ *gave* and *root* $\rightarrow$ *gave*. Figure 16.8 shows such a hypothetical graph for the sentence shown in Figure 16.7.
  - Second, from this graph extract the *maximum spanning tree* emanating from *root* – that is, the tree that: (a) starts from the *root* node, (b) contains all the sentence words, and (c) has the highest score among all possible such trees in this graph. For example, the maximum spanning tree for the graph shown in Figure 16.8 is shown with dashed lines in the same figure. It follows from its definition that the maximum spanning tree is the best parse tree that can be extracted for the given sentence.

---

[5] In addition to the basic syntactic representation we discuss in this section, Universal Dependencies includes the *enhanced dependencies* representation, which "makes some of the implicit relations between words more explicit, and augments some of the dependency labels to facilitate the disambiguation of types of arguments and modifiers." https://universal dependencies.org/u/overview/enhanced-syntax.html. These enhanced dependencies violate the tree constraint more frequently than the basic ones.

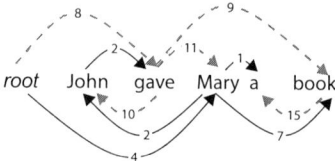

Figure 16.8 An example of a maximum spanning tree (shown with dashed lines) for a hypothetical graph containing two head predictions for each sentence word. Each edge shows a (hypothetical) prediction score; these scores are included to emphasize that the maximum spanning tree has the highest overall score of all possible spanning trees

Several algorithms have been proposed for the extraction of maximum spanning trees from graphs: Chu (1965) and Edmonds (1967) independently discovered the same algorithm, which recursively eliminates cycles from directed graphs; more recently, Eisner (1996) proposed a bottom-up algorithm that constructs well-formed trees directly.[6]

16.5   Relation Extraction

Information extraction is the NLP task that extracts structured semantic information from text. Such information includes binary relations – for example, biochemical interactions between two proteins (Krallinger et al., 2008) or *n*-ary events – that is, events with more than two arguments such as terrorist attacks, where each attack is associated with multiple arguments including the location of the attack, the identity of the attacker, number of victims, the amount of damage to physical property, and so forth. (Sundheim, 1992). Information extraction enables many important real-world applications such as discovering potential disease treatments or monitoring terrorist attacks from newswire documents.

For simplicity, in this section, we focus on (binary) relation extraction (RE) as an example of information extraction. Perhaps the most popular RE dataset at the time this book was written was TACRED (Zhang et al., 2017), which its creators described as:

"TACRED is a large-scale relation extraction dataset with 106,264 examples built over newswire and web text .... Examples in TACRED cover 41 relation types ... (e.g., `per:schools_attended` and `org:members`) or are labeled as `no_relation` if no defined relation is held."

---

[6]   We refer the interested reader to these publications for details on the respective algorithms or to (Jurafsky and Martin, 2009) for a description of the Chu (1965)/Edmonds (1967) algorithm.

> Person
>
> Billy Mays, the bearded, boisterous pitchman who, as the undisputed king of TV yell and sell,
>
> City
>
> became an unlikely pop culture icon, died at his home in Tampa, Fla, on Sunday.

> Person
>
> Pandit worked at the brokerage Morgan Stanley for about 11 years until 2005, when he and some
>
> Organization
>
> Morgan Stanley colleagues quit and later founded the hedge fund Old Lane Partners.

> Person
>
> He   received an undergraduate degree from Morgan State University in 1950 and applied for
>
> Organization
>
> admission to graduate school at the University of Maryland in College Park.

Figure 16.9 Examples of relation mentions from the TACRED corpus, from https://nlp.stanford.edu/projects/tacred. The first example is an instance of the per:city_of_death relation, which holds between a person and the city where this person died; the second example is a mention of the org:founded_by relation, which holds between an organization and the person who founded it. The last example is not a relation, according to the TACRED relation schema

At testing time, the task provides two named entities that co-occur in a sentence and requires the prediction of the relation that holds between them (or no_relation if none of the 41 TACRED relations applies). Figure 16.9 shows a few examples from this dataset.

Naively, one can repurpose any of the text classification architectures introduced in Section 16.1 for RE. For example, one could train the TN from Figure 16.3 to predict if any of the 41 TACRED relations (or none) hold between the two entities in the input sentence. While this is a reasonable baseline, it suffers from multiple limitations:

- First, using the representation of the [CLS] token to predict the specific relation that holds between two entity mentions in the sentence brings in unnecessary context (from unrelated parts of the sentence), which carries the risk of confusing the classifier. For example, consider the following sentence that contains three entity mentions (underlined): _John, who was born in England, married Mary_. There are two TACRED relations that hold in this sentence: per:country_of_birth, between _John_ and _England_, and per:spouse, between _John_ and _Mary_. However, the method introduced in Figure 16.3 would have to use the _same_ [CLS] embedding to learn _both_ relations from this sentence. This is likely to lead to confusion – for example, how is the classifier to know that the word *married* is important for the per:spouse relation, but *born* is not?

- Second, it is possible for the relation classifier to hallucinate spurious associations between the named entities themselves and the relation to be classified. For example, a classifier might learn to always predict the relation per:country_of_birth whenever it sees the country name *England* in a sentence, simply because most instances of *England* in the training data were for this relation. This overfitting on lexical artifacts has been observed for other NLP tasks (Suntwal et al., 2019; Mithun et al., 2021), and there is no reason to believe that RE is an exception. Note that it is possible that the named entities carry some useful signal for the corresponding relation – for example, from sentences such as *Angie Hicks founded Angie's List in 1995*, a classifier might be able to learn that the fact that two entities share some tokens is an indication of a strong connection between them (e.g., the relation org:founded_by holds here). However, in most cases, the surrounding context provides more generalizable evidence (e.g., the word *founded* is a strong indicator of the relation org:founded_by in this example), and, as mentioned, the danger of hallucinating meaningless associations is high.

The latter issue is solvable by replacing the named entities with a mask that indicates their semantic type and their role in the corresponding relation (Zhang et al., 2017). For example, following this masking strategy, our example sentence becomes: *PERSON-Obj founded ORGANIZATION-Subj in 1995*, which indicates that this sentence contains a relation that holds between an organization named entity as subject (or agent), and a person named entity as object (or patient).[7]

The first issue requires a more complicated solution, which involves a redesign of the RE architecture to guarantee that the classifier is exposed to the precise context necessary for the corresponding relation. To this end, Soares et al. (2019) investigated multiple strategies; we summarize two in Figures 16.10 and 16.11. Both approaches avoid using the [CLS] representation for relation classification by focusing instead on the context around the two entities. However, they diverge in the actual solutions adopted.

Figure 16.10 summarizes the RE architecture with mention pooling. The intuition behind this approach is that the contextualized representations of the entity tokens capture the information surrounding the entities, which is where the relevant signal for the underlying relation tends to be. In particular, this method feeds the input sentence through a TN (or RNN) to generate contextualized representations for all tokens.[8] Next, the representations of all

---

[7] A subtle but important implementation detail here is to add these masks – for example, PERSON-Obj and ORGANIZATION-Subj, in this example to the transformer network's vocabulary, so they are not subjected to its subword BPE tokenization.

[8] Soares et al. (2019) did not use entity masking, which is why the two figures show entities with multiple tokens.

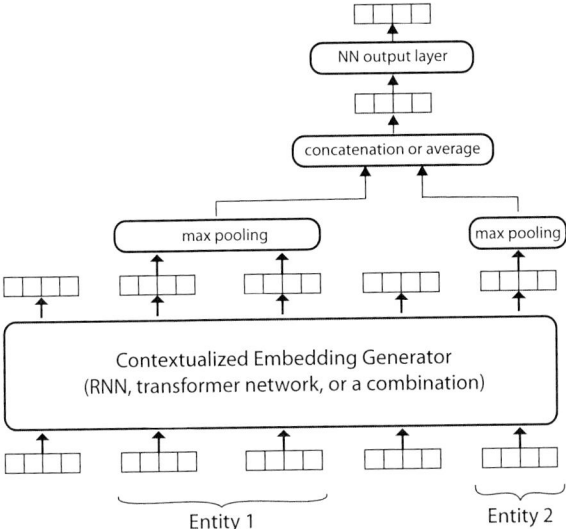

Figure 16.10 Relation extraction architecture with mention pooling. In this example, the first entity spans two tokens, while the
second entity spans one. We omit the [CLS] and [SEP] tokens for
simplicity

tokens belonging to an entity are "max pooled" into a single vector of the same
dimension. Max pooling combines an arbitrary number of vectors of the same
length by keeping the maximum value at each position. For example, feed-
ing the two vectors $[1.5, 0.9, -1.7]$ and $[0.3, -0.2, 1.3]$ into the max pooling
component produces the vector $[1.5, 0.9, 1.3]$. The advantage of max pooling
is that it reduces the representation of an entity to a single vector regardless of
how many tokens it contains. Despite the fact that this process loses information, the hope is that the strongest signal is preserved by keeping the maximum
values. Next, the max-pooled representations of the two entities are combined
(typically through concatenation, but averaging is also a common solution) and
then fed to an output layer with one neuron per relation type. This network is
trained with the usual cross-entropy loss used for multiclass classification.

The second architecture (Figure 16.11) avoids mention pooling through
further data manipulation. In particular, this approach inserts four additional
tokens into the input sentence to mark the start/end of the first/second entity.
Then, it aggregates the contextualized representations of the two entity start
tokens – that is, [E1] and [E2] – and feeds the resulting vector to an output
layer. The advantage of this strategy is that it avoids the potential lossy process
of mention pooling, while keeping the relation classifier focused on the context
surrounding the two entities.

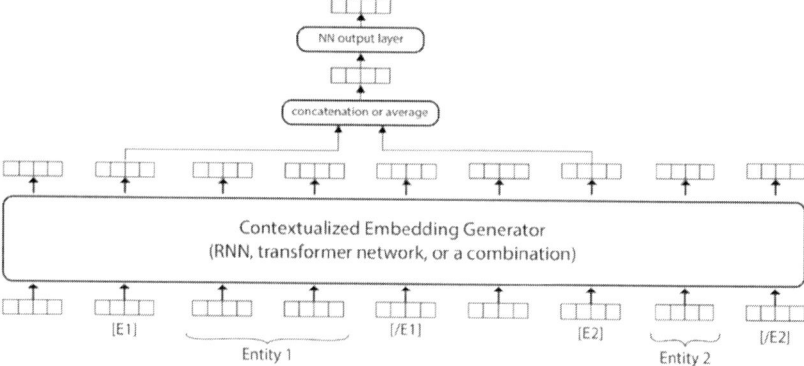

Figure 16.11 Relation extraction architecture with entity markers

In their experiments, Soares et al. (2019) reported the second architecture (Figure 16.11) performing better than the first one on four different RE datasets. However, results may change when these approaches are combined with the entity masking previously discussed in this section. For example, when entity masking is applied, both entities are represented by a single token and mention pooling is no longer necessary. In the authors' implementation, a simpler form of the architecture shown in Figure 16.10, which uses entity masking and skips mention pooling, obtained statistically similar results to the approach shown in Figure 16.11.

16.6   Question Answering

Question answering (QA) can be seen as next-generation search engine technology. That is, unlike traditional search engines, which return a list of documents in response to a keyword-based query, QA answers *natural language questions* with *short, exact answers*.[9] In addition to web or enterprise search, QA is necessary in chat bots and other dialogue systems. Today, two forms of QA dominate the research landscape. The first is *extractive QA*, also known as *reading comprehension*. As its name suggests, extractive QA approaches extract answers from a set of provided text passages. The second type of QA is *multiple-choice QA*, in which QA methods have to select the correct answer from a list of provided answers (usually with the help of supporting evidence extracted from a textual knowledge base). We detail common architectures for these two types of QA next.

---

[9] This "next-generation" search technology is closer than we think. Google announced in 2020 that TNs are used in nearly all Google queries (Raghavan, 2020).

In meteorology, precipitation is any product of the condensation of
atmospheric water vapor that falls under **gravity**. The main forms of
precipitation include drizzle, rain, sleet, snow, **graupel** and hail...
Precipitation forms as smaller droplets coalesce via collision with other rain
drops or ice crystals within a cloud. Short, intense periods of rain in
scattered locations are called "showers."

Q: What causes precipitation to fall?
A: **gravity**

Q: What is another main form of precipitation besides drizzle, rain, snow,
sleet and hail?
A: **graupel**

Q: Where do water droplets collide with ice crystals to form precipitation?
A: within a cloud

Figure 16.12 Sample passage and question-answer pairs from the
SQuAD dataset

Other legislation followed, including the Migratory Bird Conservation Act of
1929, a 1937 treaty prohibiting the hunting of right and gray whales, and the
Bald Eagle Protection Act of 1940. These later laws had a low cost to
society – the species were relatively rare – and little opposition was raised.

Q: What was the name of the 1937 treaty?
Plausible but incorrect A: **Bald Eagle Protection Act**

Figure 16.13 Example of an unanswerable question from the
SQuAD dataset

16.6.1 *Extractive Question Answering*

Figure 16.12 shows an example from SQuAD, a popular extractive QA
dataset (Rajpurkar et al., 2016, 2018). The first version of the SQuAD dataset
contained more than 100,000 questions associated with passages that contained the correct answers. The second version introduced an additional 50,000
unanswerable questions, which were associated with passages that contained
plausible but incorrect answers. Figure 16.13 shows an example of such an
unanswerable question.

Most of today's extractive QA approaches follow the architecture introduced
in (Devlin et al., 2018), which is based on Figure 12.4. In particular, extractive
QA methods concatenate the input question and supporting passage into a single sequence, where the two texts are separated by [SEP]. For example, the
input corresponding to the third question from Figure 16.12 is:[10]

[CLS] Where do water droplets collide with ice crystals to form precipitation? [SEP] In
meteorology, precipitation is ...smaller droplets coalesce via collision with other rain

[10] We omit some text from the supporting passage for brevity.

drops or ice crystals within a cloud. Short, intense periods of rain in scattered locations are called "showers." [SEP]

Note that the transformer library will handle the generation of the position and segment embeddings shown in Figure 12.4.

This network is trained to predict which tokens in the passage are the start and end of the answer span. For example, for our question, the token *within* is marked as the start of the answer span and the period immediately following *cloud* is marked as the end of the answer span. Devlin et al. (2018) implemented this using two binary classifiers: one to predict which token in the passage is the start of the answer span and a separate one to predict which token is the end. More formally, let us call C_i the contextualized embedding produced by the underlying TN for token i in the passage. Then, the probability that token i is the start of the answer span is computed as:

$$P(\text{start}|i) = \frac{e^{S \cdot C_i}}{\sum_j S \cdot C_j},\qquad(16.1)$$

where j iterates over all the tokens in the passage, and S is a "start vector" that is trained with the rest of the network parameters. Similarly, the probability that token i is the end of the answer span is computed as:

$$P(\text{end}|i) = \frac{e^{E \cdot C_i}}{\sum_j E \cdot C_j},\qquad(16.2)$$

where E is the "end vector." The S and E vectors were introduced to capture start/end information, which allows the two classifiers to share the contextualized embeddings C_i.

At prediction time, this extractive QA method selects the answer span $[i,j)$ with the maximum $P(\text{start}|i) \times P(\text{end}|j)$ with $i < j$. Note that choosing the maximum $P(\text{start}|i) \times P(\text{end}|j)$ reduces to selecting the largest $S \cdot C_i + E \cdot C_j$ because

$$P(\text{start}|i) \times P(\text{end}|j) = \frac{e^{S \cdot C_i + E \cdot C_j}}{\sum_k S \cdot C_k \times \sum_k E \cdot C_k}$$

and the denominator does not change for different i and j.

To be deployed in realistic scenarios, this QA approach must be extended to handle two additional issues. First, SQuAD provides the passages relevant to each question in the dataset. This is unlikely to happen in the real world, where extractive QA methods must be deployed on collections of documents containing up to trillions of documents.[11] Thus, in order to deploy a relatively slow TN-based architecture, an extractive QA system must first retrieve a small

---

[11] Google indexes more than 100 trillion pages: https://www.google.com/search/how searchworks.

Q: Differential heating of air can be harnessed for what?

Answer choices:
(A) **electricity production**
(B) erosion prevention
(C) transfer of electrons
(D) reduce acidity of food

Annotated supporting facts:
F1: Differential heating of air produces wind.
F2: Wind is used for producing electricity.

Figure 16.14 Example of a multiple-choice question from the QASC dataset, and the necessary facts to answer it

number of relevant passages from the entire collection to reduce its runtime. Document (and passage) retrieval is beyond the scope of this book. We refer the reader interested in learning about information retrieval (IR) to the excellent introduction by Schütze et al. (2008), and the reader eager to start coding to the Lucene software library, which implements most of the concepts discussed in this book.[12] Note that, similar to NLP, the field of IR is also in the process of adopting deep learning. For example, Karpukhin et al. (2020) proposed a neural retrieval architecture that outperforms Lucene for the retrieval of passages for QA.

Second, an extractive QA system must robustly handle unanswerable questions. Luckily, the second version of SQuAD (Rajpurkar et al., 2018) introduced training data for this situation, and the TN can be fairly easily adapted to identify such questions. For example, Devlin et al. (2018) took advantage of the [CLS] token to detect unanswerable questions. That is, for questions that have no answer in the supporting passage, they considered the [CLS] token to be both the start and end of the answer span. At testing time, if the $S \cdot C_{[CLS]} + E \cdot C_{[CLS]}$ score is larger than the equivalent score for all $i < j$ combinations, the question is marked as having no answer.

16.6.2 Multiple-Choice Question Answering

In contrast to extractive QA, multiple-choice QA presents a list of candidate answers for a given question (akin to multiple-choice questions in school exams), from which the QA system must select the correct one. Figure 16.14 shows an example of a multiple-choice question from the Question Answering via Sentence Composition (QASC) dataset, which contains 10,000 multiple-choice questions from actual US elementary and middle-school science exams (Khot et al., 2020). This seemingly simple change in the task setting introduces three complications:

[12] https://lucene.apache.org

- The answer choices are provided without the context necessary to choose the right one. For example, the QASC dataset comes with a separate corpus of more than 17 million science facts extracted from textbooks and the web. The first task of the multiple-choice QA system is to identify a small number of facts that support a given candidate answer (Figure 16.14 shows an example of two facts necessary to identify answer [A] as the correct one). Similar to extractive QA, these facts tend to be retrieved using information retrieval techniques, which are beyond the goal of this book (Schütze et al., 2008). To account for lexical differences between question, candidate answers, and supporting facts (a phenomenon beautifully called "bridging the lexical chasm" by Berger et al. [2000]), more complex techniques choose supporting facts whose representations (or embeddings) are most similar to the representations of the given question and candidate answer (Yadav et al., 2019a).

- Importantly, many multiple-choice QA tasks require the composition of several supporting facts to answer a question. For example, to understand that the differential heating of air can be used to produce electricity, one needs to aggregate the information provided by the two distinct facts shown in Figure 16.14 – that is, that differential heating produces wind, and that wind can be used to generate electricity. Note that in this case, the linking concept (wind), does not appear anywhere in the question or the candidate answer. This compositional task is often referred to *multi-hop QA*, or, perhaps too ambitiously, as *multi-hop reasoning*. Methods to address fact composition fall beyond the scope of this introductory book.[13]

- Last, the text used to describe the candidate answers does not need to match the text in the background knowledge base. For example, the correct answer in Figure 16.14 (*electricity production*) does not match exactly the text used in the final supporting fact (*producing electricity*). For this reason, the architectures used for answer classification in multiple-choice QA have to encode all three texts available – that is, question, candidate answer, and supporting facts. For example, to classify one candidate answer using a transformer architecture, a typical architecture uses the following input: [CLS] <QA text> [SEP] <fact text>, where <QA text> concatenates the question and candidate answer text, <fact text> concatenates the text for all supporting facts retrieved and composed using one of the aforementioned methods, and the [CLS] representation is used to train a binary classifier that indicates if the given candidate answer is correct or not (Yadav et al., 2019b). The same method is applied to all candidate answers; the candidate answer classified as correct with the highest score is selected as the method's output.

---

[13] Intuitively, methods that maximize the coverage of the concepts mentioned in the question and candidate answer while at the same time minimize the overlap between the distinct facts tend to perform well (Yadav et al., 2019b).

16.7 Machine Translation

Machine translation (MT) is arguably the most impactful NLP application. Today's wide deployment of MT (either through search engines that translate the web pages they ingest, or as a standalone application) allows hundreds of cultures unprecedented access to knowledge and art in other languages. This was unthinkable just a few decades ago.[14]

Neural networks have drastically simplified the design and improved the performance of MT approaches. For example, before the advent of neural networks in the NLP space, statistical machine translation (SMT) minimally relied on two components: an *alignment model*, which learned how to translate words or phrases from the source language into the target language, and a *language model*, which verified that the generated text in the target language is coherent (Koehn, 2009). More complex directions infused syntax in this approach – for example, by learning how to change word order to obey the syntax of the target language. For example, such an MT approach would learn how to change from the English subject-verb-object word order to the Japanese subject-object-verb order (Williams et al., 2016). In contrast, neural MT methods unified all these disparate components in the simpler encoder-decoder architecture we introduced in Chapters 14 (theory) and 15 (implementation).

The vast majority of today's commercial MT systems rely on the encoder-decoder architecture we discussed in detail in Chapters 14 and 15 (or variations of it). One important aspect we did not mention yet is that, similar to other NLP tasks, neural MT approaches benefit from pretraining. However, BERT's monolingual masked language model pretraining objective, which we discussed in Section 12.3.1, is not a good fit for MT: while it may capture language modeling information, it obviously cannot learn how to align words/phrases between source and target languages. This requires a new pretraining objective.

One transformer variant that is pretrained with such an objective is the **Text-to-Text Transfer Transformer** (T5) (Raffel et al., 2020). In a departure from BERT, T5 is trained on 18 distinct NLP problems, all of which are formulated as text-to-text transfer (hence the name). Figure 16.15 shows three examples of such tasks. Other tasks include: English-to-Romanian translation, English-to-French translation, reading comprehension of English texts (Khashabi et al., 2018), evaluating the acceptability (grammaticality) of English sentences (Warstadt et al., 2018), and so forth. Xue et al. (2020) expanded the same idea to 101 languages. What is critical is that all these problems are implemented in the same encoder-decoder framework sharing the same vocabulary, model, and loss function (cross-entropy over the decoded tokens). For example, when the network is exposed to an English-to-German training sentence, it maximizes the prediction probabilities of the corresponding

---

[14] For example, as of July 2022, Google Translate covered 133 languages.

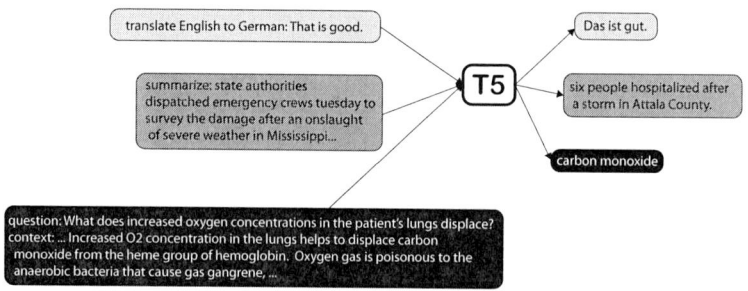

Figure 16.15 Examples of 3 of the 18 NLP problems that T5 trains on, all of which are formulated as text-to-text transfer. The three tasks are: English-to-German translation, summarization, and question answering

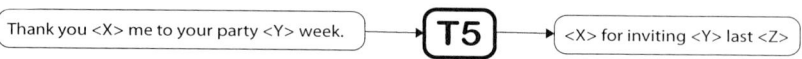

Figure 16.16 Example data point for T5 pretraining

German tokens – for example, *Das ist gut.* in Figure 16.15; when it is exposed to a data point coming from a summarization dataset, it maximizes the probabilities of the tokens in the summary. As Figure 16.15 shows, to distinguish between the different NLP problems, T5 adds a prefix to each input, which is unique per task – for example, *translate English to German*. This training procedure is clearly beneficial for downstream MT tasks that cover the same languages: it captures cross-language alignment information through the translation problems as well as language modeling information through all tasks.

To further improve its performance, the T5 authors also included an unsupervised pretraining step, which trains on data generated automatically from English texts.[15] This data-generation procedure randomly masks contiguous spans of texts with sentinel tokens. For each manipulated input text, the corresponding output to be decoded contains all the masked spans delimited by the sentinel tokens used to replace them, plus a final sentinel token ($<Z>$). Figure 16.16 shows an example of such a pretraining data point that was generated automatically. These data are used to pretrain the T5 model before it is exposed to the 18 NLP problems.

Importantly, all these $18 + 1$ pretraining tasks share a single vocabulary of 32,000 subword tokens, which was acquired from all the available training data across four languages (English, German, French, and Romanian). Because of

---

[15] Since T5 is meant to be fine-tuned for downstream tasks, one could argue that the training procedure summarized in Figure 16.15 is pretraining and that step is pre-pretraining.

this, the resulting T5 model "can only process a predetermined, fixed set of languages" (Raffel et al., 2020).

Other than these training and pretraining procedures, the T5 architecture follows very closely the "vanilla" TN architecture we introduced in Chapters 12 and 14. The one difference that is important to mention is that T5 abandons the *absolute* position embeddings used by the original TN (Section 12.1.1) and replaces them with *relative* position embeddings that capture the distance between tokens rather than the absolute token position in the sentence. This makes intuitive sense for many NLP applications. For example, in the case of syntactic parsing, the exact position of a token in the sentence provides minimal information about its likelihood of being an object, but the fact that it immediately follows the verb does. Because positions operate between pairs of tokens, T5 embeds position information in the self-attention mechanism we introduced in Section 12.1.2. In particular: (a) each offset value is mapped to a scalar parameter that is learned with the rest of the network parameters, and (b) this parameter is "added to the corresponding score used for computing the attention weights" between the key and the query (Raffel et al., 2020). This allows T5 to consider the offsets between any two tokens in the text to process without increasing the computational overhead of the network in a considerable way.

16.8   Summary

This chapter described the implementation of several common NLP applications: text classification, POS tagging, named entity recognition, syntactic dependency parsing, RE, QA and MT. We introduced robust methods that achieve (near) state-of-the-art results by today's standards and can be implemented using the neural architectures we introduced in the previous chapters as building blocks.

The methods discussed in this chapter are sufficient for the ideal supervised ML scenario, in which one has to train a ML method that addresses one task and sufficient and meaningful training data are available. However, we did not discuss in this book other real-world situations. For example, how to train the machine when only minimal training data for the task at hand is provided (*semi-supervised learning*)? Or none exists (*unsupervised learning*)? What to do when there are training data, but they differ from the data the ML system will be exposed to in its real-world deployment (*domain adaptation*)? How to train multiple tasks together so they benefit from each other (*multi-task learning*)? Unfortunately, because this book has to conclude at some point, we leave all these issues unaddressed. Nevertheless, we hope that the book has provided the necessary foundation, enabling the reader to autonomously tackle these and other research problems. Good luck!

Appendix A   Overview of the Python Language and Key Libraries

This appendix is a review of some features of Python, NumPy, and PyTorch that were used in this book. It is not meant to be a comprehensive tutorial on computer programming, Python as a programming language, or any of the libraries used. Several excellent tutorials are available for free online including these:

- The Python Tutorial – https://docs.python.org/tutorial
- NumPy quickstart – https://numpy.org/doc/stable/user/quickstart.html
- NumPy: The absolute basics for beginners – https://numpy.org/doc/stable/user/absolute_beginners.html
- Introduction to PyTorch – https://pytorch.org/tutorials/beginner/basics/intro.html
- The Hugging Face Course – https://huggingface.co/course

All the examples in this appendix have been tested using Python 3.9, NumPy 1.22, and PyTorch 1.11.

A.1   Python

Python is a high-level, multi-paradigm programming language that emphasizes code readability. Over time, it has become widely used in data science and deep learning, mostly because of libraries such as NumPy, SciPy, pandas, and PyTorch.

We assume some familiarity with computer programming, so we skip explanations of Booleans, integers, and floating point numbers, which are defined as in other programming languages. Instead, we review some of the features that make Python so convenient to work with:

  (i) Containers,
 (ii) String formatting,
(iii) Functions and parameter handling,
(iv) Classes and objects, and
 (v) Context managers.

A.1.1   Containers

Containers are data structures that store collections of objects. Python provides some common built-in containers, which are divided in two main categories: sequences and mappings. Note that libraries such as NumPy provide more specialized containers like arrays.

Sequences

Sequences are containers that are iterable, have a length, and make their elements accessible through integer indices. Examples of Python's built-in sequences are list, tuple, and str (an abbreviation for "string").

In Python, both list and tuple are heterogeneous collections of objects, which means they can have elements of different types. The main difference between them is that a list is mutable, whereas a tuple is immutable. That is, one can assign different objects to a particular position in a list, or one can add or remove elements from the list, changing the list size. However, none of these operations are supported by tuple:

```
[1]:   # a list is surrounded by square brackets
       x = [1, 2, 3]
       # a tuple is surrounded by parenthesis
       y = (1, 2, 3)
```

```
[2]:   # confirm type
       print(x, type(x))
       # assign 99 to the first position
       x[0] = 99
       print(x)
       # add 100 to the end of the list, changing its size
       x.append(100)
       print(x)
       # remove first element of the list
       x.pop(0)
       print(x)
```

```
[1, 2, 3] <class 'list'>
[99, 2, 3]
[99, 2, 3, 100]
[2, 3, 100]
```

```
[3]:   # confirm type
       print(y, type(y))
```

```
# try to assign 99 to the first position
y[0] = 99
```

```
(1, 2, 3) <class 'tuple'>
```

```
---------------------------------------------------------------
TypeError                           Traceback (most recent call last)
Input In [3], in <cell line: 4>()
      2 print(y, type(y))
      3 # assign 99 to the first position
----> 4 y[0] = 99

TypeError: 'tuple' object does not support item assignment
```

Sequences have a length that can be queried using the built-in `len()` function.

```
[4]: print(len(x))
     print(len(y))
```

```
3
3
```

Sequences can also be concatenated – that is, you can create a new sequence by merging two sequences using the plus operator. Sometimes we want to create a new sequence by repeating an existing one several times. This can be achieved with concatenation, but Python provides a shortcut with the star operator. This is analogous to the addition and multiplication of numbers:

```
[5]: # concatenation
     print(x + x + x)
     # repetition
     print(x * 3)
```

```
[2, 3, 100, 2, 3, 100, 2, 3, 100]
[2, 3, 100, 2, 3, 100, 2, 3, 100]
```

Python strings are immutable sequences of characters, and support the same immutable sequence operations we have seen so far:

```
[6]:  s = "this is a string"
      print(type(s))
      print(len(s))
      print(s + s + s)
      print('-' * 50)
```

```
<class 'str'>
16
this is a stringthis is a stringthis is a string
--------------------------------------------------
                    '
```

Individual elements of a Python sequence can be accessed using integer indexes that represent the element's position in the sequence. These indices are zero-based – the first position has index 0, not 1:

```
[7]:  # print list
      print(x)
      # print second element
      print(x[1])
```

```
[2, 3, 100]
3
```

Python sequences support more advanced indexing functionality that makes working with sequences very convenient. One of these features is the use of negative indexes, which enable access to elements starting from the end of the sequence. They are equivalent to subtracting the index from the sequence length. For example, an index of -1 in a sequence of length 5 is equivalent to $5 - 1 = 4$:

```
[8]:  # print list
      print(x)
      # print last element
      print(x[-1])
      # print penultimate element
      print(x[-2])
```

```
[2, 3, 100]
100
3
```

Another useful feature is slices that select a subsequence. A slice is denoted by a *start* position (inclusive) and a *stop* position (exclusive; one after the last desired position), separated by a colon.[1] If you omit the start, it will default to zero, and similarly, if you omit the stop, it will default to the length of the sequence:

```
[9]:  print(s)
      print('5 to 8:', s[5:9])
      print('0 to 8:', s[:9])
      print('5 to the end:', s[5:])
      print('6th from the end, to the end:', s[-6:])
      print('6th from the end, to 3rd from the end:', s[-6:-3])
```

```
this is a string
5 to 8: is a
0 to 8: this is a
5 to the end: is a string
6th from the end, to the end: string
6th from the end, to 3rd from the end: str
```

Optionally, a third number can be provided to specify a *step* (or *stride*), which allows us to skip some of the elements between *start* and *stop*, as well as to specify the direction of traversal:

```
[10]:  x = [0, 1, 2, 3, 4, 5, 6, 7, 8, 9, 10]
       # even numbers
       print(x[::2])
       # odd numbers
       print(x[1::2])
       # reverse
       print(x[::-1])
```

```
[0, 2, 4, 6, 8, 10]
[1, 3, 5, 7, 9]
[10, 9, 8, 7, 6, 5, 4, 3, 2, 1, 0]
```

Sometimes we need to confirm that a sequence contains a particular element. This can be done using the in keyword. Note, however, that this performs a linear search. That is, each element in the sequence is checked in order until either the element of interest is found or the whole sequence is consumed.

---

[1] A rationale for this convention can be found in this famous manuscript by E. W. Dijkstra: www.cs.utexas.edu/users/EWD/ewd08xx/EWD831.PDF.

Mappings

Mappings are collections that associate *keys* to *values*. The primary example in Python is a `dict`, for *dictionary*:

```
[14]: d = {'x': 1, 'y': 2, 'z': 3}
      print(d, type(d))
```

```
{'x': 1, 'y': 2, 'z': 3} <class 'dict'>
```

Dictionaries have methods to access their *keys*, *values*, and key-value pairs (referred to as *items*):

```
[15]: print(d.keys())
      print(d.values())
      print(d.items())
```

```
dict_keys(['x', 'y', 'z'])
dict_values([1, 2, 3])
dict_items([('x', 1), ('y', 2), ('z', 3)])
```

One can iterate over the keys of a dictionary with the in keyword (i.e., without explicitly using the keys() method):

```
[16]: for k in d:
          print(k, d[k])
```

```
x 1
y 2
z 3
```

Any immutable object can be used as a *key* in a dict. This includes numbers, strings, and tuples, but not, for example, lists:

```
[15]: d[0] = True
      d['s'] = True
      d[(1, 2)] = True
      print(d)
```

```
{'x': 1, 'y': 2, 'z': 3, 0: True, 's': True, (1, 2): True}
```

```
[16]: x = [1, 2]
      d[x] = True
```

```
---------------------------------------------------------------
TypeError                    Traceback (most recent call last)
Input In [16], in <cell line: 2>()
      1 x = [1, 2]
----> 2 d[x] = True

TypeError: unhashable type: 'list'
```

Items stored in a dict can be deleted using the del keyword.

```
[17]: del d['x']
      print(d)
```

```
{'y': 2, 'z': 3, 0: True, 's': True, (1, 2): True}
```

The in keyword is used to check if a *key* is contained in a mapping. The length of a dictionary is the number of keys it contains. Note that, unlike sequences, verifying if a key is contained in a dictionary has constant runtime (rather than linear in the container size, as for sequences):

```
[18]: print(d['z'])
      print('z' in d)
      print('x' not in d)
```

```
3
True
True
```

A.1.2   String Formatting

Throughout the book, we have used *f-strings* to format strings. They were introduced in Python 3.6 as a convenient alternative to string formatting. To use f-strings, simply begin a string with f before the opening quotation mark. Inside the string, Python expressions surrounded by curly brackets will be evaluated and the resulting value will be included in the string:

```
[1]: n = 0.125
     print(f'number: {n}')
     print(f'addition: {n + 1}')
     print(f'number with two decimal places: {n:.2f}')
     print(f'percentage with two decimal places: {n:.2%}')
```

```
number: 0.125
addition: 1.125
number with two decimal places: 0.12
percentage with two decimal places: 12.50%
```

In these examples, we show a few ways of formatting floating point numbers. The first example uses the value of the variable n directly. The second example shows that an expression can be used inside the curly brackets (in this case, an addition). The third and fourth examples show the use of formatting specifications. That is, the Python expression to be formatted is followed by a colon and the specification to apply. These examples show how to restrict the number of decimal digits (to only two in this case), and how to print the floating point number as a percentage. More alternatives are available using the Format Specification Mini-Language.[2]

Two particularly useful string formatting options are the addition of commas to large numbers (by using a comma as the format specification), and including the name of the variable being printed (by appending an equals sign to the variable name):

```
[2]: n = 1_000_000
     print(f'number: {n}')
     print(f'number with commas: {n:,}')
     print(f'variable name: {n=}')
```

```
number: 1000000
number with commas: 1,000,000
variable name: n=1000000
```

A.1.3   Functions

Functions in Python are *first-class objects*. That is, they can be assigned to variables, they can be passed as arguments to other functions, and they can be returned as the resulting value of a function call. The following example shows how to define a function that takes two arguments and returns their sum:

[2] https://docs.python.org/3/library/string.html#formatspec.

```
[24]:  def add(x, y):
           return x + y

       add(2, 3)
```

[24]: 5

As mentioned, one of the main characteristics of first-class objects such as functions is that they can be assigned to variables. Next, the function add is assigned to the variable x, which can now be used as the function itself:

```
[25]:  x = add
       x(2, 3)
```

[25]: 5

Here we see an example of the function add being passed to another function, call_func:

```
[26]:  def call(f, x, y):
           return f(x, y)

       call(add, 2, 3)
```

[26]: 5

Functions can also be returned from other functions. In the next example, we have a function that takes a number x as input and returns a function that adds x to its arguments:

```
[29]:  def make_adder(x):
           def adder(y):
               return x + y
           return adder

       f = make_adder(5)
       f(3)
```

[29]: 8

Sometimes we need to define simple, short-lived functions. Python provides a shortcut to define anonymous functions using the `lambda` keyword. These functions consist of a single expression, and the resulting value is returned automatically, without using the `return` keyword. Lambda expressions are commonly used when dealing with libraries that expect functions as arguments, as is the case with the `map()` method of pandas DataFrames.

```
[27]:  f = lambda x, y: x + y
       f(2, 3)
```

```
[27]: 5
```

One aspect of Python functions that makes them very convenient is their parameter handling mechanism. So far we have seen examples of positional arguments, where the values passed to the function are assigned to the function parameters based on their position. Python also supports named arguments, which allow us to assign arguments to parameters by name instead of by position:

```
[31]:  add(y=5, x=8)
```

```
[31]: 13
```

Further, Python supports the unpacking of sequences to form function arguments. This is done through the star operator. In the example that follows, the `args` sequence is unpacked, such that 2 is assigned to x, and 3 is assigned to y:

```
[28]:  args = [2, 3]
       add(*args)
```

```
[28]: 5
```

Similarly, dictionaries can be unpacked with the double-star operator. As one would expect, in this case, the keys of the dictionary must match the function's parameter names:

```
[29]:  kwargs = {'x': 2, 'y': 3}
       add(**kwargs)
```

[29]: 5

The same * operator can be used during a function declaration to indicate that the function allows a variable number of positional arguments.

```
[30]: def f(*args):
          print(args)

      f(1, 2, 3, 4)
```

(1, 2, 3, 4)

Similarly, the use of the ** operator during a function declaration indicates that the function allows a variable number of named arguments.

```
[31]: def f(**kwargs):
          print(kwargs)

      f(x=1, y=2, z=3)
```

{'x': 1, 'y': 2, 'z': 3}

Last, Python functions support default parameter values. This is very useful for designing application programming interfaces (APIs) that expose many options to the user, but that have reasonable default values. This feature is used extensively by NumPy, PyTorch, and the Hugging Face libraries:

```
[36]: def f(x='a', y='b', z='c'):
          print(f'{x=} {y=} {z=}')

      f(y=99)
```

x='a' y=99 z='c'

Built-in Functions
Python includes many built-in functions.[3] Here we mention some of the most common. We have already seen the len() function that returns the length of a sequence. Another important function is range(), which is used to create a sequence of numbers.

---

[3] A complete list of Python's built-in functions is available here: https://docs.python.org/3/library/functions.html.

```
[48]:   for i in range(5):
            print(i)
```

```
0
1
2
3
4
```

The function enumerate() receives an iterable as input and returns another iterable in which each element is a tuple containing an item's position in the sequence together with the actual item:

```
[49]:   s = 'text'
        for i, c in enumerate(s):
            print(i, c)
```

```
0 t
1 e
2 x
3 t
```

The function zip() can be used to iterate over several iterables in parallel:

```
[50]:   xs = ['a', 'b', 'c']
        ys = [1, 2, 3]
        for x, y in zip(xs, ys):
            print(x, y)
```

```
a 1
b 2
c 3
```

One useful trick involving the zip() function and the star operator for unpacking arguments is illustrated next. Many times when dealing with data we have a list of tuples, and in each tuple we have a feature vector and its corresponding label. We often need to transform this list into two lists: one with all the feature vectors and another with all the labels. By using the star operator, we unpack the list, passing the tuples directly to the zip() function, which then iterates over them in parallel, returning the first elements of all the tuples in a single sequence, followed by the second elements of all the tuples, and so on (if the tuples have more than two elements):

```
[51]:  data = [('a', 1), ('b', 2), ('c', 3)]
       names, values = zip(*data)
       print(names)
       print(values)
```

```
('a', 'b', 'c')
(1, 2, 3)
```

A.1.4   Classes and Objects

Python classes bundle together data and code, which is Python's way to support *object-oriented programming*.[4] Understanding Python classes is important for our purposes because both the PyTorch and Hugging Face libraries use classes extensively to provide components that can be employed to build new functionality.

In the next example, we create a class called Parent to illustrate how classes are defined. The methods with names surrounded by double underscores are usually referred to as *dunder methods*. These dunder methods are used to implement special functionality, allowing us to implement objects that can be used like sequences, mappings, and callables, among other categories of objects.[5] In general, one should not call dunder methods directly.

In the next example, we use the __init__() method as the *constructor* of the class. It receives a parameter x and stores it as an *attribute* of the object being created. We also define a method called __call__() that allows us to treat this object as a *callable* – that is, we can treat it as a function. Finally, we define a regular method called value() that simply returns the attribute x:

```
[41]:  class Parent:
           def __init__(self, x):
               self.x = x

           def __call__(self, n):
               return self.x + n

           def value(self):
               return self.x
```

Next, we instantiate a new object of class Parent, implicitly calling the __init__() method, and assign it to the variable called parent. Then we use the object as a callable, passing the argument 10, implicitly calling the

---

[4] https://en.wikipedia.org/wiki/Object-oriented_programming
[5] The dunder methods used to give Python objects special functionality can be found here: https://docs.python.org/3/reference/datamodel.html.

__call__() method. We also call it explicitly, to show that the method is there and the behavior of calling it implicitly or explicitly is exactly the same. Finally, we call the regular method called value():

```
[42]:   parent = Parent(5)
        print(parent(10))
        print(parent.__call__(10))
        print(parent.value())
```

```
15
15
5
```

Python classes support *inheritance*, which allow us to arrange classes in a hierarchy where more basic functionality can be implemented in *superclasses*, and more specialized in *subclasses*.

The following example defines a class called Child that inherits from Parent. The __init__() method of Child overrides (or hides) the method of Parent, but we can still access the method of the superclass using the super() function. We have seen this several times when we implemented PyTorch modules throughout the book. We have also overridden the value() method, and we have provided access to the parent's value() method in the child's parent_value() method. Note that we didn't override the __call__() method, which means that the Child class inherits it from the parent class (Parent):

```
[44]:   class Child(Parent):
            def __init__(self, x, y):
                super().__init__(x)
                self.y = y

            def value(self):
                return self.x + self.y

            def parent_value(self):
                return super().value()
```

```
[45]:   child = Child(5, 8)
        print(child(10))
        print(child.value())
        print(child.parent_value())
```

```
15
13
5
```

A.1.5   Context Managers

The with statement is used in Python to manage external resources with *context managers*. The classic example is reading from a file. When we read from a file, we have to open it before we start and we must close it after we are finished. By using the with statement, we ensure that the file is first opened and then later closed automatically when the contained block of code exits. We have seen the with statement before when we used the torch.no_grad() context manager to disable gradient calculation.

Next, we show an implementation of a context manager that opens a file and closes it when we are finished reading or writing it. This is for illustration purposes only, since Python's open() function already provides this functionality. Further, you will seldom need to implement your own context managers. Note that our context manager is implemented using the __enter__() and __exit__() dunder methods. The value returned by __enter__() is bound to the variable that follows the as keyword, which would be the file itself in this example. The __exit__() method is executed when we leave the with block:

```
[47]:  class FileContextManager:
           def __init__(self, filename, mode):
               self.file = open(filename, mode)

           def __enter__(self):
               return self.file

           def __exit__(self, exc_type, exc_value, traceback):
               self.file.close()

       with FileContextManager('test.txt', 'r') as f:
           print(f.read())
```

```
this is the file's content
```

A.2   NumPy

NumPy is a widely used numerical library for Python that provides support for multidimensional arrays, as well as a comprehensive set of mathematical functions that operate on these arrays. In this section, we will review some key features of NumPy, including:

(i) Creating arrays and some key array attributes,
(ii) Vectorized operations on arrays,
(iii) Indexing,
(iv) Broadcasting, and
(v) A few of NumPy's built-in methods.

A.2.1   Arrays

NumPy provides functions to create arrays from existing data, arrays with a particular shape, or arrays defined by numerical ranges. We cover a few of them in what follows.[6]

The array() function can be used to create a NumPy array from an existing Python list:

[1]:
```
import numpy as np

x = np.array([1, 2, 3, 4, 5])
print(x)
```

[1 2 3 4 5]

Multidimensional arrays can be created by nesting Python lists. Next, we create a two-dimensional array with two rows and three columns:

[2]:
```
x = np.array([[1, 2, 3], [4, 5, 6]])
print(x)
```

[[1 2 3]
 [4 5 6]]

The function arange() returns an array of numbers in an interval. It provides an interface similar to Python's range() function:

[3]:
```
x = np.arange(10)
print(x)
```

---

[6]   The documentation for all array creation functions is available here: https://numpy.org/
doc/stable/reference/routines.array-creation.html.

```
[0 1 2 3 4 5 6 7 8 9]
```

Another useful array creation function is `linspace()`, which creates an array of evenly spaced numbers over an interval. Here, we show how to create an array with five numbers between 2 and 3:

```
[4]: x = np.linspace(2, 3, num=5)
     print(x)
```

```
[2.   2.25 2.5  2.75 3.  ]
```

NumPy provides functions to create multidimensional arrays of a particular shape. Here, we show the `zeros()` and `ones()` functions that create arrays filled with zeros and ones, respectively:

```
[5]: x = np.zeros(shape=(2, 3))
     print(x)
```

```
[[0. 0. 0.]
 [0. 0. 0.]]
```

```
[6]: x = np.ones(shape=(2, 3))
     print(x)
```

```
[[1. 1. 1.]
 [1. 1. 1.]]
```

The shape of a NumPy array, represented as a tuple, can be retrieved using its `shape` attribute:

```
[7]: print(x.shape)
```

```
(2, 3)
```

Importantly, unlike Python lists, all elements in NumPy arrays must be of the same type. Several types are supported, mostly based on the types of the C programming language.[7] In our examples, we mostly use the int64 and float64 types. (We will also use Boolean arrays when we discuss array indexing.)

The type of the array elements is stored in its dtype attribute:

[8]:
```
print(x.dtype)
```

```
float64
```

When we create an array from a list, the dtype is inferred from the elements of the list. Next, we show examples of arrays created with int and float values:

[9]:
```
x = np.array([[1, 2, 3], [4, 5, 6]])
print(x)
print('shape:', x.shape)
print('dtype:', x.dtype)
```

```
[[1 2 3]
 [4 5 6]]
shape: (2, 3)
dtype: int64
```

[10]:
```
x = np.array([[1.0, 2.0, 3.0], [4.0, 5.0, 6.0]])
print(x)
print('shape:', x.shape)
print('dtype:', x.dtype)
```

```
[[1. 2. 3.]
 [4. 5. 6.]]
shape: (2, 3)
dtype: float64
```

The dtype can also be explicitly specified when the array is created using the dtype parameter of the array function:

---

[7] The documentation for the dtypes supported by NumPy is available here: https://numpy.org/doc/stable/reference/arrays.scalars.html.

```
[11]: x = np.array([[1, 2, 3], [4, 5, 6]], dtype=np.float64)
      print(x)
      print('shape', x.shape)
      print('dtype', x.dtype)
```

```
[[1. 2. 3.]
 [4. 5. 6.]]
shape (2, 3)
dtype float64
```

```
[12]: x = np.zeros(shape=(2, 3), dtype=np.int64)
      print(x)
      print('shape', x.shape)
      print('dtype', x.dtype)
```

```
[[0 0 0]
 [0 0 0]]
shape (2, 3)
dtype int64
```

A.2.2   Vectorized Operations

NumPy's efficiency comes from its highly optimized functions and operations over arrays. In general, using Python loops to iterate over the individual elements of a NumPy array should be avoided. Instead, operations should be applied to the array itself, allowing NumPy to efficiently operate over the elements of the array.

Next, we show that adding and multiplying an array by a scalar is equivalent to applying that operation between the scalar and every element of the array, in bulk:

```
[13]: x = np.arange(5)
      print(x)
      print(x + 10)
      print(x * 5)
```

```
[0 1 2 3 4]
[10 11 12 13 14]
[ 0  5 10 15 20]
```

NumPy also supports element-wise operations between arrays of the same shape. Next, we show the addition and multiplication of two arrays of the same size:

```
[14]:  x = np.array([2, 4, 6, 8])
       y = np.array([1, 2, 3, 4])
       print(x + y)
       print(x * y)
```

```
[ 3  6  9 12]
[ 2  8 18 32]
```

An operation that is very useful in linear algebra, and thus in ML, is matrix multiplication, including the dot product. Matrix multiplication is performed using the *at* operator. In the next example, we show the dot product of the x and y arrays defined earlier:

```
[15]:  print(x @ y)
```

```
60
```

The same operator can be used to multiply a matrix and a one-dimensional array, as long as their shapes are compatible. Next, we show the multiplication of a (2×3) matrix and a three-dimensional vector:

```
[16]:  mat = np.array([[1, -2, 4], [3, 0, -5]])
       vec = np.array([3, 2, 1])
       print(mat @ vec)
```

```
[3 4]
```

The next example shows the use of the *at* operator to multiply a (2×3) matrix and a (3×2) matrix, resulting in a (2×2) matrix:

```
[17]:  mat1 = np.array([[1, -2, 4], [3, 0, -5]])
       mat2 = np.array([[2, 1], [3, 4], [0, -2]])
       print(mat1 @ mat2)
```

```
[[ -4 -15]
 [  6  13]]
```

A.2.3   Indexing

NumPy supports a sophisticated mechanism to access array elements and slices. Not only do NumPy arrays support indexing using integers, just like Python lists, but they also support arrays of integers and Booleans to access several elements at once.

Here, we create a (3×5) array initialized with random values, which we will use for the indexing examples that follow:

```
[18]:  x = np.random.rand(3, 5)
       x
```

```
[18]:  array([[0.87108457, 0.893787  , 0.02692156, 0.72924685, 0.
       →01412141],
              [0.90573701, 0.40813536, 0.85640188, 0.11093727, 0.
       →06586925],
              [0.27893989, 0.8061504 , 0.67575313, 0.97175433, 0.
       →52530966]])
```

A single element of a multidimensional array can be accessed by specifying the coordinates in each dimension, separated by commas:

```
[19]:  x[1, 1]
```

```
[19]:  0.40813536461556266
```

Python's slice syntax is also supported for each dimension. Here, we retrieve all the elements corresponding to the second row in the array:

```
[20]:  x[1, :]
```

```
[20]:  array([0.90573701, 0.40813536, 0.85640188, 0.11093727, 0.
       →06586925])
```

In the next example, we retrieve all the elements corresponding to the second column in the array:

```
[21]:  x[:,1]
```

```
[21]:  array([0.893787  , 0.40813536, 0.8061504 ])
```

Here, we retrieve a smaller matrix by discarding the first row and the first column:

[22]: ```
x[1:, 1:]
```

[22]: ```
array([[0.40813536, 0.85640188, 0.11093727, 0.06586925],
       [0.8061504 , 0.67575313, 0.97175433, 0.52530966]])
```

Arrays of integers can also be used to index an array, retrieving the elements corresponding to the positions in the index array, in that order:

[23]: ```
x = np.arange(10)
indices = [2, 4, 1]
print(x)
print(x[indices])
```

```
[0 1 2 3 4 5 6 7 8 9]
[2 4 1]
```

Another useful way to retrieve elements from an array is Boolean indices. In this case, an array of Booleans of the same shape as the array of values must be provided, and the elements corresponding to the True values will be returned. In the example that follows, we retrieve the even values of the array by computing an array of Booleans that indicates which values divided by two have a remainder of zero. Then we use this Boolean array to index the array of values, retrieving the even elements:

[24]: ```
x = np.array([1, 2, 3, 4, 5])
print(x)
print(x % 2)
print(x % 2 == 0)
mask = x % 2 == 0
print(x[mask])
```

```
[1 2 3 4 5]
[1 0 1 0 1]
[False  True False  True False]
[2 4]
```

A.2.4   Broadcasting

So far, we have shown arithmetic operations over arrays of the same shape. Arithmetic operations can also be applied to arrays of different shapes, as long as their shapes are compatible. In these situations, the smaller array is repeated over the bigger array in a process called *broadcasting*.[8]

Here we define a two-dimensional array x of shape (2 × 3) and a one-dimensional array with two elements:

```
[25]:  x = np.array([[1, 2, 3], [4, 5, 6]])
       y = np.array([10, 20])
```

When we try to add them together, we get an exception stating that the operands of the addition cannot be broadcasted together:

```
[26]:  x + y
```

```
---------------------------------------------------------------
ValueError                         Traceback (most recent call last)
Input In [26], in <cell line: 1>()
----> 1 x + y

ValueError: operands could not be broadcast together with shapes␣
↪(2,3) (2,)
```

The first condition for compatible shapes is that they must have the same number of dimensions. The second condition is that the corresponding sizes for each dimension must be equal, or one of them must be 1.

We now use the expand_dims() function to add a new dimension to our array, so that it now has the shape (2 × 1):

```
[27]:  y = np.expand_dims(y, axis=1)
       print(y)
       print(y.shape)
```

```
[[10]
 [20]]
(2, 1)
```

Now we have two arrays, each with two dimensions. The first one is of shape (2 × 3), and the second one is of shape (2 × 1). These shapes are compatible according to the rules of broadcasting. When we add them together, the smaller array will be repeated over the columns of the bigger array:

---

[8] The documentation for broadcasting is available here: https://numpy.org/doc/stable/user/basics.broadcasting.html.

```
[28]: x + y
```

```
[28]: array([[11, 12, 13],
             [24, 25, 26]])
```

A.2.5   A Few Built-in NumPy Methods

The NumPy library contains a vast collection of functions intended to create, manipulate, and apply mathematical operations on arrays. In this subsection, we show just a few functions to illustrate their use.[9]

In the classification examples implemented throughout the book, we often had to find the value corresponding to the highest predicted score. This can be done using the argmax() function, which returns the index corresponding to the highest value of the array. To exemplify the argmax() function, we first create an array and shuffle its values:

```
[29]: x = np.arange(10)
      print(x)
      np.random.shuffle(x)
      print(x)
```

```
[0 1 2 3 4 5 6 7 8 9]
[7 3 6 0 5 8 4 9 2 1]
```

Then we use the argmax() function to find the index of the maximum value, and we print both the index and the value:

```
[30]: i = np.argmax(x)
      print('argmax', i, 'max', x[i])
```

```
argmax 7 max 9
```

Sometimes the maximum value is not sufficient, and we must retrieve the top-k values instead (e.g., when looking for the top-k most similar words to a given word in Chapter 9). The argsort() function returns an array of indices that corresponds to the sorted values. This array of indices can be used to index the value array, retrieving the sorted values:

---

[9] A comprehensive list of NumPy functions is available here: https://numpy.org/doc/stable/reference/routines.html.

[31]:
```
indices = np.argsort(x)
print(indices)
print(x[indices])
```

```
[3 9 8 1 6 4 2 0 5 7]
[0 1 2 3 4 5 6 7 8 9]
```

NumPy arrays contain several methods that can be used instead of the functions implemented at the library level. Here, we use as an example the `min()` method that returns the minimum value in the array. To show its usage, we first create a new (2×3) array populated with random integers between 0 and 100:

[32]:
```
x = np.random.randint(100, size=(2, 3))
print(x)
```

```
[[79 12 24]
 [58 66  1]]
```

Next, we use the `min()` method without arguments to retrieve the smallest value in the array. Then we also show how to restrict the application of the function to a particular direction using the `axis` parameter, retrieving the smallest value per column and per row. This way of operating over a particular dimension is common in NumPy:

[33]:
```
print(x.min())
print(x.min(axis=0))
print(x.min(axis=1))
```

```
1
[58 12  1]
[12  1]
```

A.3   PyTorch

PyTorch is an ML library used extensively in NLP and computer vision. The main data structure PyTorch provides is the *tensor*, which shares many features with NumPy arrays. The most important functionality provided by tensors is the support for GPUs, which is crucial for the efficient training of neural networks.

In this section, we provide a brief description of how to create tensors and modules (the building blocks of PyTorch neural networks).

A.3.1   PyTorch Tensors

Similar to NumPy arrays, tensors can be created directly from Python lists, and the types are inferred from the list elements:

```
[1]: import torch

     torch.tensor([[1, 2], [3, 4], [5, 6]])
```

```
[1]: tensor([[1, 2],
             [3, 4],
             [5, 6]])
```

Also, dtypes can be specified explicitly when creating a tensor:

```
[2]: torch.tensor([[1, 2], [3, 4], [5, 6]],
                  dtype=torch.float32)
```

```
[2]: tensor([[1., 2.],
             [3., 4.],
             [5., 6.]])
```

Tensors can also be created from NumPy arrays. This is useful since NumPy arrays are used extensively by many other libraries such as pandas or SciPy:

```
[3]: import numpy as np

     x = np.array([[1, 2, 3], [4, 5, 6]])
     t = torch.from_numpy(x)
     print(t)
```

```
     tensor([[1, 2, 3],
             [4, 5, 6]])
```

Tensors can also be converted back to Python lists or NumPy arrays:

```
[4]: t.tolist()
```

```
[4]: [[1, 2, 3], [4, 5, 6]]
```

```
[5]:  t.numpy()
```

```
[5]:  array([[1, 2, 3],
             [4, 5, 6]])
```

A.3.2   Modules

A module is a class that inherits from PyTorch's nn.Module. Modules are used to implement either complete neural models or individual layers in complex neural networks. For example, the layers provided by PyTorch such as nn.Linear also inherit from nn.Module.[10]

Next, we create a module called FeedForward, which connects two linear layers with a nonlinearity (ReLU) between them. The two main methods that we need to implement are __init__() and forward(). It is very important that we call nn.Module's constructor from our constructor, since it sets up several important components required for our modules to interact correctly with the rest of PyTorch. The forward() method implements the *forward pass* of our module:

```
[6]:  from torch import nn

      class FeedForward(nn.Module):
          def __init__(self, in_size, hidden_size, out_size):
              super().__init__()
              self.fc1 = nn.Linear(in_size, hidden_size)
              self.relu = nn.ReLU()
              self.fc2 = nn.Linear(hidden_size, out_size)

          def forward(self, x):
              x = self.fc1(x)
              x = self.relu(x)
              x = self.fc2(x)
              return x
```

```
[7]:  ff = FeedForward(4, 3, 2)
      ff
```

---

[10] The documentation for the different layers implemented in PyTorch is available here: https://pytorch.org/docs/stable/nn.html.

```
[7]: FeedForward(
       (fc1): Linear(in_features=4, out_features=3, bias=True)
       (relu): ReLU()
       (fc2): Linear(in_features=3, out_features=2, bias=True)
     )
```

The correct way to use a class that inherits from nn.Module is through its __call__() method, which is achieved by using the object as if it were a function. This, in turn, calls the forward() method. Note, for example, that our module's forward() method does not call the forward() method of the linear layers. Instead, we use the included modules as functions – for example, self.fc1(x). Similarly, we do not call the forward() method of our module in the next example.

Importantly, a layer's input size refers to the *second* dimension of the input provided to the module. The first dimension is the number of examples that are being passed to the model – that is, the batch size. This is true for all PyTorch layers, since PyTorch is designed to support batching. It is important to keep this in mind when passing a single example to a PyTorch layer, since it must be packaged as a batch of size one. The example that follows shows how to use our example module on a batch that contains two examples of size four:

```
[8]: t = torch.tensor(
         [[1, 2, 3, 4], [5, 6, 7, 8]],
         dtype=torch.float32,
     )

     ff(t)
```

```
[8]: tensor([[ 0.7461, -1.0864],
             [ 1.3010, -1.8234]], grad_fn=<AddmmBackward0>)
```

A.3.3   GPU Usage in PyTorch

As mentioned previously, one of the main features of PyTorch is its support of GPUs. We can use the cuda.is_available() function to check if PyTorch has detected a GPU that it can use..

```
[9]: torch.cuda.is_available()
```

```
[9]: True
```

We can indicate to PyTorch that we intend to perform a computation on the CPU or the GPU by using a `device` object, which can be created as follows:.

```
[10]: device = torch.device('cpu')
      device
```

```
[10]: device(type='cpu')
```

```
[11]: device = torch.device('cuda')
      device
```

```
[11]: device(type='cuda')
```

PyTorch tensors and modules provide a `to()` method that can be used to copy the data to the specified device. Note that both the module and the tensor data need to be on the same device in order to interact with each other:

```
[12]: ff = ff.to(device)
      t = t.to(device)
      ff(t)
```

```
[12]: tensor([[ 0.0553, -0.2621],
              [-0.2935,  0.0889]], device='cuda:0',␣
          ↪grad_fn=<AddmmBackward0>)
```

Appendix B   Character Encodings: ASCII and Unicode

Every NLP practitioner needs to understand how computers represent text. In this appendix, we will discuss different representations of text and try to demystify the concepts involved. In particular, we will discuss the difference between text and bytes, what encoding/decoding means in this context, and text normalization. This is a vast topic and we will not cover it completely. We aim to explain the fundamentals, in particular the ones most needed for NLP.

B.1   How Do Computers Represent Text?

Computers represent text the same way that they represent images, sounds, video, and everything else: as numbers. *Character encodings* establish a mapping between characters and unique numbers that identifies them. Note that character encodings are not exclusive to computers. For example, Morse code is a character encoding that predates computers, but that illustrates the usefulness of encoding text into a representation that can be transmitted over long distances.[1]

In the United States, ASCII character encoding became an early standard for encoding text in computers.[2] The ASCII coding consists of 127 characters. Of these, 96 are *printable* characters (i.e., letters, digits, punctuation, and other symbols), and the rest are *control* characters, which were not meant to be printed but to control devices such as printers. As mentioned, each of these characters needs to be represented as a number by the computer. We list the numbers corresponding to ASCII control characters in Table B.1 and those for printable characters in Table B.2.

Computers represent numbers as sequences of ones and zeros, with each binary digit in the sequence referred to as a *bit*.[3] Internally, computers manipulate groups of bits at a time. For mostly historical reasons we will not explore here, computers group sequences of eight bits together; this is referred to as a

---

[1] As electric pulses over wires.

[2] ASCII is short for American Standard Code for Information Interchange.

[3] The term *bit* is a portmanteau for *binary digit*.

Table B.1 ASCII control characters

Number	Description	Number	Description
0	Null	16	Data Line Escape
1	Start of Heading	17	Device Control 1
2	Start of Text	18	Device Control 2
3	End of Text	19	Device Control 3
4	End of Transmission	20	Device Control 4
5	Enquiry	21	Negative Acknowledgment
6	Acknowledgment	22	Synchronous Idle
7	Bell	23	End of Transmit Block
8	Backspace	24	Cancel
9	Horizontal Tab	25	End of Medium
10	Line Feed	26	Substitute
11	Vertical Tab	27	Escape
12	Form Feed	28	File Separator
13	Carriage Return	29	Group Separator
14	Shift Out	30	Record Separator
15	Shift In	31	Unit Separator

byte. Because a byte has eight bits, it can encode $2^8 = 256$ values. Thus, since ASCII has 127 characters, any ASCII character can be stored in a byte.

As Table B.2 shows, the ASCII encoding is tailored for languages that rely on the Latin alphabet, and not for others. In order to remedy this limitation, several standards emerged that used ASCII as a base but added characters, taking advantage of the fact that ASCII defines only 127 out of 256 possible values in a byte. ISO-8859-1, commonly referred to as Latin-1, is an example of an encoding that adds characters to ASCII in order to support additional languages spoken in Europe and parts of Africa. Windows-1252 is another of these encodings, which extends ISO-8859-1 and was popularized by Microsoft. Other regions of the world developed their own standards to suit their needs, such as the Japanese Industrial Standard (JIS) in Japan.[4]

As the Internet and the world became more connected, the need to share documents across the many regions of the world became more pressing, and the different encodings became problematic. Further, the various ASCII extensions are still woefully insufficient to encode the variety of the world's alphabets. This led to a push for a universal encoding for all writing systems: in 1991, the Unicode Foundation published the first version of the Unicode standard. The Unicode standard can be thought of as a large table that assigns a unique identifier to each character, much like ASCII, but with tens of thousands of characters instead of a couple of hundred. Each of these numerical identifiers is

---

[4] https://en.wikipedia.org/wiki/JIS_encoding.

Table B.2 ASCII printable characters

Number	Character	Number	Character	Number	Character	Number	Character
32	(space)	56	8	80	P	104	h
33	!	57	9	81	Q	105	i
34	"	58	:	82	R	106	j
35	#	59	;	83	S	107	k
36	$	60	<	84	T	108	l
37	%	61	=	85	U	109	m
38	&	62	>	86	V	110	n
39	'	63	?	87	W	111	o
40	(	64	@	88	X	112	p
41	)	65	A	89	Y	113	q
42	*	66	B	90	Z	114	r
43	+	67	C	91	[	115	s
44	,	68	D	92	\	116	t
45	-	69	E	93	]	117	u
46	.	70	F	94	^	118	v
47	/	71	G	95	_	119	w
48	0	72	H	96	`	120	x
49	1	73	I	97	a	121	y
50	2	74	J	98	b	122	z
51	3	75	K	99	c	123	{
52	4	76	L	100	d	124		
53	5	77	M	101	e	125	}
54	6	78	N	102	f	126	~
55	7	79	O	103	g	127	(delete)

referred to as a *code point*.[5] Note that not all of these code points correspond to what one would typically consider a character. Some of these, called *combining characters*, must be combined with another code point to form a character. This means that there can be a sequence of code points that together form a single character. For example, diacritical marks such as accents are commonly combined with Latin letters to form characters in many European languages – for example, á in Spanish.

Importantly, Unicode code points do not fit into a single byte the way ASCII characters do. Instead, we need a way of transforming a code point into multiple bytes. The most convenient encoding for this is UTF-8, which encodes a code point into a sequence of bytes of variable length, between one and four bytes, depending on its value. One nice property of UTF-8 is that ASCII is a subset of UTF-8, as the first 127 Unicode code points correspond to the ASCII characters. In practice, this means that any file encoded with ASCII can be

---

[5] There are 143,859 characters as of Unicode 13.0. For backward compatibility, the first 128 Unicode code points correspond to the ASCII characters.

decoded with UTF-8. Further, the algorithm for the UTF-8 encoding uses as few bytes as possible to encode the code points, which helps reduce the size of the represented texts.

Last, the Unicode standard also establishes other character properties such as names and numeric values. For example, the character 1/2 has a numeric value of 0.5.

```
>>> s = '1/2'
>>> unicodedata.numeric(s)
0.5
```

The Unicode standard also provides rules for text normalization, collation, and even rendering.

B.2   How to Encode/Decode Characters in Python

Now that we have a better understanding of what character encodings are and why they exist, let's see how they work in Python. We'll start with a string that has an acute accent in the first character:

```
>>> s = 'ábaco'
```

Calling the encode method on a string returns a bytes Python object, which is a sequence of integers between 0 and 255 (inclusive). Python prints bytes objects as strings, except that it prepends a b to it, and it prints bytes that do not map to ASCII characters using a backslash followed by the letter x and two hexadecimal digits (e.g., \xc3 and \xa1):[6]

```
>>> s.encode('utf8')
b'\xc3\xa1baco'
>>> s.encode('latin1')
b'\xe1baco'
```

Note that UTF-8 requires two bytes to encode the character á and Latin-1 requires only one (because of its extension to the ASCII standard with new characters that fit within a byte).

If you encode a string and decode it with the *same encoding*, the content will be preserved:

```
>>> s.encode('utf8').decode('utf8')
'ábaco'
```

However, if you encode with one encoding and decode with another, the message gets corrupted or the decoding simply fails.

---

[6] Recall that one hexadecimal digit takes values between 0 and 15, and a hexadecimal number of two digits takes values between 0 and 255, similar to a byte. For example, the hexadecimal digit a corresponds to the decimal number 10, and the hexadecimal number a1 corresponds to the decimal number $10 \times 16 + 1 = 161$.

```
>>> s.encode('utf8').decode('latin1')
'Ã¡baco'
>>> s.encode('latin1').decode('utf8')
Traceback (most recent call last):
  File "<stdin>", line 1, in <module>
UnicodeDecodeError: 'utf-8' codec can't decode byte 0xe1
in position 0: invalid continuation byte
```

To address this, the `errors` flag can be provided, which tells Python how to handle failures. The default value is `strict` (problems cause a crash), but other options include `ignore` (skip problem characters), and `replace` (replace problem characters with an unknown token).

```
>>> s.encode('latin1').decode('utf8', errors='ignore')
'baco'
>>> s.encode('latin1').decode('utf8', errors='replace')
' baco'
```

An alternative is to find out the correct encoding of the file to be processed using command-line tools such as `chardet`:[7]

```
$ chardet readme.txt
readme.txt: utf-8 with confidence 0.99
```

B.3   Text Normalization

Being able to compare strings is critical to many NLP applications. For this, we need to be able to compare sequences of code points and determine if they are equivalent. Unicode allows several sequences of code points to render identically, so they *appear* to be the same even though the underlying representation is not. For example, consider the character *á*, which can be represented either directly as that character or as *a* combined with an acute accent combining character. This is shown in the following snippet:

```
>>> print(s1)
á
>>> print(s2)
á
>>> s1 == s2
False
>>> import unicodedata
>>> len(s1)
1
>>> unicodedata.name(s1)
'LATIN SMALL LETTER A WITH ACUTE'
>>> len(s2)
2
>>> unicodedata.name(s2[0])
'LATIN SMALL LETTER A'
>>> unicodedata.name(s2[1])
'COMBINING ACUTE ACCENT'
```

---

[7] https://github.com/chardet/chardet.

Table B.3 The four normalization forms in Unicode

Name	Description
Normalization Form D (NFD)	Canonical decomposition
Normalization Form C (NFC)	Canonical decomposition, followed by canonical composition
Normalization Form KD (NFKD)	Compatibility decomposition
Normalization Form KC (NFKC)	Compatibility decomposition, followed by canonical composition

B.3.1   Unicode Normalization Forms

Unicode defines two types of equivalence between characters: *canonical equivalence* and *compatibility equivalence*. According to the Unicode Standard Annex 15[8]:

Canonical equivalence is a fundamental equivalency between characters or sequences of characters which represent the same abstract character, and which when correctly displayed should always have the same visual appearance and behavior.
...
   Compatibility equivalence is a weaker type of equivalence between characters or sequences of characters which represent the same abstract character (or sequence of abstract characters), but which may have distinct visual appearances or behaviors.

To make use of these types of equivalence, Unicode defines four different forms of normalization, based on composition and decomposition. Essentially, composition can be thought of as replacing sequences of characters that combine with a single composite character that is visually equivalent. Decomposition performs the opposite operation: composite characters are broken down into the combining characters that could be used to create their visual equivalent, with a consistent ordering. These normalization forms are listed in Table B.3.

Normalization Form C (NFC) is the one most commonly used, at least for American, European, and Korean languages.[9] To apply this normalization in Python we can use the `unicodedata` module, as shown next.

```
>>> import unicodedata
>>> unicodedata.normalize('NFC', s)
'ábaco'
```

Decomposition is useful when we want to strip diacritics from their characters. To do this, we first decompose the characters in a string, then remove all combining characters, and, finally, combine the remaining characters:

[8] https://unicode.org/reports/tr15/.
[9] www.w3.org/wiki/I18N/CanonicalNormalization.

```
def remove_diacritics(text):
    s = unicodedata.normalize('NFD', text)
    s = ''.join(c for c in s if unicodedata.combining(c))
    return unicodedata.normalize('NFC', s)
```

Sometimes, depending on our application, canonical equivalence may not be sufficient. Unicode provides several variants of the Greek letter π – for example, GREEK SMALL LETTER PI, GREEK PI SYMBOL, DOUBLE-STRUCK SMALL PI, MATHEMATICAL BOLD SMALL PI. Sometimes these distinctions are meaningful, but sometimes we just need to know if it is the Greek letter π in any of its variants. In these situations, compatibility equivalence is more appropriate:

```
>>> c1 = unicodedata.lookup('GREEK SMALL LETTER PI')
>>> c2 = unicodedata.lookup('GREEK PI SYMBOL')
>>> c1 == c2
False
>>> nfc1 = unicodedata.normalize('NFC', c1)
>>> nfc2 = unicodedata.normalize('NFC', c2)
>>> nfc1 == nfc2
False
>>> nfkc1 = unicodedata.normalize('NFKC', c1)
>>> nfkc2 = unicodedata.normalize('NFKC', c2)
>>> nfkc1 == nfkc2
True
```

B.3.2   Casefolding

Another string equivalence some applications need is case-insensitive equivalence. For example, when building a classifier over social media texts, we may prefer a case-insensitive classifier, as case is inconsistently used in such informal texts. This is usually done by converting all strings to lowercase before comparing them, or using them in downstream components. This process is called *casefolding*. This is sufficient when dealing with ASCII or Latin-1 characters. However, when we consider other writing systems, things get more complicated. Unicode's casefolding is similar to transforming text to lowercase, but it adds transformations that make the resulting strings more suitable for case-insensitive analyses:

```
>>> s1 = 'groß'
>>> s2 = 'gross'
>>> s1.lower()
'groß'
>>> s1.casefold()
'gross'
>>> s1.lower() == s2.lower()
False
>>> s1.casefold() == s2.casefold()
True
```

References

Alammar, Jay. 2018. The illustrated transformer. `http://jalammar.github.io/illustrated-transformer/`. Accessed: 2023-06-19.

Artetxe, Mikel, Labaka, Gorka, Agirre, Eneko, and Cho, Kyunghyun. 2018. Unsupervised neural machine translation. *Proceedings of the International Conference for Learning Representations (ICLR)*.

Ba, Jimmy Lei, Kiros, Jamie Ryan, and Hinton, Geoffrey E. 2016. Layer normalization. *arXiv preprint arXiv:1607.06450*.

Bahdanau, Dzmitry, Cho, Kyunghyun, and Bengio, Yoshua. 2015. Neural machine translation by jointly learning to align and translate. *CoRR*. **abs/1409.0473**.

Baldridge, Jason. 2013. *Someone needs to do a talk on "The Unreasonable Effectiveness of Bags of Words" #nlproc #computervision*. Tweeted by @jasonbaldridge on April 22, 2013.

Ballesteros, Miguel, Dyer, Chris, Goldberg, Yoav, and Smith, Noah A. 2017. Greedy transition-based dependency parsing with stack LSTMs. *Computational Linguistics*, **43**(2), 311–347.

Banerjee, Satanjeev, and Lavie, Alon. 2005. METEOR: An automatic metric for MT evaluation with improved correlation with human judgments. Pages 65–72 of *Proceedings of the ACL Workshop on Intrinsic and Extrinsic Evaluation Measures for Machine Translation and/or Summarization*.

Baum, Leonard E., and Eagon, John Alonzo. 1967. An inequality with applications to statistical estimation for probabilistic functions of Markov processes and to a model for ecology. *Bulletin of the American Mathematical Society*, **73**(3), 360–363.

Baum, Leonard E., and Petrie, Ted. 1966. Statistical inference for probabilistic functions of finite state Markov chains. *The Annals of Mathematical Statistics*, **37**(6), 1554–1563.

Bellman, Richard. 1954. The theory of dynamic programming. *Bulletin of the American Mathematical Society*, **60**(6), 503–515.

Bellman, Richard. 1957. *Dynamic programming*. Princeton, NJ: Princeton University Press.

Beltagy, Iz, Peters, Matthew E., and Cohan, Arman. 2020. Longformer: The long-document transformer. *arXiv preprint arXiv:2004.05150*.

Berger, Adam, Caruana, Rich, Cohn, David, Freitag, Dayne, and Mittal, Vibhu. 2000. Bridging the lexical chasm: Statistical approaches to answer-finding. Pages 192–199 of *Proceedings of the 23rd Annual International ACM SIGIR Conference on Research and Development in Information Retrieval*.

Block, Hans-Dieter. 1962. The perceptron: A model for brain functioning. *Reviews of Modern Physics*, **34**(1), 123–135.

Bojanowski, Piotr, Grave, Edouard, Joulin, Armand, and Mikolov, Tomas. 2017. Enriching word vectors with subword information. *Transactions of the Association for Computational Linguistics*, **5**, 135–146.

Bojar, Ondřej, Chatterjee, Rajen, Federmann, Christian, et al. 2016. Findings of the 2016 conference on machine translation. Pages 131–198 of *Proceedings of the First Conference on Machine Translation: Volume 2, shared task papers*. Berlin: Association for Computational Linguistics.

Bolukbasi, Tolga, Chang, Kai-Wei, Zou, James Y., Saligrama, Venkatesh, and Kalai, Adam T. 2016. Man is to computer programmer as woman is to homemaker? Debiasing word embeddings. *Advances in Neural Information Processing Systems*, **29**, 4349–4357.

Brown, Peter F., Della Pietra, Vincent J., Desouza, Peter V., Lai, Jennifer C., and Mercer, Robert L. 1992. Class-based n-gram models of natural language. *Computational Linguistics*, **18**(4), 467–480.

Bunescu, Razvan, and Mooney, Raymond. 2005. A shortest path dependency kernel for relation extraction. Pages 724–731 of *Proceedings of Human Language Technology Conference and Conference on Empirical Methods in Natural Language Processing*.

Chambers, Nathanael, Cer, Daniel, Grenager, Trond, et al. 2007. Learning alignments and leveraging natural logic. Pages 165–170 of *Proceedings of the ACL-PASCAL Workshop on Textual Entailment and Paraphrasing*.

Charniak, Eugene, Knight, Kevin, and Yamada, Kenji. 2003. Syntax-based language models for statistical machine translation. *Proceedings of Machine Translation Summit IX: Papers*.

Cho, Kyunghyun, Van Merriënboer, Bart, Gulcehre, Caglar, et al. 2014. Learning phrase representations using RNN encoder-decoder for statistical machine translation. *arXiv preprint arXiv:1406.1078*.

Chu, Yoeng-Jin. 1965. On the shortest arborescence of a directed graph. *Scientia Sinica*, **14**, 1396–1400.

Cireşan, Dan Claudiu, Meier, Ueli, Gambardella, Luca Maria, and Schmidhuber, Jürgen. 2010. Deep, big, simple neural nets for handwritten digit recognition. *Neural Computation*, **22**(12), 3207–3220.

Collins, Michael. 1996. A new statistical parser based on bigram lexical dependencies. *arXiv preprint cmp-lg/9605012*.

Collins, Michael. 2002. Discriminative training methods for hidden Markov models: Theory and experiments with perceptron algorithms. Pages 1–8 of *Proceedings of the 2002 Conference on Empirical Methods in Natural Language Processing (EMNLP 2002)*.

Collins, Michael, and Roark, Brian. 2004. Incremental parsing with the perceptron algorithm. Pages 111–118 of *Proceedings of the 42nd Annual Meeting of the Association for Computational Linguistics (ACL-04)*.

Conneau, Alexis, Khandelwal, Kartikay, Goyal, Naman, et al. Unsupervised cross-lingual representation learning at scale. *arXiv preprint arXiv:1911.02116*.

Cortes, Corinna, and Vapnik, Vladimir. 1995. Support-vector networks. *Machine Learning*, **20**(3), 273–297.

Cox, David R. 1958. The regression analysis of binary sequences. *Journal of the Royal Statistical Society: Series B (Methodological)*, **20**(2), 215–232.

Cramer, Jan Salomon. 2002. The origins of logistic regression. Tinbergen Institute Working Paper No. 2002-119/4, Available at SSRN: https://ssrn.com/abstract=360 300 or http://dx.doi.org/10.2139/ssrn.360300

Crammer, Koby, Dekel, Ofer, Keshet, Joseph, Shalev-Shwartz, Shai, and Singer, Yoram. 2006. Online passive-aggressive algorithms. *Journal of Machine Learning Research*, **7**, 551–585.

Crammer, Koby, and Singer, Yoram. 2003. Ultraconservative online algorithms for multiclass problems. *Journal of Machine Learning Research*, **3**(Jan.), 951–991.

Crystal, David. 1997. *The Cambridge encyclopedia of language (2nd ed.)*. Cambridge: Cambridge University Press.

Cui, Hang, Sun, Renxu, Li, Keya, Kan, Min-Yen, and Chua, Tat-Seng. 2005. Question answering passage retrieval using dependency relations. Pages 400–407 of *Proceedings of the 28th Annual International ACM SIGIR Conference on Research and Development in Information Retrieval*.

De Marneffe, Marie-Catherine, and Manning, Christopher D. 2008. *Stanford typed dependencies manual*. Technical report. Stanford, CA: Stanford University.

Dean, Jeffrey, Corrado, Greg S., Monga, Rajat, et al. 2012. Large scale distributed deep networks. In *Proceedings of the Conference on Neural Information Processing Systems*.

Deerwester, Scott, Dumais, Susan T., Furnas, George W., Landauer, Thomas K., and Harshman, Richard. 1990. Indexing by latent semantic analysis. *Journal of the American Society for Information Science*, **41**(6), 391–407.

Devlin, Jacob, Chang, Ming-Wei, Lee, Kenton, and Toutanova, Kristina. 2018. Bert: Pre-training of deep bidirectional transformers for language understanding. *arXiv preprint arXiv:1810.04805*.

Dickerson, Desiree. 2019. How I overcame impostor syndrome after leaving academia. *Nature*, **574**(7780), 588–589.

Dietterich, Thomas G. 2000. Ensemble methods in machine learning. Pages 1–15 of *International workshop on multiple classifier systems*. Berlin: Springer.

Domingos, Pedro. 2015. *The master algorithm: How the quest for the ultimate learning machine will remake our world*. New York: Basic Books.

Donaldson, Julai, and Scheffler, Axel. 2008. *Where's my mom?* London: Dial Books.

Dozat, Timothy, and Manning, Christopher D. 2016. Deep biaffine attention for neural dependency parsing. *arXiv preprint arXiv:1611.01734*.

Dreyfus, Hubert L. 1992. *What computers still can't do: A critique of artificial reason*. Cambridge, MA: MIT Press.

Dreyfus, Stuart. 1962. The numerical solution of variational problems. *Journal of Mathematical Analysis and Applications*, **5**(1), 30–45.

Dreyfus, Stuart. 1990. Artificial neural networks, back propagation, and the Kelley–Bryson gradient procedure. *Journal of Guidance, Control, and Dynamics*, **13**(5), 926–928.

Dubey, Shiv Ram, Singh, Satish Kumar, and Chaudhuri, Bidyut Baran. 2021. A comprehensive survey and performance analysis of activation functions in deep learning. *arXiv preprint arXiv:2109.14545*.

Duchi, John, Hazan, Elad, and Singer, Yoram. 2011. Adaptive subgradient methods for online learning and stochastic optimization. *Journal of Machine Learning Research*, **12**(7), 2121–2159.

Duda, Richard O., Hart, Peter E., et al. 1973. *Pattern classification and scene analysis*. Vol. 3. New York: Wiley.

Dyer, Chris, Ballesteros, Miguel, Ling, Wang, Matthews, Austin, and Smith, Noah A. 2015. Transition-based dependency parsing with stack long short-term memory. *arXiv preprint arXiv:1505.08075.*

Edmonds, Jack. 1967. Optimum branchings. *Journal of Research of the National Bureau of Standards B*, **71**(4), 233–240.

Eisner, Jason M. 1996. Three new probabilistic models for dependency parsing: An exploration. In *COLING 1996 Volume 1: The 16th International Conference on Computational Linguistics.*

Elman, Jeffrey L. 1990. Finding structure in time. *Cognitive Science*, **14**(2), 179–211.

Fernández-González, Daniel, and Gómez-Rodríguez, Carlos. 2019. Left-to-right dependency parsing with pointer networks. Pages 710–716 of *Proceedings of the 2019 Conference of the North American Chapter of the Association for Computational Linguistics: Human language technologies, volume 1 (long and short papers).* Minneapolis, MN: Association for Computational Linguistics.

Ferrer-i Cancho, Ramon, and McCowan, Brenda. 2009. A law of word meaning in dolphin whistle types. *Entropy*, **11**, 688–701.

Finkel, Jenny Rose, Grenager, Trond, and Manning, Christopher D. 2005. Incorporating non-local information into information extraction systems by Gibbs sampling. Pages 363–370 of *Proceedings of the 43rd annual meeting of the Association for Computational Linguistics (ACL'05).*

Firth, John R. 1957. A synopsis of linguistic theory, 1930–1955. *Studies in Linguistic Analysis*, 10–32.

Fowler, Henry Watson. 1994. *A dictionary of modern English usage.* Hertfordshire: Wordsworth Editions.

Francis, Winthrop Nelson. 1964. *A standard sample of present-day English for use with digital computers.* Ann Arbor: University of Michigan Press.

Galley, Michel, and Manning, Christopher D. 2009. Quadratic-time dependency parsing for machine translation. Pages 773–781 of *Proceedings of the Joint Conference of the 47th Annual Meeting of the ACL and the 4th International Joint Conference on Natural Language Processing of the AFNLP.*

Garg, Nikhil, Schiebinger, Londa, Jurafsky, Dan, and Zou, James. 2018. Word embeddings quantify 100 years of gender and ethnic stereotypes. *Proceedings of the National Academy of Sciences*, **115**(16), E3635–E3644.

Glorot, Xavier, and Bengio, Yoshua. 2010. Understanding the difficulty of training deep feedforward neural networks. Pages 249–256 of *Proceedings of the Thirteenth International Conference on Artificial Intelligence and Statistics. JMLR Workshop and Conference Proceedings.*

Glorot, Xavier, Bordes, Antoine, and Bengio, Yoshua. 2011. Deep sparse rectifier neural networks. Pages 315–323 of *Proceedings of the Fourteenth International Conference on Artificial Intelligence and Statistics. JMLR Workshop and Conference Proceedings.*

Goodfellow, Ian, Bengio, Yoshua, Courville, Aaron, and Bengio, Yoshua. 2016. *Deep learning.* Cambridge, MA: MIT Press.

Greene, Stephan, and Resnik, Philip. 2009. More than words: Syntactic packaging and implicit sentiment. Pages 503–511 of *Proceedings of Human Language Technologies: The 2009 Annual Conference of the North American Chapter of the Association for Computational Linguistics.*

Griffiths, Dawn. 2009. *Head first statistics.* Sebastopol, CA: O'Reilly.

Hajic, Jan, Ciaramita, Massimiliano, Johansson, Richard, et al. 2009. The CoNLL-2009 shared task: Syntactic and semantic dependencies in multiple languages. Pages 1–18 of *Proceedings of the Thirteenth Conference on Computational Natural Language Learning (CoNLL 2009): Shared Task.*

Harris, Zellig S. 1954. Distributional structure. *Word*, **10**(2–3), 146–162.

He, Kaiming, Zhang, Xiangyu, Ren, Shaoqing, and Sun, Jian. 2015. Delving deep into rectifiers: Surpassing human-level performance on imagenet classification. Pages 1026–1034 of *Proceedings of the IEEE International Conference on Computer Vision.*

He, Pengcheng, Liu, Xiaodong, Gao, Jianfeng, and Chen, Weizhu. 2020. Deberta: Decoding-enhanced bert with disentangled attention. *arXiv preprint arXiv:2006. 03654.*

Hochreiter, Sepp, and Schmidhuber, Jürgen. 1997. Long short-term memory. *Neural Computation*, **9**(8), 1735–1780.

Hornik, Kurt. 1991. Approximation capabilities of multilayer feedforward networks. *Neural Networks*, **4**(2), 251–257.

Ioffe, Sergey, and Szegedy, Christian. 2015. Batch normalization: Accelerating deep network training by reducing internal covariate shift. Pages 448–456 of *International Conference on Machine Learning*. PMLR.

Ivakhnenko, Alekseĭ Grigorévich, and Lapa, Valentin Grigorévich. 1966. *Cybernetic predicting devices*. Technical report. Purdue University School of Electrical Engineering.

Iyyer, Mohit, Manjunatha, Varun, Boyd-Graber, Jordan, and Daumé III, Hal. 2015. Deep unordered composition rivals syntactic methods for text classification. Pages 1681–1691 of *Proceedings of the 53rd Annual Meeting of the Association for Computational Linguistics and the 7th International Joint Conference on Natural Language Processing (Volume 1: Long papers).*

Jarrett, Kevin, Kavukcuoglu, Koray, Ranzato, Marc'Aurelio, and LeCun, Yann. 2009. What is the best multi-stage architecture for object recognition? Pages 2146–2153 of *2009 IEEE 12th International Conference on Computer Vision*. IEEE.

Jiao, Xiaoqi, Yin, Yichun, Shang, Lifeng, et al. 2019. TinyBERT: Distilling BERT for natural language understanding. *arXiv preprint arXiv:1909.10351.*

Johnson, Justin. 2021. EECS 442: Computer Vision – How to Use Colab. `https://web.eecs.umich.edu/~justincj/teaching/eecs442/WI2021/colab.html`.

Jurafsky, Daniel, and Martin, James H. 2009. *Speech and language processing 2nd ed.* Upper Saddle River, NJ: Prentice Hall.

Jurafsky, Daniel, and Martin, James H. 2022. *Speech and language processing (third edition draft)*. `https://web.stanford.edu/~jurafsky/slp3`.

Kahneman, Daniel. 2011. *Thinking, fast and slow*. New York: Macmillan.

Kahneman, Daniel, Sibony, Olivier, and Sustein, Cass R. 2021. *Noise: A flaw in human judgment*. New York: Little, Brown Spark; Hachette Book Group.

Kalchbrenner, Nal, and Blunsom, Phil. 2013. Recurrent continuous translation models. Pages 1700–1709 of *Proceedings of the 2013 Conference on Empirical Methods in Natural Language Processing.*

Karpukhin, Vladimir, Oguz, Barlas, Min, Sewon, et al. 2020. Dense passage retrieval for open-domain question answering. Pages 6769–6781 of *Proceedings of the 2020 Conference on Empirical Methods in Natural Language Processing (EMNLP)*. Online: Association for Computational Linguistics.

Kelley, Henry J. 1960. Gradient theory of optimal flight paths. *ARS Journal*, **30**(10), 947–954.

Khandelwal, Urvashi, He, He, Qi, Peng, and Jurafsky, Dan. 2018. Sharp nearby, fuzzy far away: How neural language models use context. *arXiv preprint arXiv:1805.04623*.

Khashabi, Daniel, Chaturvedi, Snigdha, Roth, Michael, Upadhyay, Shyam, and Roth, Dan. 2018. Looking beyond the surface: A challenge set for reading comprehension over multiple sentences. Pages 252–262 of *Proceedings of the 2018 Conference of the North American Chapter of the Association for Computational Linguistics: Human Language Technologies, volume 1 (long papers)*.

Khot, Tushar, Clark, Peter, Guerquin, Michal, Jansen, Peter, and Sabharwal, Ashish. 2020. Qasc: A dataset for question answering via sentence composition. Pages 8082–8090 of *Proceedings of the AAAI Conference on Artificial Intelligence*.

Kiat, Lim Swee. 2021. Attention primer. `https://github.com/greentfrapp/attention-primer/blob/master/1_counting-letters/README.md`.

Kingma, Diederik P., and Ba, Jimmy. 2015. Adam: A method for stochastic optimization. Pages 1–13 of *Proceedings of the 3rd International Conference for Learning Representations (ICLR)*.

Kiperwasser, Eliyahu, and Goldberg, Yoav. 2016. Simple and accurate dependency parsing using bidirectional LSTM feature representations. *Transactions of the Association for Computational Linguistics*, **4**, 313–327.

Kitaev, Nikita, Kaiser, Lukasz, and Levskaya, Anselm. 2020. Reformer: The efficient transformer. In *International Conference on Learning Representations*.

Knight, Kevin. 2009. Bayesian inference with tears. www.isi.edu/natural-language/people/bayes-with-tears.pdf.

Koehn, Philipp. 2009. *Statistical machine translation*. Cambridge: Cambridge University Press.

Krallinger, Martin, Leitner, Florian, Rodriguez-Penagos, Carlos, and Valencia, Alfonso. 2008. Overview of the protein–protein interaction annotation extraction task of BioCreative II. *Genome Biology*, **9**(2), 1–19.

Krizhevsky, Alex, Sutskever, Ilya, and Hinton, Geoffrey E. 2012. Imagenet classification with deep convolutional neural networks. Pages 1–9 of *Proceedings of the Conference on Advances in Neural Information Processing Systems*.

Kruskal, Joseph B. 1983. An overview of sequence comparison: Time warps, string edits, and macromolecules. *SIAM Review*, **25**(2), 201–237.

Lafferty, John, McCallum, Andrew, and Pereira, Fernando C. N. 2001. Conditional random fields: Probabilistic models for segmenting and labeling sequence data. Pages 282–289 of *Proceedings of the Eighteenth International Conference on Machine Learning*.

Lample, Guillaume, Ballesteros, Miguel, Subramanian, Sandeep, Kawakami, Kazuya, and Dyer, Chris. 2016. Neural architectures for named entity recognition. Pages 260–270 of *Proceedings of the 2016 Conference of the North American Chapter of the Association for Computational Linguistics: Human Language Technologies*. San Diego, CA: Association for Computational Linguistics.

Lefkowitz, Melanie. 2019. Professor's perceptron paved the way for AI: 60 years too soon. https://news.cornell.edu/stories/2019/09/professors-perceptron-paved-way-ai-60-years-too-soon.

Leshno, Moshe, Lin, Vladimir Ya, Pinkus, Allan, and Schocken, Shimon. 1993. Multilayer feedforward networks with a nonpolynomial activation function can approximate any function. *Neural Networks*, **6**(6), 861–867.

Levy, Omer, and Goldberg, Yoav. 2014. Dependency-based word embeddings. Pages 302–308 of *Proceedings of the 52nd Annual Meeting of the Association for Computational Linguistics (Volume 2: Short papers)*.

Levy, Omer, Goldberg, Yoav, and Dagan, Ido. 2015. Improving distributional similarity with lessons learned from word embeddings. *Transactions of the Association for Computational Linguistics*, **3**, 211–225.

Li, Xin, and Roth, Dan. 2002. Learning question classifiers. In *COLING 2002: The 19th International Conference on Computational Linguistics*.

Lin, Dekang. 1997. Using syntactic dependency as local context to resolve word sense ambiguity. Pages 64–71 of *35th Annual Meeting of the Association for Computational Linguistics and 8th Conference of the European Chapter of the Association for Computational Linguistics*.

Ling, Xiao, and Weld, Daniel S. 2012. Fine-grained entity recognition. In *Twenty-Sixth AAAI Conference on Artificial Intelligence*.

Linnainmaa, Seppo. 1970. The representation of the cumulative rounding error of an algorithm as a Taylor expansion of the local rounding errors. PhD thesis, master's thesis (in Finnish), University of Helsinki.

Liu, Yinhan, Ott, Myle, Goyal, Naman, et al. 2019. RoBERTa: A robustly optimized BERT pretraining approach. *arXiv preprint arXiv:1907.11692*.

Luong, Minh-Thang, Pham, Hieu, and Manning, Christopher D. 2015. Effective approaches to attention-based neural machine translation. *arXiv preprint arXiv:1508.04025*.

Maas, Andrew L., Daly, Raymond E., Pham, Peter T., et al. 2011. Learning word vectors for sentiment analysis. Pages 142–150 of *Proceedings of the 49th Annual Meeting of the Association for Computational Linguistics: Human Language Technologies*. Portland, OR: Association for Computational Linguistics.

Maas, Andrew L, Hannun, Awni Y., Ng, Andrew Y., et al. 2013. Rectifier nonlinearities improve neural network acoustic models. In *Proceedings of ICML*. Atlanta, GA.

Manning, Christopher D. 2011. Part-of-speech tagging from 97% to 100%: Is it time for some linguistics? Pages 171–189 of *International Conference on Intelligent Text Processing and Computational Linguistics*. Tokyo: Springer.

Manning, Christopher D. 2015. Computational linguistics and deep learning. *Computational Linguistics*, **41**(4), 701–707.

Marcinkiewicz, Mary Ann. 1994. Building a large annotated corpus of English: The Penn Treebank. Using large corpora. Pages 313–330 of *Proceedings of the Association for Computational Linguistics*.

Marcus, Gary, and Davis, Ernest. 2019. *Rebooting AI: Building artificial intelligence we can trust*. New York: Penguin Random House.

Matilal, Bimal Krishna. 1990. *The word and the world: India's contribution to the study of language*. Delhi: Oxford University Press.

McCulloch, Warren S., and Pitts, Walter. 1943. A logical calculus of the ideas immanent in nervous activity. *The Bulletin of Mathematical Biophysics*, **5**(4), 115–133.

McDonald, Ryan, Pereira, Fernando, Ribarov, Kiril, and Hajic, Jan. 2005. Non-projective dependency parsing using spanning tree algorithms. Pages 523–530 of

Proceedings of Human Language Technology Conference and Conference on Empirical Methods in Natural Language Processing.

Mehrabi, Ninareh, Morstatter, Fred, Saxena, Nripsuta, Lerman, Kristina, and Galstyan, Aram. 2021. A survey on bias and fairness in machine learning. *ACM Computing Surveys (CSUR)*, **54**(6), 1–35.

Mikolov, Tomas, Sutskever, Ilya, Chen, Kai, et al. 2013. Distributed representations of words and phrases and their compositionality. Pages 3111–3119 of *Advances in neural information processing systems*, vol. 26. New York: Curran Associates, Inc.

Minsky, Marvin, and Papert, Seymour. 1969. *Perceptron: An introduction to computational geometry.* Cambridge, MA: MIT Press.

Mintz, Mike, Bills, Steven, Snow, Rion, and Jurafsky, Dan. 2009. Distant supervision for relation extraction without labeled data. Pages 1003–1011 of *Proceedings of the Joint Conference of the 47th Annual Meeting of the ACL and the 4th International Joint Conference on Natural Language Processing of the AFNLP.*

Mithun, Mitch, Suntwal, Sandeep, and Surdeanu, Mihai. 2021. Data and model distillation as a solution for domain-transferable fact verification. In *Proceedings of the 2021 Conference of the North American Chapter of the Association for Computational Linguistics: Human Language Technologies.*

Moschitti, Alessandro. 2006. Efficient convolution kernels for dependency and constituent syntactic trees. Pages 318–329 of *Proceedings of European Conference on Machine Learning.* Berlin: Springer.

Mrkšić, Nikola, Ó Séaghdha, Diarmuid, Thomson, Blaise, et al. 2016. Counterfitting word vectors to linguistic constraints. Pages 142–148 of *Proceedings of the 2016 Conference of the North American Chapter of the Association for Computational Linguistics: Human Language Technologies.* San Diego, CA: Association for Computational Linguistics.

Nair, Vinod, and Hinton, Geoffrey E. 2010. Rectified linear units improve restricted Boltzmann machines. Pages 807–814 of *Proceedings of the 27th International Conference on Machine Learning*, Haifa, Israel.

Nesterov, Yurii. 1983. A method for unconstrained convex minimization problem with the rate of convergence $O(1/k^2)$. *Doklady ANSSSR*, **269**, 543–547.

Ng, Andrew. 2019. Stanford University's CS229: Machine learning. https://cs229.stanford.edu/syllabus-fall2020.html.

Nielsen, Michael. 2019. Neural networks and deep learning. http://neuralnetworksanddeeplearning.com.

Nivre, Joakim. 2003 (Apr.). An efficient algorithm for projective dependency parsing. Pages 149–160 of *Proceedings of the Eighth International Conference on Parsing Technologies.*

Nivre, Joakim, De Marneffe, Marie-Catherine, Ginter, Filip, et al. 2016. Universal dependencies v1: A multilingual treebank collection. Pages 1659–1666 of *Proceedings of the Tenth International Conference on Language Resources and Evaluation (LREC'16).*

Nivre, Joakim, de Marneffe, Marie-Catherine, Ginter, Filip, et al. 2020. Universal dependencies v2: An evergrowing multilingual treebank collection. *arXiv preprint arXiv:2004.10643.*

Novikoff, Albert B. 1963. *On convergence proofs for perceptrons.* Technical report. Stanford Research Institute.

Olah, Christopher. 2015. Understanding LSTM networks. http://colah.github.io/posts/2015-08-Understanding-LSTMs.

Ontanón, Santiago, Ainslie, Joshua, Cvicek, Vaclav, and Fisher, Zachary. 2021. Making transformers solve compositional tasks. *arXiv preprint arXiv:2108.04378*.

Pang, Bo, and Lee, Lillian 2008. Opinion mining and sentiment analysis. *Foundations and Trends® in Information Retrieval*, **2**(1–2), 1–135.

Papineni, Kishore, Roukos, Salim, Ward, Todd, and Zhu, Wei-Jing. 2002. Bleu: A method for automatic evaluation of machine translation. Pages 311–318 of *Proceedings of the 40th Annual Meeting of the Association for Computational Linguistics*.

Pearl, Raymond, and Reed, Lowell J. 1920. On the rate of growth of the population of the United States since 1790 and its mathematical representation. *Proceedings of the National Academy of Sciences*, **6**(6), 275–288.

Pennington, Jeffrey, Socher, Richard, and Manning, Christopher D. 2014. GloVe: Global Vectors for Word Representation. Pages 1532–1543 of *Empirical Methods in Natural Language Processing (EMNLP)*.

Petrov, Slav, Das, Dipanjan, and McDonald, Ryan. 2011. A universal part-of-speech tagset. *arXiv preprint arXiv:1104.2086*.

Polyak, Boris T. 1964. Some methods of speeding up the convergence of iteration methods. *USSR Computational Mathematics and Mathematical Physics*, **4**(5), 1–17.

Qian, Ning. 1999. On the momentum term in gradient descent learning algorithms. *Neural Networks*, **12**(1), 145–151.

Raffel, Colin, Shazeer, Noam, Roberts, Adam, et al. 2020. Exploring the limits of transfer learning with a unified text-to-text transformer. *J. Mach. Learn. Res.*, **21**(140), 1–67.

Raghavan, Prabhakar. 2020. How AI is powering a more helpful Google. https://blog.google/products/search/search-on

Raina, Rajat, Madhavan, Anand, and Ng, Andrew Y. 2009. Large-scale deep unsupervised learning using graphics processors. Pages 873–880 of *Proceedings of the 26th Annual International Conference on Machine Learning*.

Rajpurkar, Pranav, Jia, Robin, and Liang, Percy. 2018. Know what you don't know: Unanswerable questions for SQuAD. Pages 784–789 of *Proceedings of the 56th Annual Meeting of the Association for Computational Linguistics (Volume 2: Short papers)*. Melbourne: Association for Computational Linguistics.

Rajpurkar, Pranav, Zhang, Jian, Lopyrev, Konstantin, and Liang, Percy. 2016. SQuAD: 100,000+ questions for machine comprehension of text. Pages 2383–2392 of *Proceedings of the 2016 Conference on Empirical Methods in Natural Language Processing*. Austin, TX: Association for Computational Linguistics.

Ramshaw, Lance A., and Marcus, Mitchell P. 1999. Text chunking using transformation-based learning. Pages 157–176 of *Natural Language Processing Using Very Large Corpora*. Dordrecht: Springer.

Ratinov, Lev, and Roth, Dan. 2009. Design challenges and misconceptions in named entity recognition. Pages 147–155 of *Proceedings of the Thirteenth Conference on Computational Natural Language Learning (CoNLL-2009)*.

Reed, Lowell Jacob, and Berkson, Joseph. 1929. The application of the logistic function to experimental data. *Journal of Physical Chemistry*, **33**(5), 760–779.

Resnik, Philip Stuart. 1993. *Selection and information: A class-based approach to lexical relationships*. Philadelphia: University of Pennsylvania.

Rosenblatt, Frank. 1958. The perceptron: A probabilistic model for information storage and organization in the brain. *Psychological Review*, **65**(6), 386–408.

Rumelhart, David E., Hinton, Geoffrey E., and Williams, Ronald J. 1985. *Learning internal representations by error propagation*. Technical report. California University San Diego La Jolla Institute for Cognitive Science.

Rumelhart, David E., Hinton, Geoffrey E., and Williams, Ronald J. 1986. Learning representations by back-propagating errors. *Nature*, **323**(6088), 533–536.

Sanh, Victor, Debut, Lysandre, Chaumond, Julien, and Wolf, Thomas. 2019. DistilBERT, a distilled version of BERT: Smaller, faster, cheaper and lighter. *arXiv preprint arXiv:1910.01108*.

Schmidhuber, Jürgen. 1992. Learning complex, extended sequences using the principle of history compression. *Neural Computation*, **4**(2), 234–242.

Schütze, Hinrich. 1992. Word space. *Advances in Neural Information Processing Systems*, **5**, 895–902.

Schütze, Hinrich, Manning, Christopher D., and Raghavan, Prabhakar. 2008. Introduction to information retrieval. Page 260 of *Proceedings of the International Communication of Association for Computing Machinery Conference*.

Sellam, Thibault, Das, Dipanjan, and Parikh, Ankur P. 2020. BLEURT: Learning robust metrics for text generation. Pages 7881–7892 of *Proceedings of the 58th Annual Meeting of the Association for Computational Linguistics*.

Sennrich, Rico, Haddow, Barry, and Birch, Alexandra. 2015. Neural machine translation of rare words with subword units. *arXiv preprint arXiv:1508.07909*.

Shannon, Claude Elwood. 1948. A mathematical theory of communication. *The Bell System Technical Journal*, **27**(3), 379–423.

Shwartz, Vered. 2019. A systematic comparison of English noun compound representations. Pages 92–103 of *Proceedings of the Joint Workshop on Multiword Expressions and WordNet (MWE-WN 2019)*. Florence: Association for Computational Linguistics.

Smith, Larry, Tanabe, Lorraine K., Kuo, Cheng-Ju, et al. 2008. Overview of BioCreative II gene mention recognition. *Genome Biology*, **9**(2), 1–19.

Soares, Livio Baldini, FitzGerald, Nicholas, Ling, Jeffrey, and Kwiatkowski, Tom. 2019. Matching the blanks: Distributional similarity for relation learning. *arXiv preprint arXiv:1906.03158*.

Socher, Richard, Perelygin, Alex, Wu, Jean, et al. 2013. Recursive deep models for semantic compositionality over a sentiment treebank. Pages 1631–1642 of *Proceedings of the 2013 Conference on Empirical Methods in Natural Language Processing*.

Srivastava, Nitish, Hinton, Geoffrey, Krizhevsky, Alex, Sutskever, Ilya, and Salakhutdinov, Ruslan. 2014. Dropout: A simple way to prevent neural networks from overfitting. *Journal of Machine Learning Research*, **15**(56), 1929–1958.

Sundheim, Beth M. 1992. *Overview of the fourth message understanding evaluation and conference*. Technical report. Naval Command Control and Ocean Surveillance Center, RDT & E Division, San Diego, CA.

Suntwal, Sandeep, Paul, Mithun, Sharp, Rebecca, and Surdeanu, Mihai. 2019. On the importance of delexicalization for fact verification. Pages 3413–3418 of *Proceedings of the 2019 Conference on Empirical Methods in Natural Language Processing and the 9th International Joint Conference on Natural Language Processing (EMNLP-IJCNLP)*. Hong Kong: Association for Computational Linguistics.

Surdeanu, Mihai, Ciaramita, Massimiliano, and Zaragoza, Hugo. 2011. Learning to rank answers to Non-factoid questions from web collections. *Computational Linguistics*, **37**(2), 351–383.

Surdeanu, Mihai, Johansson, Richard, Meyers, Adam, Màrquez, Lluís, and Nivre, Joakim. 2008. The CoNLL 2008 shared task on joint parsing of syntactic and semantic dependencies. Pages 159–177 of *CoNLL 2008: Proceedings of the Twelfth Conference on Computational Natural Language Learning*.

Sutskever, Ilya, Martens, James, Dahl, George, and Hinton, Geoffrey. 2013. On the importance of initialization and momentum in deep learning. Pages 1139–1147 of *International Conference on Machine Learning*. PMLR.

Sutskever, Ilya, Vinyals, Oriol, and Le, Quoc V. 2014. Sequence to sequence learning with neural networks. *Advances in Neural Information Processing Systems*, **27**.

Swanson, Don R. 1986. Undiscovered public knowledge. *Library Quarterly*, **56**(2), 103–118.

Taulé, Mariona, Martí, M. Antònia, and Recasens, Marta. 2008. AnCora: Multilevel annotated corpora for Catalan and Spanish. In *Proceedings of the Sixth International Conference on Language Resources and Evaluation (LREC'08)*. Marrakech: European Language Resources Association (ELRA).

Tikhonov, Andrey Nikolayevich. 1943. On the stability of inverse problems. *Doklady Akademii Nauk SSSR*, **39**(5), 195–198.

Tjong Kim Sang, Erik F. 2002. Introduction to the CoNLL-2002 shared task: Language-independent named entity recognition. In *COLING-02: The 6th Conference on Natural Language Learning 2002 (CoNLL-2002)*.

Tjong Kim Sang, Erik F., and De Meulder, Fien. 2003. Introduction to the CoNLL-2003 shared task: Language-independent named entity recognition. Pages 142–147 of *Proceedings of the Seventh Conference on Natural Language Learning at HLT-NAACL 2003*.

Vacareanu, Robert, Barbosa, George Caique Gouveia, Valenzuela-Escarcega, Marco A., and Surdeanu, Mihai. 2020a. Parsing as tagging. Pages 5225–5231 of *Proceedings of the 12th Language Resources and Evaluation Conference*.

Vacareanu, Robert, Valenzuela-Escarcega, Marco A., Sharp, Rebecca, and Surdeanu, Mihai. 2020b. An unsupervised method for learning representations of multi-word expressions for semantic classification. In *The 28th International Conference on Computational Linguistics in Barcelona (COLING 2020)*.

Valenzuela-Escárcega, Marco A., Babur, Ozgun, Hahn-Powell, Gus, et al. 2018. Large-scale automated machine reading discovers new cancer driving mechanisms. *Database: Journal of Biological Databases and Curation*.

Valenzuela-Escárcega, Marco A., Hahn-Powell, Gus, Surdeanu, Mihai, and Hicks, Thomas. 2015. A domain-independent rule-based framework for event extraction. Pages 127–132 of *Proceedings of ACL-IJCNLP 2015 System Demonstrations*.

Vardakas, Konstantinos Z., Tsopanakis, Grigorios, Poulopoulou, Alexandra, and Falagas, Mathew E. 2015. An analysis of factors contributing to PubMed's growth. *Journal of Informetrics*, **9**(3), 592–617.

Vaswani, Ashish, Shazeer, Noam, Parmar, Niki, et al. 2017. Attention is all you need. Pages 5998–6008 of *Advances in Neural Information Processing Systems*.

Verhulst, Pierre-François. 1838. Notice sur la loi que la population suit dans son accroissement. *Correspondence mathematique et physique*, **10**, 113–126.

Verhulst, Pierre-François. 1845. La loi d'accroissement de la population. *Nouveaux Memories de l'Académie Royale des Sciences et Belles-Lettres de Bruxelles*, **18**, 14–54.

Viterbi, Andrew J. 2006. A personal history of the Viterbi algorithm. *IEEE Signal Processing Magazine*, **23**(4), 120–142.

Vulić, Ivan, and Mrkšić, Nikola. 2018. Specialising word vectors for lexical entailment. Pages 1134–1145 of *Proceedings of the 2018 Conference of the North American Chapter of the Association for Computational Linguistics: Human Language Technologies, volume 1 (Long papers)*. New Orleans, LA: Association for Computational Linguistics.

Wang, Sinong, Li, Belinda Z., Khabsa, Madian, Fang, Han, and Ma, Hao. 2020. Linformer: Self-attention with linear complexity. *arXiv preprint arXiv:2006.04768*.

Warstadt, Alex, Singh, Amanpreet, and Bowman, Samuel R. 2018. Neural network acceptability judgments. *arXiv preprint arXiv:1805.12471*.

Werbos, Paul J. 2010. Applications of advances in nonlinear sensitivity analysis. Pages 762–770 of *System modeling and optimization: Proceedings of the 10th IFIP Conference. New York City*, August 31–September 4, 1981. Berlin: Springer.

Werbos, Paul J. 1990. Backpropagation through time: What it does and how to do it. *Proceedings of the IEEE*, **78**(10), 1550–1560.

Wikipedia. Garden-path sentence. https://en.wikipedia.org/wiki/Garden-path_sentence.

Williams, Philip, Sennrich, Rico, Post, Matt, and Koehn, Philipp. 2016. *Syntax-based statistical machine translation*. Morgan & Claypool (Synthesis Lectures on Human Language Technologies, edited by Graeme Hirst). San Rafael, CA: Morgan & Claypool.

Wilson, D Randall, and Martinez, Tony R. 2003. The general inefficiency of batch training for gradient descent learning. *Neural networks*, **16**(10), 1429–1451.

Xue, Linting, Constant, Noah, Roberts, Adam, et al. 2020. mT5: A massively multilingual pre-trained text-to-text transformer. *arXiv preprint arXiv:2010.11934*.

Yadav, Vikas, Bethard, Steven, and Surdeanu, Mihai. 2019a (6). Alignment over heterogeneous embeddings for question answering. Pages 2681–2691 of *Proceedings of the 2019 Conference of the North American Chapter of the Association for Computational Linguistics: Human Language Technologies, volume 1 (Long and short papers)*.

Yadav, Vikas, Bethard, Steven, and Surdeanu, Mihai. 2019b (Nov.). Quick and (not so) dirty: Unsupervised selection of justification sentences for multi-hop question answering. In *Proceedings of the 2019 Conference on Empirical Methods in Natural Language Processing and the 9th International Joint Conference on Natural Language Processing, (Long papers)*.

Yarowsky, David. 1992. Word-sense disambiguation using statistical models of Roget's categories trained on large corpora. In *COLING 1992 Volume 2: The 14th International Conference on Computational Linguistics*.

Young, Ed. 2009. *The Real Wisdom of the Crowds*. www.nationalgeographic.com/science/article/the-real-wisdom-of-the-crowds.

Young, Tom, Hazarika, Devamanyu, Poria, Soujanya, and Cambria, Erik. 2018. Recent trends in deep learning based natural language processing. *IEEE Computational Intelligence Magazine*, **13**(3), 55–75.

Yu, Mo, and Dredze, Mark. 2015. Learning composition models for phrase embeddings. *Transactions of the Association for Computational Linguistics*, **3**, 227–242.

Yule, G. Udney. 1925. The growth of population and the factors which control it. *Journal of the Royal Statistical Society*, **38**, 1–59.

Zaheer, Manzil, Guruganesh, Guru, Dubey, Kumar Avinava, Ainslie, et al. 2020. Big Bird: Transformers for longer sequences. Pages 17283–17297 of Hugo Larochelle, Marc Aurelio Ranzato, Raia Hadsell, Maria-Florina Balcan, and Hao Lin (eds), *Advances in Neural Information Processing Systems*, vol. 33. Curran Associates, Inc.

Zeiler, Matthew D. 2012. Adadelta: An adaptive learning rate method. *arXiv preprint arXiv:1212.5701*.

Zhang, Xiang, Zhao, Junbo, and LeCun, Yann. 2015. Character-level convolutional networks for text classification. *Advances in Neural Information Processing Systems*, **28**, 1–9.

Zhang, Yuhao, Zhong, Victor, Chen, Danqi, Angeli, Gabor, and Manning, Christopher D. 2017. Position-aware attention and supervised data improve slot filling. Pages 35–45 of *Proceedings of the 2017 Conference on Empirical Methods in Natural Language Processing (EMNLP 2017)*.

Zipf, George Kingsley. 1932. *Selected studies of the principle of relative frequency in language*. Cambridge, MA: Harvard University Press.

Index

activation functions
  hyperbolic tangent, 94
  Leaky ReLU, 96
  logistic, 30
  rectified linear unit, 94
  ReLU, 94
  sigmoid, 30
  softmax, 41, 76
  tanh, 94
applications
  dependency parsing, 255
    Universal dependency types, 255
  machine translation, 229, 269
    fine-tuning, 237
    greedy decoding, 235
  named entity recognition, 252
    annotation schemas, 253
  part-of-speech tagging, 147
    recurrent neural networks, 171, 249
    transformer networks, 204, 249
    Universal tags, 249
  question answering, 264
    extractive, 265
    multiple-choice, 267
  relation extraction, 260
  text classification, 11
    bag of words, 12, 109
    distributional representations, 140, 246
    recurrent neural networks, 247
    transformer networks, 196, 248
ASCII character encoding, 301
ASCII control characters, 301
ASCII printable characters, 301
avoiding overflow, 58

backpropagation
  equations, 79
binary_classification_report, 56

bit, 301
byte, 302

character encodings, 301
chardet, 305
classification task
  binary, 10
  multiclass, 10
classifier, 9
common rules of computation for derivatives, 40
CoNLL-U format, 165
cosine similarity, 18
cost functions
  binary cross-entropy, 43, 98
  cross-entropy, 43, 99
  mean squared error, 80, 97
  negative log likelihood, 34
  regularization, 99
CountVectorizer, 52
  fit, 52
  fit_transform, 52
  transform, 52
curse of dimensionality, 117

dataset partitions
  development, 11
  testing, 11
  training, 11
datasets
  AG News, 62, 108
  AnCora, 165
  Large Movie Review Dataset, 50
  WMT 2016, 229
deep learning, 1
distributional hypothesis, 117
distributional representations
  co-occurrence vectors, 118
  low-rank approximations, 120

321